T0202891

Undergraduate Lecture Notes in Physics

Undergraduate Lecture Notes in Physics (ULNP) publishes authoritative texts covering topics throughout pure and applied physics. Each title in the series is suitable as a basis for undergraduate instruction, typically containing practice problems, worked examples, chapter summaries, and suggestions for further reading.

ULNP titles must provide at least one of the following:

- An exceptionally clear and concise treatment of a standard undergraduate subject.
- A solid undergraduate-level introduction to a graduate, advanced, or non-standard subject.
- A novel perspective or an unusual approach to teaching a subject.

ULNP especially encourages new, original, and idiosyncratic approaches to physics teaching at the undergraduate level.

The purpose of ULNP is to provide intriguing, absorbing books that will continue to be the reader's preferred reference throughout their academic career.

Series editors

More information about this series at http://www.springer.com/series/8917

Carl S. Helrich

Analytical Mechanics

Carl S. Helrich
Goshen, IN
USA

Additional material to this book can be downloaded from http://extras.springer.com.

ISSN 2192-4791 ISSN 2192-4805 (electronic)
Undergraduate Lecture Notes in Physics
ISBN 978-3-319-44490-1 ISBN 978-3-319-44491-8 (eBook)
DOI 10.1007/978-3-319-44491-8

Library of Congress Control Number: 2016948228

Printed on acid-free paper

This Springer imprint is published by Springer Nature
The registered company is Springer International Publishing AG
The registered company address is: Gewerbestrasse 11, 6330 Cham, Switzerland

To my wife,
for her patience and understanding

Preface

This textbook presents what Joseph-Louis Lagrange called *Analytical Mechanics*. Historically this was a great advance beyond the methods of Euclidean geometry employed by Isaac Newton in the *Philosophiae Naturalis Principia Mathematica*. With the methods of Lagrange and Leonhard Euler, we could actually perform calculations. Lagrange and Euler used the calculus and did not require the formidable expertise in the use of geometry that Newton possessed.

The step introduced by William Rowan Hamilton simplified the formulation. Hamilton's ideas also represent a great step forward in our understanding of the meaning of Analytical Mechanics. This, coupled with the simplification added by Carl Gustav Jacobi, provided us with a pathway to the more modern uses of Analytical Mechanics including applications to astrophysics, complex systems, and chaos.

Our approach will introduce a modern version of what was done in the 18th and 19th centuries. We will follow essentially the historical development because the ideas unfold most logically if we do so. We will, however, pay more attention to the development of Analytical Mechanics as a valuable tool than to a historical study.

Our final step will be the relativistic formulation of Analytical Mechanics. That is an absolute necessity in any complete study of Analytical Mechanics.

Logically we begin this text with a chapter on the history of mechanics. Many texts include brief historical comments or even added pages outlining individual contributions. That is certainly an improvement on the anecdotes that our professors often passed on to us without citation. Those anecdotes piqued our interest and added flavor. But they lacked a continuity of thought and that all-important accuracy that we prize. Analytical Mechanics is the oldest of the sciences. And the history stretches from the beginnings of philosophy in Miletus in 600 BCE to the advances in scientific thought introduced in the Prussian Academy and in Great Britain. I have sincerely endeavored to shorten this, as any serious student will easily recognize. But I still worry about the length.

Because my own understanding of science has been greatly enriched by studies in history, I cannot recommend that a professor ignore the first chapter completely. The student should understand something of the interesting and tortured individual Newton was. And we cannot really comprehend the origins of the ideas that gave birth to Analytical Mechanics without encountering the work of Pierre Maupertuis, Johann Bernoulli,[1] Euler, and Lagrange. The sections of Hamilton and Jacobi may be held until after the students have gained an appreciation for the methods of Analytical Mechanics. But those sections will be of interest to students as they encounter the chapter on the Hamilton-Jacobi approach. They should see the simplicity of what Jacobi brought and his great respect for the ideas of Hamilton. Then to emphasize the importance these ideas, I include an outline of Erwin Schrödinger's original published derivation of his wave equation from the Hamilton-Jacobi equation. With the caveat surrounding a second variation, the quantum theory is buried in the theory of Hamilton and Jacobi.

In Chap. 1, I have not included the historical events leading to Albert Einstein's development of the Special Theory of Relativity in 1905. Some of this I have placed in the final chapter. The historical importance of Einstein's contributions is more easily understood by a reader who has a general grasp of the classical theory of fields, which is not our primary topic.

Beyond the history, the primary part of the text, in which I present the basis and applications of Analytical Mechanics, begins with Chap. 2 on Lagrangian Mechanics. There the issue is the Euler-Lagrange equation and the variational problem, which is solved by the Euler-Lagrange equation. This I follow by a chapter on Hamiltonian Mechanics, which, through the Legendre transformation, is a logical next step from the approach of Euler and Lagrange. The canonical equations were actually obtained by Hamilton in his papers of 1834 and 1835 with another goal in mind. But the procedure was the Legendre transformation. With these chapters, we have Analytical Mechanics essentially in place.

Then I introduce the Hamilton-Jacobi approach. I do not present a method to be memorized and applied because doing so obscures the logic and the simplicity. I follow in spirit, but not in precise detail, the ideas of Jacobi. The generator of a canonical transformation will take center stage, as it did for him. The final method does not follow a head-down approach, but one with finesse.

In all texts there are final chapters. And all courses are of finite duration. Therefore, there will always be parts of the student's experience that will become lost in the fuzziness of the final days. In this text, those final chapters contain studies of complex systems, chaos, and relativistic mechanics. Each of these chapters deals with subjects of entire courses at many institutions. I have written the chapters on complex systems and on chaos as introductions to these very interesting topics. They may then be treated as windows opening onto studies that may occupy

---

[1]Johann was the original name given by his parents. Jean or John appears sometimes, depending upon whether the author is French or English. Johann Bernoulli was born and died in Basel, Switzerland.

the students' interests completely at a later time. They may even provide interest for the last weeks of a semester. But the final chapter on special relativity is not of the same character.

I elected deliberately to make the final chapter on the Special Theory of Relativity an almost self-contained unit. The reader who is not completely familiar with the theory of classical fields will be able to pass over a portion of the chapter in which we develop the field strength tensor and electromagnetic force. However, the approach to relativistic mechanics and finally to relativistic Analytical Mechanics should be considered carefully by the serious reader. There I have followed some of the classic sources, such as Wolfgang Pauli, Peter G. Bergmann, and Wolgang Rindler. The principal product of this work is the Hamiltonian and the canonical equations for relativistic motion in the electromagnetic field. We required the nonrelativistic approximation to these results for our treatment of this motion in a previous chapter.

I cannot expect that all students will be stirred, as some of mine have been, when they see the connections among the ideas common to theoretical physics. But I hope they are.

I am deeply indebted to generations of students who have gone through this intellectual adventure with me during the past forty years. I am thankful that I have been part of their intellectual pathways and for the questions with which they continued to press me. They have seen me grow in understanding and love for the ideas I try to express here.

I am also very thankful to my teachers who introduced me to the beauty and power of Analytical Mechanics. Isaac Greber particularly stands out. He presented us with remarkably inspired and almost impossibly difficult problems to which I devoted all of my energies on many cold winter nights in Cleveland. Isaac has been a friend and an inspiration.

Goshen, IN, USA
May 2016

Carl S. Helrich

Contents

Chapter 1
History

> We must gather and group appearances until the scientific imagination discerns their hidden laws, and unity arises from variety; and then from unity we must deduce variety, and force the discovered law to utter its revelations about the future.
>
> William Rowan Hamilton

1.1 Introduction

Analytical Mechanics is the theory on which we base our understanding of motion. The histories of the development of some parts of classical physics span limited time frames and can be neatly treated based on what we accept as the experimental and mathematical philosophy of physical science that emerged after the work of Johannes Kepler and Isaac Newton. But the history of classical mechanics cannot be confined in this fashion. The philosophical steps taken by Kepler and Newton were a landmark on the path leading to the theory of mechanics. But the time frame for these steps was short. We have always observed motion around us and in the cosmos above us. And at some point we began to ask questions about the motion. The history of our study of motion must go back to the original scientific thinkers. Those thinkers were the ancient Greeks.

What we may identify as ancient Greek thought resulted in the ideas of Plato and Aristotle and in the academy at Athens, which was closed by the Eastern Roman emperor Justinian I in 529 CE. A number of the Academy's thinkers found new homes among the Arabs and brought with them Greek manuscripts. Then the revelation to Muhammad in 610 CE resulted in the beginning of Islam.[1] This changed the basis on

[1] Islam is the monotheistic Abrahamic religion based on the Qur'an, which is considered to be the exact word of God by adherents to Islam.

C.S. Helrich, *Analytical Mechanics*, Undergraduate Lecture Notes in Physics, DOI 10.1007/978-3-319-44491-8_1

which scientific and philosophical thought was developed among the Muslim Arabs. Neither Muslim[2] science nor the religion of Islam developed in isolation from one another. The considerable contributions of Muslim thinkers to astronomy, cosmology, and mathematics and the educational system that included the first universities cannot be considered independently of the development of Islam. It was this intellectual culture that came to western Europe with the Muslim conquest of Spain and the formation of *Al Andalus* (Muslim Spain).

The great interest among the Latin speaking intellectuals of western Europe in Greek and Arabic science and mathematics, which was introduced into Al Andalus, resulted in an extended period of translation of the Arabic texts into Latin. These included the works of Greek thinkers that had been corrected by Muslim scientists and mathematicians, as well as original Muslim texts. Coincidental with this translation universities were also founded across Europe. The entire intellectual landscape changed in western Europe. We may be tempted to call this the first, or scientific, renaissance in Europe.

With the fall of Al Andalus in 1492, and other economic and geopolitical changes, Muslim intellectual dominance waned and scientific progress became centered in western Europe. In 1543 the work of Nicolaus Copernicus on the heliocentric universe was published. At the beginning of the 17th century Kepler changed astronomy into astrophysics and in 1687 Isaac Newton published *Philosophiae Naturalis Principia Mathematica* which contained his formulation of the laws of mechanics.

In the 18th century, at the hands of Pierre Louis Maupertuis, Leonhard Euler and Joseph-Louis Lagrange, Newton's laws became *Analytical Mechanics*, which we could actually use in calculations. The final and most beautiful form of the Analytical Mechanics came from the work of William Rowan Hamilton (1835) and Carl Gustav Jacobi (1837). The formulation of Jacobi leads to the modern applications of Analytical Mechanics to astrophysics and to chaos theory. Here we also find the basis of the Schrödinger Equation and quantum mechanics. And, with Albert Einstein and Hermann Minkowski, we began to understand space and time as a continuum and Analytical Mechanics took on a form applicable to high energies.

In this chapter we will separate the periods outlined here into sections. We shall attempt to provide sufficient detail for the reader to understand the development of the ideas as they relate to the history of Analytical Mechanics. At the same time we shall attempt to be as brief as feasible.

1.2 Ancient Greece

The roots of the ideas and concepts that we now accept as fundamental originated with the Presocratic philosophers, or simply Presocratics, of Greece. The Presocratics were Greek thinkers of the 6th and 5th centuries BCE (Before the Common Era).

[2]A Muslim is an adherent to the religion of Islam.

In the 19th century Hermann Diels introduced the term Presocratic to contrast these philosophers, whose interests were primarily in cosmology and physical speculation, from those after Socrates (ca. 47–399 BCE). Socrates' primary interest was in moral problems [14].

Identifying the Presocratics as philosophers rather than simply as thinkers is also somewhat problematic [14]. We shall, however, identify them as philosophers in part because of tradition, but also because they rejected explanations in terms of the gods or of the supernatural [[69], p. 8]. This is not to imply anything about personal belief of the Presocratics or of any other philosophers. It is rather an acceptance of the fact that the scientist or philosopher must seek understanding that does not explicitly invoke the supernatural.

Greek thought did not originate in isolation. The Babylonians were astronomers and mathematicians, certainly surpassing the Egyptians. The Babylonians were interested in more than arithmetical calculations. They possessed as well an algebra with the ability to handle quadratic equations. Observations by the Babylonians of the appearances and disappearances of Venus date to ca. 1600 BCE. The Greek astronomer Claudius Ptolemy had access to the complete Babylonian records of eclipses and used the first year of the reign of Nabonassar (747 BCE) as his base line [[69], pp. 6, 7].

Because they were a society based on agriculture and needed to know the planting seasons, the Egyptians developed what is considered to be the first intelligent calendar in human history. This was far superior to the civil calendars of the individual Greek city states, and to the lunar calendar of Babylon. Greek astronomers, in late antiquity used the Egyptian calendar in preference to the Greek and Babylonian [[69], pp. 65, 6].

But we are interested in the science of mechanics and not in the development of ancient science. We will then, unapologetically, base our discussion on the Greeks.

Some of the philosophers identified as Presocratic were contemporaries of Socrates whose thought differs from that of Socrates. And what we know of the Presocratic philosophers comes almost exclusively from the writings of Aristotle. These nuances in interpretation become more important with the emergence of the more coherent theologies of Judaism, Christianity, and Islam.

1.2.1 Milesians

Miletus is a city in modern day Turkey on the coast of the Mediterranean Sea. We identify a trio of Greek thinkers, who lived in this region, as Milesians.

Thales (born early 6th century BCE) is considered to be the first of the Presocratic thinkers. According to Aristotle, Thales proclaimed that "the world is alive and full of gods" [[69], p. 9]. He did not mean that the Olympian gods controlled the world. More probably he meant that divinity is present in all that is [[91], p. 7].

For example, Thales considered that magnetic lodestone possessed a soul, which he thought of as a unifying principle [14]. When he identified the first principle from which all things come as water it seems that he meant water was the source of all material and that all material in some way contained water. But this water was not just H_2O [14]. Rather he was speaking if of the possibility that all matter could be based on a unifying principle. Aristotle interpreted this as related to the material cause of things [[69], p. 18].

Tradition holds that Anaximander was Thales' student [14]. At least he also lived in Miletus around 560 BCE [[91], p. 8]. Anaximander had a more abstract, although similar, concept for the unifying principle of the universe. He spoke of *to apeiron*, the Boundless, which filled all of the cosmos. The Boundless is manifested as matter, motion, energy, law and perhaps even purpose [[91], p. 9]. It is the beginning and end of all there is. In the words of Aristotle this is the Divine.

Simplicius, a 6th century CE commentator on Aristotle's Physics cites a rather poetic claim by Anaximander that the Boundless is

> The source from which existing things derive their existence is also that to which they return at their destruction, for they pay penalty and retribution to each other for their injustice according to the assessment of Time [[14], [91], p. 9].

It seems that here Anaximander is referring to a system that is ruled by the justice of the ordering of Time. This contrasts to the capricious chaos of the disordered world subject to the whims of Olympian gods [14].

Anaximenes also thought in terms of an underlying substance, which he identified as air. Air can take on different colors, temperatures, tastes and smells. And he added a basic theory of how air is transformed. Air condenses to form water and boils to reform as a gas. And air also can become solid. These are based on observation and not on qualitative concepts alone [[14], [69], p. 22].

In the Milesians we see the first steps toward a unifying idea or concept for the cosmos. And we see the beginning of observation in order to decide on the mechanism underlying change. This is the beginning of the transition from being to becoming.

1.2.2 Beyond Miletus

If we go North of Miletus, but remain in modern Turkey, we come to the ancient Greek city of Colophon. Until recently Xenophanes of Colophon was considered to be only a minor poet [14]. We now realize that he was an important thinker who explored human knowledge and limitations and influenced later philosophers as they explored human thought and the possibility that human's may have a god's eye view.

Xenophanes pointed out that humans depict their god's in a familiar (human) form. But then he claimed that there is a single greatest god.

One god greatest among gods and men,
Resembling mortals neither in body nor in thought.
... whole [he] sees, whole [he] thinks, and whole [he] hears,
but completely without toil he agitates all things by the
thought of his mind. [Xenophanes, quoted in [14]]

This greatest god was indifferent to the affairs of humans. Xenophanes asserts that we as humans are epistemologically autonomous and that we must rely on our own capacity for inquiry. We are limited in our certainty beyond our direct experience [14].

Here we find the beginnings of what we may even consider to be modern scientific thought.

Ephesus was in the same part of what is now Turkey as Colophon. Heraclitus of Ephesus is as difficult to comprehend as we may wish. Socrates is said to have remarked regarding Heraclitus, "What I understood is noble, and also, I think, what I did not understand." [[91], p. 11]. Plato accused him of being incoherent and Aristotle claimed that he denied the law of non-contradiction [14].

Heraclitus was interested in process and in *logos* rather than substance or matter. It is not clear exactly what Heraclitus meant when he spoke of logos. He seemed to mean a lawful, divine principle that governed the universe, the cosmos. He also claimed it was possible, but not easy, for humans to understand logos [14].

This seems to be the *Dharma* of the Indians to the East, and the *Wisdom* of the Hebrews to the South. In Proverbs 8 Wisdom says she was constantly with God in the beginning [[91], p. 10].

From Heraclitus, therefore, we have a cosmos which is changing, but in the change is lawful order. Logos provides a unity of purpose [[91], p. 12].

Samos is just off of the coast of modern Turkey where ancient Miletus, Colophon and Ephesus were located. Pythagoras was born there around 570 BCE. He was not an isolated mathematician. At one point he and about 300 followers settled in the Greek colony of Croton in southern Italy to escape the tyranny of Polycrates in Samos [[91], p. 24]. These Pythagoreans formed a commune, which was of a religious nature and accepted women as full members [[91], p. 12].

The Pythagoreans seem to have invented the concept of mathematical proof [[91], p. 13]. Their primary interest, however, was in numbers. The Milesians held that the primary things were material substances or abstract substances as the Boundless of Anaximander. But the Pythagoreans found that the principle of all things was in numbers [[69], p. 25]. The ratios of the musical harmonies provided one of their favorite illustrations. This they passed on to astronomical concepts when they spoke of the motion of the cosmos in terms of a music of the spheres [[69], p. 27].

The Pythagoreans also conducted experiments, such as their empirical investigations in acoustics. For example they observed the relationship between the amount of water in a jar and the note produced when the jar is struck as well as the relationship between string lengths and the notes they produced. Plato refers to some of these experiments in acoustics. Plato's evidence is all the more convincing because he strongly disapproved of these methods [[69], p. 31].

The Pythagoreans sought to give knowledge an actual mathematical basis. But this went beyond a mere description. The formal structure of observed phenomena was expressible in terms of numbers. For many Pythagoreans things, concrete material objects, were also numbers.

These ideas place the Pythagoreans far ahead in the use of mathematics and a mathematically based structure for physics, even though we cannot accept many of their concepts.

The Milesians dealt with the concept of change and Heraclitus had already raised the difficulties. Then Parmenides, from the Greek colony Elea on the West coast of Italy and South of Naples, denied that change can occur at all based on a skepticism regarding human perception.

What we have of Parmenides (ca. 515–ca. 460 BCE), besides Plato's remarks and acknowledgement, is a single philosophical poem. But the poem is so important to philosophical thought that it has been preserved by many [[69], p. 21]. He divided the poem into two parts: the *Way of Truth* and the *Way of Seeming*, which he claimed was deceitful. He deals with *being*, which is and cannot not be, and *not-being*. Nothing that is can come from not-being. Being may be existence itself, the totality of what is or what can be spoken about [[69], p. 38].

If we are to accept Parmenides we must deny validity to human perception based on human senses. This may not come as a difficult point to the modern physicist, who has become accustomed to the fact that we must rely on instruments that surpass our senses. But it was a difficult point to the Greek philosophers of nature in the late 5th century BCE. The problem was to find a way for change to exist while accepting Parmenides' dictum that what exists cannot arise from that which does not exist [[69], p. 39].

Empedocles (ca. 490–ca. 430 BCE) of Acragas, an ancient Greek city on Sicily, accepted Parmenides' admonition that we could not rely on the senses. But he did not accept Parmenides' insistence that reason alone was the way to truth. He conceded the fact that the senses are poor instruments for measurement. But so too is the human mind. Parmenides was correct in asserting that being cannot arise from not-being. But this does not mean that what exists is unique. The *roots* earth, air, fire and water have existed eternally, but change is produced by their mixing. This mixing occurs under the influence of the opposing forces of *Love* and *Strife* [[69], p. 39].

Here we have the origins of the concept we now know as *element* and even the concept of proportion. This was a truly inspired guess. He only applied this idea in a limited way. For example he claimed that bone was composed of fire, water and earth in the ratio 4:2:2 [[69], pp. 40–42].

Anaxagoras (ca. 510–ca. 428 BCE) was born in the Ionian region of the northern most Greek settlements in what is now western Turkey. But he lived most of his life in Athens. There he was a friend and teacher of Pericles, the Athenian statesman and general. And he was prosecuted on the charge of impiety, which seems to have been motivated by a political faction seeking to discredit Pericles [[69], p. 43].

Anaxagoras dealt with the same problem of change imposed by Parmenides and his solution reflected the ideas of Empedocles. His mixtures were, however, based not only on the four roots, but included mixtures of what had previously been considered qualities rather than things, "...no thing comes to be or passes away, but is mixed together and dissociated from the things that are." [14]. Qualities included hot and cold or wet and dry. No natural substance was more elemental or more fundamental than any other [[69], p. 44–45].

Anaxagoras also attempted to come to grips with what we may call dynamics. He attributed the motion of his ingredients to an external intelligent force. The same force that governed the rotation of the heavens governs as well the events on earth, including life and death. We are not able to determine the truth because of the limitations of our senses, but appearances are a view into the unseen [14].

These ideas did not later satisfy Plato and Aristotle because of their lack of a teleology [14]. Nevertheless, they hold a kernel of Newton's claim that what holds the moon in orbit causes the apple to fall.

1.2.3 Atomism

The idea of atomism was first suggested by Leucippus (5th century BCE) of Miletus and then developed by Democritus of Abdera, which is on the Eastern tip of the Greek peninsula and northern banks of the Aegean Sea. Democritus was born in about 460 BCE, which is shortly after Socrates was born in Athens. So Leucippus, Democritus, along with Diogenes of Apollonia (active after 440 BCE) end the Presocratics. We shall not indulge in the controversy over whether or not Leucippus was a real person.

The Greek atomic theory cannot be equated to modern atomic theory, nor is modern atomic theory a natural extension of Greek atomic theory. The basic postulate that only atoms and the void are real is a familiar concept in modern theory, with the caveat that a quantum vacuum is not necessarily a complete void.

But the Greek atom is strictly indivisible, which is what *Atomon* means. Aristotle illustrated the modes of differences in atoms in terms of shape of individual atoms, arrangement of atoms with one another, and positions of atoms. Greek atoms were in continuous motion and underwent collisions with one another sometimes hooking or becoming barbed together, or if their shapes correspond they may otherwise cohere. This coherence produces compound bodies that may have different effects on the senses. That is they may have different colors and tastes [[69], pp. 45–46]. We must be prepared to be very flexible if we wish to build the atomic theory of John Dalton upon this basis,

Democritus (460–370 BCE) was not a complete atomic realist, at least not in the modern sense. He claimed that "There are two forms of knowing, one genuine and the other bastard. To the bastard belong all these: sight, hearing, smell, taste, touch. The other, the genuine, has been separated from this" [14]. Democritus was also identified in antiquity as one who taught that life should be guided by a cheerful nature. His thoughts were not confined to what we may identify as the basis of physical science.

Diogenes (ca. 412–323 BCE) brings us back to the cosmic concepts of the Milesians. All things arise from one substance and are the same thing, and there is an intelligence that guides the cosmos. This single substance he identifies with air, as Anaximenes did, and he claims that "which possesses intelligence is what human beings call air." For Diogenes air is soul and intelligence [14].

1.2.4 Plato

Cicero (107–143 BCE) wrote that "Socrates called down philosophy from the skies." Socrates turned from an interest in physics to one in morals and ethics. Athens became the intellectual center of Greece. Plato and Aristotle founded schools in Athens, which attracted philosophers and scientists from all of Greece [[69], p. 66].

Plato was a student of Socrates and Aristotle was a student of Plato. As Socrates marks a change in the direction of Greek philosophy so Plato and the Aristotle begin a purpose for philosophy and a synthesis of developments before them. This was primarily because of the importance of education in Greece. Plato's primary goal was to educate people to govern Athens [[69], p. 67], [[91], p. 40].

Plato wrote in dialogue form, a skillful use of literary art. It was not required that the ideas be true. They are the thoughts and opinions of the speaker. Plato made no pretense of presenting physical truths. In the *Timaeus*, Plato's dialogue on the cosmos and physics, what Timaeus (the speaker) expresses is to be taken as a likely story. This is not the place to learn physics for the sake of the science.

In Plato's writing we can find, however, indications of the science to come. Plato has Timaeus describe how the Demiurge, the architect of the world, created the universe out of chaos. Timaeus says that the Demiurge, the god, took everything that was in chaotic motion and brought it into order, which the god judged to be in every way better [[128], p. 3, cf. Genesis 1:2]. Plato also claimed that "God ever geometrizes." Here there is at least an indication of the importance of mathematical order in the cosmos. Without that order there can be no science.

We find the reason for the study of science perhaps clearly expressed by Plato in the *Republic*. In Book VII of the *Republic* Plato has Socrates describing the training of the philosopher kings, who were to be the guardians of the ideal state. Socrates claims that astronomy should be one of the basic subjects to be studied. In the dialogue Glaucon misunderstands the reason for this. He assumes this is based on the practical requirements of agriculture and navigation. Socrates finds this amusing and wonders if Glaucon is afraid that people may consider that he is proposing useless subjects for study. The dialogue eventually comes to the real issue. The purpose of the guardian's education is to lead the guardian away from the visible world to the intelligible, to make their souls cultivate reason rather than sensation [[69], pp. 67–68].

For Plato reality was in the *Forms* or *Ideas*. The theory of Ideas is the central concept in the dialogues. Justice, Love, Beauty and Good are examples of these Forms. The Forms are the anchor in the life of a good person. They are the basis of the laws of the just republic. These Forms are all there is to know [[91], p. 39].

1.2.5 Aristotle

No thinker has had the effect on science as Aristotle (384−322 BCE). From the 4th century BCE until the 17th century CE Aristotle's ideas were dominant in western science. Aristotle's influence over such a vast period of time creates problems in attempting to distinguish his ideas from the ideas of others who contributed to Aristotelianism. But our interest here is limited to what we can understand of Aristotle's contributions to the science of mechanics. So we shall be selective in our treatment.

It seemed to Aristotle natural that humans would philosophize because the natural world presented wonders and puzzles. He did not consider it worthwhile to spend effort being skeptical about existence before setting to work. The observable world does not appear as systematically deceptive. We have sense organs, which provide us with data that must be considered reliable. We must not, however, consider that our observations are infallible. In this way Aristotle's approach resembles that of the modern day scientist [107].

Cause and effect were the basis of understanding. Aristotle begins with the phenomena as they appear. And to this he adds what others have considered regarding these phenomena. This is termed the *endoxic method*, which, in principle, resembles our citation of previous work on a subject. These endoxa, as Aristotle termed them, he calls reputable or credible opinions. He does not accept all endoxa without questioning their veracity, just as modern scientists may question the conclusions others draw from the data [107].

Aristotle based his arguments on syllogism, which consists of two premises and a conclusion. For example, all As are Bs and all Bs are Cs, *therefore* all As are Cs. There may be three primary premises: axioms, definitions and hypotheses. Axioms are unquestionably true, definitions provide the meanings of terms and hypotheses refer to the assumptions that certain properties result from the definitions [[69], p. 99].

For Aristotle knowledge existed when we know the cause of a particular attribute and that the attribute could not have been otherwise. There is then a universal connection between the subject and the attribute. Aristotle, however, did not consider that the work of the physicist was to be logical proof. The physicist should try to discover the causes themselves. These form the middle term in the syllogism [[69], p. 101]. For example 'all vines are broad-leaved trees' is the cause of the fact that 'all vines are deciduous.'

We cannot, however, draw an unbroken line from Aristotle to what we may call the science of mechanics. Although causality is central to Aristotle's thought, he considered there to be four causes, while the science of mechanics considers only the cause of change in motion of a material body. In the Aristotelian picture this is closest to the *efficient cause*, which initiates the change taking place. But the *final cause* is the reason the efficient cause exists. The final cause moves us closer to a divine actuality [[91], p. 46]. That is, Aristotle's causality contained a teleology.[3]

In addition to the efficient and final causes there were the *formal* and *material* causes. And Aristotle spoke of change in a larger sense than we may speak of a change in the motion of a material body. These engage the human in the process of change. For example if someone forms a bowl of silver the formal cause is the concept or design that the bowl will take. And the material cause is the silver itself [[91], p. 46].

The concept of motion was not foreign to Aristotle. The stars, the sun, and the moon move relative to anyone on the earth. Massive bodies move downward toward the earth and fire and air move upward. Rivers flow downhill and animals move of their own accord. These motions to Aristotle were natural. But there was also another motion. If a body is not moved by itself, in a natural fashion, but is moved by something else, the motion is violent [[91], p. 48].

Aristotle's concept of causality applied also to violent motion, which would have occurred in the making of the silver bowl. But the hopelessness of trying to reduce causality to the consideration of efficient cause alone is evident.

Aristotle claimed that knowledge comes only from reason. It is logic that creates scientific knowledge. While modern scientists may not agree entirely with this position, it is not the syllogistic logic, nor is it necessarily the concept of four causes that stand in the way of drawing a direct line from Aristotle to the science of mechanics. The difficulty rather seems to lie in Aristotle's axioms and his metaphysics. The logic must begin at some point and there must be some basis for choosing that point.

An example is the shape of the heaven, which Aristotle claimed must be spherical, since that is the shape most appropriate to its substance as well as the shape which is primary [[91], p. 51]. The natural motion of the heavens, as Aristotle knew, was circular, which must be natural for the stars, planets, sun and moon. The natural motion of terrestrial matter, *earth*, *air*, *fire* and *water*, is upwards or downwards. Therefore the heaven must contain a fifth element with a natural motion which is circular. This was the *aether*, which permeates the heaven. Beyond the fact that this fifth element fits well with Greek religious beliefs in the divinity of the heavens, Aristotle's primary reason for proposing the aether was to solve the serious physical problem of natural circular motion [[69], pp. 110, 111]. But Aristotle's axiomatic claim that the shape of the heaven is spherical leads logically to problems that must be solved by unforeseen devices.

Before Aristotle there existed no part of natural philosophy that could have been called dynamics. Aristotle at least began to consider motion of inanimate bodies. In addition to his concept of natural motion for terrestrial matter he considered the details of the motion of bodies. He suggested that the speed of a body in a medium will be inversely proportional to the density of the medium. This suggestion led him

[3]Teleology implies that an occurence may be explained by referring to some purpose or final goal. For example, rational human conduct is understood in terms of human desires and William Paley (1802) proposed that biological organisms were designed by an Intelligent Being for a specific function.

to deny the existence of the void, which was a part of the atomic picture. If the void existed the speed of a body would be infinite, which is absurd. He claimed as well that the speed of a body is directly proportional to the force applied to produce the motion and inversely proportional to the weight of the body, although he recognized that this rule did not always apply [[69], p. 113].

Here we have statements producing exact algebraic relations that we now know are not correct. In general terms, however, these rules do describe the behavior of moving bodies. We may then claim that the difficulty lay more in the fact that Aristotle did not pursue the logic far enough. But therein lies also the strength of Aristotle's science. It was logical and clearly accounted for the world as we all knew it. The fact that Aristotelian physics was flawed was revealed by a more mathematically based picture and careful measurement.

1.2.6 Reflection on Greek Science

In our brief outline of ancient Greek thought we can see that the philosophers, or scientists if we choose, encountered the same sorts of problems and proposed some of the same solutions that are familiar to scientists in the 21st century. And we have not yet solved some of the primary problems of perception and of measurement that they encountered. In their search for fundamental underlying principles we may also see our own search. We encounter, particularly in the quantum theory, the power of mathematics as, perhaps, even more than simply a descriptive medium. This the Pythagoreans understood rather well, even though we may deny the validity of some of their spiritual ideas. The study of Analytical Mechanics, which lies before us, will also bring us an appreciation for the Platonic idea of form, particularly as we encounter the variational calculus and eventually the methods of Hamilton and Jacobi.

1.3 Islamic Science

Islamic scientific thought was of great importance for the scientific revolution in western Europe. The history of Islamic science has also suffered unwarranted neglect making Muslim contributions unfamiliar to many modern students of physics. We have, therefore, elected to consider this part of our historical treatment in more detail than some more familiar contributions of western Europe to the development of mechanics.

We also primarily consider the Islamic contributions to astronomy, since this has the greatest bearing on the development of analytical mechanics.

1.3.1 The Rise of Islamic Science

As we pointed out the closing of the Academy in Athens (529 CE) resulted in many Greek philosophers finding refuge in Arab courts and that the revelation to Muhammad (610 CE) was almost coincidental with the closing of the Academy. Both of these events influenced the development science in the Arab world. A date for the origin of scientific thought in Muslim society cannot, however, be determined with any certainty. It is clear that these origins preceded the period during which the great Muslim scientist Abu Musa ibn Hayyan (Latin Geber) lived (ca. 721–ca. 815 CE). But we only have accurate records after the founding of Baghdad (726 CE) [[84], p. 12]. The serious Islamic encounter with Greek, Persian and Indian scientific thought seems to have been associated with the reforms initiated by Abd al-Malik ibn Marwan (ca. 647–705 CE), who was the Umayyad Caliph[4] from 685–705 [[6], vol 1, p. 20]. Al-Malik made Arabic the official language in the financial administration of government, which had previously been conducted in Greek and Persian [[6], vol 4, p. 646]. This reform had a domino-like effect in the civil service.

The mathematical procedures used in accounting were not common knowledge. A financial officer needed also to have knowledge of surveying to understand estate properties, which in turn required an understanding of geometry and trigonometry [[84], p. 10]. Complications were also imposed by the usage of both solar and lunar calendars, the understanding of which required some rudiments of astronomy [[108], p. 54].

Changing the official language of financial administration to Arabic meant that fluency in Greek and Persian no longer provided access to governmental positions. And the government was the greatest employer. The selection of persons to fill these positions now became based on understanding of mathematics, science and medicine. The result was what could be considered a healthy competition to be recognized for technical competence. George Saliba argues forcefully and coherently that the reforms of al-Malik were the seeds that resulted in the rise of Islamic science. Specifically these reforms required that the known sources of this technical knowledge, which were primarily Greek, but included Persian and Indian sources, had to be translated into Arabic. And provisions needed to be made for mathematical and scientific education in Arabic [[108], pp. 58–72].

1.3.2 Expansion and Contact

The 8th century also saw a geographical expansion of Islam, which afforded greater contact with other cultures and intellectual resources. That these could be understood

[4]A Caliph is the head of an Islamic form of government, who is considered a political and religious successor to the Prophet Muhammad.

The Sunni branch of Islam requires that the Caliph should be elected. The Shia branch requires that the Caliph should be an Imam chosen by God. [Wikipedia].

and accommodated implied that an Islamic scientific tradition was already present [[84], p. 10].

Included in this expansion was the invasion of the Hispanic peninsula by Arab and Berber forces from North Africa beginning in 711. The region which comprised parts of present day Spain, Portugal, Gibraltar, and France became *Al-Andalus*. We may, almost arbitrarily, claim that Muslim rule in Spain ended on January 2, 1492, when Abd Allah Muhammad XI surrendered the Emirate of Grenada to Queen Isabella I of Castile. Historically, however, the Islamic and Christian peoples had become irretrievably associated [[6], vol 20, pp. 1086, 1087, 1093], [[6], vol 3, p. 820]. So this date is primarily of textbook importance.

1.3.3 Islam and Greek Astronomy

The translation of the Greek texts was also not a benign procedure. The modernization of the mathematics resulted in new mathematical sources, such as the text on algebra by Muhammad b. Musa al-Khwarizmi, and a complete set of trigonometric functions. Critical translation of the Greek texts also revealed inconsistencies, which were then corrected in the Arabic translations. Particularly there were striking deviations between Ptolemy's geometrical model of the universe that appeared in the *Almagest* and his Aristotelian-based account of the celestial spheres that appeared in the *Planetary Hypotheses*.

While the *Almagest* produced results that were in agreement with the observed phenomena, these were obtained by geometrical constructs that violated the Aristotelian cosmological requirement that all heavenly motion must be spherical with the center of heaviness, which is the earth, located at the fixed center of the universe. According to Aristotle motion in the cosmos was the natural motion of the spheres carrying the ethereal bodies: the planets and the stars [[108], p.79, pp. 54, 67, 79, 88, 90–91], [[91], p. 48].

These inconsistencies demanded, at the very least, a search for a consistent picture. This search, which began in earnest in the 9th century and continued into the 16th century, accepted the basic tenets of Aristotelian physics. The metaphysics of Aristotle, however, provided difficulties for Islamic thought. For example Aristotle held that God was a primary cause, but eternally absorbed in self-contemplation, and the eternal and rational world did not involve God. These ideas were in direct conflict with the corresponding concepts in the *Qur'an*. The difficulty, then, was not only in the inconsistency between Ptolemy and Aristotle. There were other profound difficulties that Islamic scientists and philosophers found in Aristotle as well [[84] pp. 22, 32–33].

Islam simply cannot conceive of nature as something which is independent and self-supporting. The order science finds in nature is a result of the fact that God's laws are orderly and the study of nature, then, leads to an understanding of God and is a form of worship [[84], pp. 6, 8].

In practical terms the search for an Islamic astronomy was the search for a mathematically based structure describing the universe and based only on circular motion about a central point. This required new mathematical theorems beyond those of the Greeks and resulted finally in a complete overhaul of Ptolemaic astronomy [[84] p. 50]. Here we find, as well, the sources of some of the mathematical constructs required by Nicolaus Copernicus [see e.g. [108] pp. 217–219; [109] p. 236; [86], vol 3, pp. 1108–9].

1.3.4 Islamic Astronomy

Theological considerations were central in all of Islamic thought. Therefore the formulation of Islamic astronomy, which was based in part on Ptolemaic astronomy and in part on Aristotelian cosmology, created a tension. The history of Islamic astronomy and scientific thought in general is the story of how Islamic scientists dealt with this tension [[84], p. 23].

The earliest Islamic cosmology was based on the *Qur'an* and the words of the Prophet supplemented by scientific observations and models. This is known as the *Radiant* Cosmology Drop = and is spiritual and physical. It was a counterweight to Aristotelian cosmology, which allows us to see the influence of Islam on Muslim philosophical cosmology in the 8th century [[84], pp. 81–84].

Al-Biruni (Abu al-Rayhan Muhammad Ibn Ahmad al-Biruni) (973–1048), a man of powerful intellect, provided empirical evidence that the Qur'anic view of nature affected the way nature was perceived and studied by Muslim scientists. Al-Biruni rejected the Greek picture of an eternal universe and produced a cosmology which included creation out of nothing, the equivalent of the Christian *creatio ex nihilo*. And into this scheme he introduced the central doctrine of Hindu cosmology that the rate at which time passes is not uniform. As a consequence the laws of nature are variable. Al-Biruni also questioned some central claims of Aristotelian physics based his own observations, such as the increase of the volume of water upon freezing [[6], vol 3, p. 711], [[84], pp. 89–91].

There were philosophical conflicts in the 11th and 12th centuries. Specifically, Abu Hamid al-Ghazali (1058–1111), who was a theologian and philosopher and professor at Baghdad, attacked (Neoplatonic) Aristotelian philosophy in his book *The Incoherence of the Philosophers*.

Al-Ghazali had a physical and psychological crisis after which he spent ten years in seclusion cultivating mysticism. After mastering the logic of the neoplatonists he used this against them, supposedly in defense of Islam. He wrote in opposition to the neoplatonist philosophy of Ibn Sina (d. 1037) (Latin Avicenna) [[6], vol 10, p. 387]. The philosopher and theologian of Al Ansalus, Ibn Rushd (Latin Averroes) (1126–1189), responded to al-Ghazali [[84], pp. 86–88]. Al-Ghazali held that natural phenomena resulted from God's will, while Ibn Rushd claimed they were a consequence of the natural laws God had established. The difference is subtle, but important. The philosophical conflicts among Ibn Sina, Ibn Rushd and al-Ghazali were between Islam and Aristotelian philosophy and not between Islam and science.

Muslim scientists were distressed by the liberties Ptolemy had taken with Aristotle and were developing a new astronomy. But these liberties were of less concern to Muslim theologians, who advocated a more relaxed understanding of Aristotle, such as the approach Ptolemy had taken [[84], pp. 86–88], [[108], pp. 127, 234]. And the fact that Islamic science continued to flourish after al-Ghazali's death certainly indicates that his work did not sound the death knell for Islamic science.

The Illumination school (*Ishraqi*) was founded by Shihab al-Din Suhrawardi (d. 1187). This school gave a prominent role to intuition. The philosophy was based on an unveiling of truth [[84], p. 89].

The most advanced developments in astronomy were produced at Maragha in Western Iran during what is known as the Golden Age of Islamic astronomy. This extended from the middle of the 13th to the middle of the 14th centuries. The four astronomers whose names stand out in this period are Mu'ayyad al-Din al-Urdi (d. 1266), Nasir al-Din al-Tusi (d. 1274), Qutb al-Din al-Shirazi (d. 1311) and Ibn al-Shatir (d. 1375) [[84], p. 50]. These people comprised what has been called the Maragha School [101]. Serious attempts to correct the Ptolemaic modeling, including by the Maragha School, began in the 11th century [[108], p. 150].

The specific problem lay in the constructs that Ptolemy had chosen to describe the motion of specific cosmic bodies around the (stationary) earth. The sun could not move on a sphere around the earth and still produce the seasons. The solution could be to place the sun on an eccentric sphere (center displaced from the earth) or to allow it to move on an epicycle with center on a sphere concentric with the earth. The results were identical in either case, and Ptolemy resorted to the epicyclical resolution. This, of course, did violence to the requirements of Aristotelian astronomy, even though it preserved the phenomena observed [[108], pp. 136–139]. The motion Ptolemy proposed for the moon, to account for the observed sizes of the moon, involved multiple spheres. Some of these did not move uniformly as required by Aristotelian physics [[108], pp. 141–144]. And the epicycles and eccentric motion required to describe the observations of the (then) five known planets resulted in an impossible situation. One proposed motion destroyed the other. In realizing the magnitude of the difficulties, Ptolemy pointed out that his hypotheses should not be harshly judged because they were human constructs, and not the equivalent of the divine [[118], p. 600].

The device Ptolemy had employed was to place the epicycle on a sphere, which rolled on another moving sphere. Neither of these spheres was centered on the earth, and the rolling sphere did not move uniformly about its own center. The motion was, however, uniform about a point termed the *equant*. Owen Gingerich remarks that in this Ptolemy was "devilishly clever", although he did have to ignore Aristotle's restrictions on motion. The problem apparently was that Aristotle simply did not have detailed observations [[33], p. 36].

If we focus our attention on those astronomers responsible for the development of new mathematical theorems, which were important in the description of the motion of the cosmic bodies, al-Urdi and al-Tusi stand out. With a geometrical construct using uniformly moving spheres, al-Urdi was able to obtain the same motion of the planet. This is known as Urdi's Lemma, which was subsequently used by numerous astronomers, including al-Shirazi, al-Shatir, al-Qushji (d. 1474) and Copernicus (d. 1543) [[108], pp. 141–154].

The nonuniform oscillatory motion that was solved by Ptolemy using the rolling spheres described here was solved by al-Tusi using two uniformly moving rolling spheres. The construct is known as the Tusi couple [[108], pp. 155–159]. This was used by Copernicus for his own model of the orbit of Mercury and a double Tusi couple was used by al-Shirazi in his solution for Mercury's orbit [[113], pp. 488, 503], [[108], pp. 160–161].

1.3.5 Decline of Islamic Science

There was finally a decline in Islamic science and Muslim civilization. There is no complete agreement on the cause or causes of the decline of either the civilization or of the vigor of science within that civilization. They are certainly related. But no serious historian will pretend that there was a single cause. Attempts to determine a precise time at which the decline began can be countered by firm evidence of creativity in scientific contributions by Muslim scientists after that date.

There were military and political upheavals that cannot be neglected. The Crusades, which began in 1095 and continued for two hundred years, were possibly the most important of these events. Although the result was the military triumph of Islam in the Middle East, what emerged was a less tolerant Muslim culture. What had been an enlightened and urbane culture, superior to that of western Europe in intellectual depth and breadth, emerged as a culture that was becoming religiously intolerant, less secular and less intellectually vigorous [[6], vol 6, pp. 833–34].

The Crusades also had a profound effect on the public attitude in Europe toward Islam. For example Dante Alighieri (1265–1321) placed the Islamic philosophers and scientists Ibn Sina and Ibn Rushd in limbo in the first circle of Hell with Caesar, Aristotle, Plato and Cicero. And then he placed the Prophet among the "sowers of scandal and schism." [[84], p. 114].

Toward the end of the period of the Crusades, in 1258, Baghdad, which had been an international intellectual center for five centuries, was unconditionally surrendered to the Mongol warlord Helegu. This tragedy resulted in the indiscriminate killing of over 800,000 inhabitants, destruction of all major public buildings and the uprooting of intellectual life. But Islamic science found homes in other Muslim lands, such as present day Turkey, Syria, Egypt and Iran. And the Mongols remaining converted to Islam [[84], p. 131], [[6], vol 4, p. 653].

Vasco da Gama completed a voyage from Portugal to India in 1497–1498, thereby demonstrating the possibility of opening trade between western Europe and India by a sea route. The voyages of Christopher Columbus from Spain, beginning in 1492, also opened European eyes to the possibilities of the New World as a source of raw materials. The fact that da Gama encountered difficulties in attempts to establish contact points along the route to India because of the antagonism of Muslims toward Christians probably made a direct trade route to the East more attractive [[6], vol 6, p. 1111]. At any rate the discovery of new resources and trade routes, and the colonizations that resulted, represent major international changes that we may consider as the primary geopolitical sources affecting Muslim civilization and ultimately Islamic science [cf. [108], pp. 351–3], [[84], p. 128].

1.3.6 Reflection on Islamic Science

We have only treated Islamic astronomy and, therefore cannot draw any general conclusions but Islamic science. What we see, however, is an academic discipline searching for a path trough the thickets of Greek philosophy and a new theology. The stone on which the astronomers stumbled was Ptolemy's resolution to the difficulties Aristotle did not realize existed. The resolution the Muslim astronomers proposed was mathematical. And to carry this out they had a mathematics advanced far beyond the Euclidean geometry of the Greeks.

This is not different from the approach taken later in western Europe by Copernicus and by Kepler. Both the Muslim and the western European scientists labored under the ideas of Aristotle. Although Aristotle was a careful observer, his thoughts were rooted in Greek philosophy.

1.4 Europe Encounters Islamic Science

In the Introduction we mentioned that Al-Andalus was the primary gateway for the passage of Islamic scientific thought into Europe. This passage had a dramatic effect on the intellectual life of Europe that is difficult to exaggerate. We may identify this as the transmission or translation period, which extended from the 10th to the 17th centuries. During this period the corpus of Islamic scientific and philosophical thought was translated from Arabic into Latin making it available to the intellectuals of Europe.

The translation period began when Gerbert of Aurillac, who would become Pope Sylvester II in 999, encountered the mathematics of Islam and Arabic numerals in Catalan,[5] Spain, in 967. We know of Gerbert's interest in Arabic mathematics texts from letters requesting translations of them [[84], p. 104].

The second phase of the translation period extended from the 11th to the 14th centuries. Toledo was reconquered from the Muslims in 1085 making an excellent library of Islamic works available to a wider community. Toledo already had a large population of Arabic-speaking Christians (Mozarabs) and had been a center of learning [[17], pp. 421–462].

---

[5] After 732 Catalan was no longer part of Muslim Spain.

The library attracted scholars such as Gerard of Cremona from Italy, who came to Toledo in the 1140s in search of the *Almagest*. The abundance of manuscripts there enticed Gerard to stay in Toledo, learn Arabic and translate over eight volumes. And Gonzalo Garc ía Gudiel established his own scriptorium in Toledo around 1273. The *Alfonsine Tables*, crucial for European astronomy until the late 16th century, were drawn up in Toledo by order of Alfonso X, king of Castile and Léon [[84], p. 109].

Coincident with this second phase of translations was the founding of several of the great universities of Europe. The university at Bologna was founded in 1150, that at Paris in 1200, and at Oxford in 1220. Teachers were granted the right to teach at one particular university and then could move to any location on the continent. The language of instruction was Latin. There was a distinct parallel between this emerging European university system and that in the Muslim world [[84], p. 109].

The second phase of translations also coincided with the Crusades and another translation movement in Europe of Islam's message and the life of the Prophet. Although many involved in this lived close to Muslim communities and had access to the *Qur'an* and the Hadith, some fantastic and very negative portrayals were common. This combined with the Crusades produced an increasingly negative view of Islam in Europe [[84], pp. 112, 113].

The third phase of translation, which took place between the 16th and 17th centuries, differed from the first two phases in that it was increasingly becoming based in the universities of Europe. The translations became more refined and critical than those of the first two phases as the intellectual culture of the universities developed. This phase, which has been referred to as the Golden Age of Arabic Studies in Europe also affected that university culture. Studies in Arabic became an integral part of university curricula and any humanist of this period was expected to have learned Arabic. This interest in Islamic science and culture continued into the 17th century. For example, in 1619 the chairs established in geometry and astronomy at Oxford required knowledge of the Islamic scientific tradition [27], [[84], pp. 112, 115].

1.5 Medieval Physics

The second phase of translations coincided with the late medieval period in western Europe. This period saw a flowering of scientific creativity, which was certainly related to the translations of the Arabic texts. We shall consider three people who contributed significantly to this period. Each of these people worked in the shadow of Aristotle. But each also pushed Aristotle's ideas in the direction of more modern mathematics and, arguably, experiment.

The mathematical approach that we accept as logical in physics was not part of Aristotelian physics because it was insufficient. Truth in physical science was based on logical proof that things could not be otherwise and that logic was based on syllogism. Aristotle points out that one cannot establish, using Euclidean geometry, whether a straight line is more beautiful than a curve. And the fact that rotary motion is prior to rectilinear motion can not be proven mathematically. It is prior because it is more simple and complete [[91], p. 101]. Mathematics, for Aristotle, had nothing to say about the issues in physics of beauty, simplicity and causality.

Robert Grosseteste (1174–1253)[6] proposed that under some circumstances mathematics may provide a demonstration with as much authority and explanatory power as the syllogism. In agreement with Augustine's claim that God reasons and acts mathematically, Grosseteste claimed also that mathematical structure is causally responsible for motion. Mathematics in this case meant geometry. So this must be considered a particularly radical claim. Grosseteste would only have known of the physical laws of the reflection of light from a mirror, which was discovered by Hero of Alexandria, and Archimedes law of the lever [[128], p.5], [[91], pp. 101–102].

Roger Bacon (c. 1214–c. 1292) was either Grosseteste's greatest student, or a brilliant philosopher and teacher who was acquainted with Grosseteste [[91], p. 133]. Hard evidence for details is difficult to come by. And the claim that Bacon's ideas on scientific method and the use of mathematics and experiment in science placed him far ahead of his contemporaries has been modified by more recent scholarship, which indicates that he was more a product of his time. Jeremiah Hackett simply writes that each generation must find its own Bacon [38].

Bacon was primarily interested in education, as a member of a faculty dedicated to linguistic arts. He had a deep understanding of the ancient contributions to the science of optics, where mathematics (geometry) was most readily applied. But Bacon's efforts in this should not be thought of as an exercise in modern mathematical optics. This effort was directed toward a philosophy of perception and mind. This was inspired in part by Grosseteste's ideas and those of Al-Kindi on the use of radiation force. Al-Kindi proposed that a universal force radiates from everything and was responsible for the effects observed. For Bacon this formed the basis of a universal causality and the background for his philosophy of vision and perception. This was the first attempt in the Latin world to separate the material from the spiritual. No spiritual being resided in the medium. The universal causation was material, although Bacon believed in the freedom of human will [38].

These ideas brought Bacon into conflict with his superiors in the Franciscan Order, particularly Bonaventure.[7] The issue was primarily Bacon's interest in astrology and alchemy. An order from Bonaventure compelled Bacon to cease lecturing at Oxford and to place himself under the surveillance of the order at Paris [38].

Neither Grosseteste's nor Bacon's ideas on mathematics and experiments in science set us on a clear path to the principles of mechanics. Jean Buridan[8] was credited

[6]Robert Grosseteste (c. 1175–1253) was born in humble circumstances in England, but became a statesman, scholastic philosopher, theologian, scientist and Bishop of Lincoln. He is often identified as the first Chancellor of the University of Oxford as well, although that exact role is unclear. The Franciscan Roger Bacon was one of his disciples.

[7]Saint Bonaventure, O.F.M. (1221–1274) born John of Fidanza was an Italian theologian and philosopher declared a Doctor of the Church in 1588 by Pope Sixtus V.

by the pioneering historians of science Pierre Duhem and Anneliese Maier as having an important role in the demise of Aristotelian physics. But Buridan remained Aristotelian and tried to reshape the Aristotelian picture rather than overthrow it [130].

Buridan's principal contribution was his concept of impetus. Impetus was an impressed force carried by a body in motion. This was to replace the already discredited Aristotelian idea of antiperistasis, according to which a thrown body continued to move because of the force exerted by the air moving around to the back of the body. Buridan's impetus in the body was permanent unless destroyed by resistance. However, he also claimed that the impetus was a variable quantity whose force was determined by the speed of the object and the quantity of matter it possessed. This provided an understanding of the acceleration of a body in free fall. The body gradually gained units of impetus as it fell [130].

A modern physicist would understand impetus in terms of momentum, at least since the imparted impetus provided the ability for a thrown body to continue to move. But motion represented a particularly difficult problem for the medieval scientist.

1.6 European Scientific Revolution

We may, somewhat arbitrarily, pick the publication date of Copernicus' *On the Revolutions of the Heavenly Spheres* (1543) [see [111], Chap. 8] as a beginning date for the scientific revolution in Europe. Then, in the 17th century, the science of Christian Europe began to surpass that of Islam. Kepler published his first two laws of planetary motion in his *Astronomia nova* in 1609 and the third in *Harmonices mundi* in 1619 [119], [[91], pp. 150–153], [59]. With those observational laws the spheres of Aristotle were forever broken. Then in 1687 Isaac Newton published *Philosophiae naturalis principia mathematica* and the Aristotelian claim that there was a division of motion between the cosmos and the earth was no longer tenable [[13], p. 30]. The new ideas that would form classical physics were emerging from Europe.

Copernicus was Canon of the Cathedral in Frombork, Poland. This was a responsible Church position, although he had time to pursue his work in astronomy. He knew his ideas would find resistance in the Church. And he was vulnerable. He lived with a mistress. So publication had to coaxed from him [111].

Kepler had trained for the ministry. But the seminary faculty gently moved him out when the Protestant university of Graz in Austria came looking for a professor of mathematics and astronomy. Kepler had minimal studies in both, but he was going to make a terrible minister. Kepler's sense of God's direction never completely left him as he moved from his belief that the key of the universe had been revealed to him to his discovery of the orbit of Mars based on hard data taken by Tycho Brahe. Kepler had moved astronomy from philosophy to astrophysics [59].

[8] Jean Buridan (c. 1300 – after 1358) was a French priest who proposed the concept of impetus as a force present in a freely moving body. This is considered to be the first step toward the modern concept of inertia.

1.7 Newton

I offer this work as the mathematical principles of philosophy, for the whole burden of philosophy seems to consist in this — from the phenomena of motions to investigate the forces of nature, and then from these forces to demonstrate the other phenomena.

Isaac Newton

Preface to the *Principia*, first edition

1.7.1 Introduction

We must recognize Isaac Newton as a major figure in the development of the science of mechanics. He was also an unusual person. The effort expended on trying to discern which parts of his life are definitely important and which are not would be far greater than simply providing a somewhat detailed outline of his life. We, therefore, will devote more space to his life and work than we have for others.

The understanding of motion and the concepts of mass and force that Newton encountered were also not at all the fairly simple concepts familiar to us today. To work his way through these and to arrive at the concepts he gave us was no mean task. Even the terms he gave us were not yet the terms we presently use. His concepts of space and time, although he stated them authoritatively and with conviction, were not those we presently have. Newton will always remain a major figure in the history of science. But we should not assume that history is a set of completed blocks neatly fit together. We are still without complete understanding of mass and force.

We may be thankful for the difficult and forceful personality of Newton. But we should try, in some way, to see him as he was.

1.7.2 The Person

Newton came from a complex and less than ideal background. Richard Westfall, a biographer of Newton, wrote of him "… Newton was a tortured man, an extremely neurotic personality who teetered always, at least through middle age, on the verge of breakdown" [[121], p. 53].

Newton's father Isaac Sr. was an illiterate farmer in Licolnshire, England, who married Hannah Ayscough (Askew) in April of 1642 and died in October 1642. Newton was born on Christmas day, 1642. According to Newton's own account he was very small and weak at birth and not expected to live [[121], pp. 44–45]. But Michael White, another biographer of Newton, cautions his reader regarding Newton's attempts to make his own life appear miraculous [[124], pp. 11–12]. At least he survived and spent his first three years with his mother at Woolsthorpe Manor, where he was born. Then his mother married Barnabas Smith, the aging rector of North Witham, leaving Isaac with her parents [[121], p. 49], [[124], p. 14].

Smith was of flexible religious convictions during the Second English Civil War of 1647–49. His actions meant he saved his economic position, but provoked some lines by John Milton [[121], p. 52], [[124], p.14]. And during his seven years of marriage to Hannah, Smith left young Isaac with the grandparents rather than moving him the mile and a half to the rectory. This was devastating to Newton, who had become very close to his mother. Hannah occasionally reappeared at her parents' home of an afternoon, but then was gone. The experience probably did much to mold Newton's personality [[124], pp. 15–17].

After Smith died in 1653 Hannah returned with her three Smith children to Woolsthorpe and two years later, probably at the insistence of the Ayscoughs, Newton went to grammar school at The King's School in Grantham seven miles away, where he boarded with an apothecary he remembered only as Mr. Clark [[13], p. 19]. The King's School was a grammar school, which meant instruction in Latin and Greek, but no mathematics or science. This, however, stood Newton very well. He could read and write Latin as well as he could English giving him ready access to mathematics and science texts [[121], p. 58]. And Mr. Clark's shop provided him an appreciation for the wonders of chemistry [[124], p. 23].

After the grammar school, Newton returned to Woolsthorpe. Hannah thought he should become a farmer, for which grammar school was sufficient. However Hannah's brother, the Reverend William Ayscough, together with the schoolmaster of Grantham, Mr. Stokes, recommended that Newton return to The King's School to prepare for university. The nine months at Woolsthorpe had been a nightmare and Stokes was willing to waive the tuition of forty shillings. So in 1660 Newton returned to Grantham to prepare for Cambridge [[121], p. 64], [[13], p. 19].

Newton entered Trinity College, Cambridge, in June of 1661 as a subsizar, which meant he received tuition and fees for menial service. His introversion and intellectual independence determined his course at Cambridge. The fact that Newton was a Puritan and the college was Anglican isolated him inwardly [[124], p. 49]. After a few weeks he began to approach studies at Cambridge much as he had at Grantham: on his own. He studied philosophy and first encountered mathematics. Newton also, independently, undertook experiments to test the published ideas of others, such as Descartes' ideas on optics. He bought a prism at the fair in 1663 and conducted experiments on the separation of light. And he almost blinded himself by looking at the sun through a lens [[124], pp. 58–61].

In the spring of 1665 Newton obtained a Bachelor of Arts (BA), second class. His studies outside of the curriculum resulted in the second class ranking. Then in July of 1665 the bubonic plague caused the closing of the university for about two years. Newton returned to Woolsthorpe [[124], p. 64, 85].

In Woolsthorpe Newton had what are called his *miracle years* of 1665–1666. Out of these years supposedly came the calculus, or method of fluxions, and the great synthesis of mechanics that was at last published in the *Principia* of 1687. But this is a great simplification. It is more appropriate to consider that the miracle years were the two decades from 1665–1687, and encompassed, as well, Newton's passionate pursuit of alchemy [[124], p. 86]. Newton had a fear that his ideas would be stolen and a profound belief that material should only be released after it had been thoroughly developed. White proposes these as reasons for Newton's withholding publication of his ideas for so long [[124], p.100].

Newton began alchemical studies around 1667. His mentor in this was Robert Boyle. Newton was aware of the need for secrecy, which was a common part of alchemical studies at that time. The practitioners of alchemy hid behind pseudonyms. Newton's was *Jeova Sanctus Unus* (One Holy God). We do not have a thread to unravel in his alchemical work. He kept meticulous notes. But he did not date them, as he did his notes on physical science. We can deduce periods of time only by forms of the handwriting [[124], pp.134–135, 140].

Our question as scientists and engineers of the 21st century may well be why this supposed high priest of the Enlightenment would turn to alchemy. The reason seems almost clear. Newton was aware of the primary difficulty facing mechanical and atomic theory. No deterministic mechanical theory can account for the human being as a person with free will [see [91], p. 197]. Newton's claims for alchemy exceeded those of base metals. He wrote

> For Alchemy does not trade with metals as ignorant vulgars think, which error has made them distress that noble science; but she has also material veins of whose nature God created handmaidens to conceive & bring forth its creatures …

Newton did not take religion lightly. He was a devout Puritan, but also an Arian. That is he did not accept the idea of the Trinity, considering it blasphemous. He had, however, attested to the Thirty-Nine Articles of the Anglican Church when he took his BA in 1665 and his Master of Arts (MA) in 1667. But he faced a major difficulty in accepting the Lucasian Professorship. He would be required to take holy orders and this he could not do. Remarkably Isaac Barrow, who had given up the Luasian Professorship and recommended Newton, suggested that Newton apply directly to Charles II for a special dispensation that would allow him to continue to hold the professorship without the requirement of holy orders. Barrow was then the king's chaplain and had the king's ear. The king granted the dispensation in perpetuity for the Lucasian Professorship [[124], pp. 150–151].

Newton's Arianism[9] led him to the belief that true religion preceded the ideas of the Greeks and the Romans. God was part of all nature and all knowledge and, therefore, also history. Newton conceived of this as an *alchemical history*, the understanding of which would reveal the purpose of existence, the wisdom of the ancients, and could lead to a decoding of biblical prophesy [[124], p. 154].

Newton's enthusiasm for his research was matched only by his lack of enthusiasm as a teacher. A very tiny group of students came to his first lecture and none to any after that. But he was obliged as Lucasian Professor to conduct lectures, which he had reduced to one term per year. He lectured dutifully to an empty room. He also fairly well ignored the requirement to provide written notes of ten lectures per year to the university library. Finally he gave up the pretense of lecturing completely and devoted himself entirely to his research [[124], Chap. 8].

---

[9] Arianism is a heresy that denies the divinity of Jesus. This originated with Alexandrian priest Arius (ca. 250– ca. 336).

1.7.3 Disputes

Newton became embroiled in disputes that are important to us because they reveal methods of scientific and philosophical thought that Newton was developing. They may also be interesting because they reveal something of a clash of personalities not entirely unrelated to science. Newton's principal antagonist was Robert Hooke. Hooke was once Boyle's assistant. In 1662 Boyle secured a position for Hooke as curator of experiments for the emerging Royal Society [[124], p. 175].

In this role Hooke reviewed Newton's paper of 1672 "Theory of Light and Colours," which appeared in the *Philosophical Transactions* for 19 February, 1672. Hooke's critique precipitated an acrid exchange, which exposed a difference in thinking between the two men. Newton was developing an experimental and mathematical philosophy, while Hooke's approach was quite different. Hooke referred to the results of Newton's experiments, which had demonstrated dispersion and reconstitution of white light, as a hypothesis. Newton saw this as a theory,[10] which had resulted from experimental evidence [[124], pp. 176–188].

On 7 December, 1675, Newton sent two papers to Henry (Heinrich) Oldenburg, who was first Secretary of the Royal Society. These were 'An Hypothesis Explaining the Properties of Light' and 'Discourse of Observations'. Here Newton was writing as a natural philosopher, not as a hard scientist. He had been working on the natural order of things for ten years [[124], pp. 102–103], [[124], pp. 184–185]. Both of these papers later appeared as parts of Newton's *Opticks* (1704). But Newton refused permission to publish them in 1675 [[122], pp. 102, 253].

In the course of this discussion Newton had written to Hooke

> What Descartes did was a good step. You have added much several ways, & especially in taking on the colours of thin plates into philosophical consideration. If I have seen further it is by standing on ye shoulders of giants.

That last sentence has been often quoted, missing the point completely. According to White, Newton was being truly spiteful and vicious. Hooke was so physically deformed that he had he appearance of a dwarf. The last sentence is then a double-edged sword. It was neither complementary nor indication of Newton's gratitude to his forebears [[124], p. 187].

The transformation from hypothesis to demonstrable theory was carried out on 27 April, 1676, at the Royal Society. Then the experiments were meticulously performed to validate Newton's 'Theory of Light and Colours' [[124], p. 188].

Another exchange with Hooke regarding the path of an object dropped from a tower, beginning in 1679, was also unpleasant, but brought some clarification in the greatest puzzle to Newton: universal gravitation. Hooke had wasted no time on mathematical analysis and based his ideas on supposition and reading Galileo.

---

[10] In the sciences hypothesis is a proposal that may be falsified as a result of a subsequent experiment. A theory is a substantial statement based on a set of experiments and composed in precise (often mathematical) terms. A theory may be falsified based on experiments, but not as easily as a hypothesis.

But Newton also made some woeful mistakes. For example he had first claimed that the path followed by the dropped object was a spiral. And Hooke made their communication public before the Royal Society, making Newton's errors evident [[122], pp. 148–152], [[124], pp. 195–199].

An object dropped from a tower will have an angular velocity around the earth's axis of rotation and is attracted to the center of the earth. Some thirteen years earlier Newton had obtained the mathematical description of circular motion in an inverse square-law force. But Kepler had shown the planetary orbits were elliptical. And Hooke had surmised this to be the path followed by the dropped object. Therefore Newton applied his calculus to show that the orbit around an inverse square force is an ellipse with the force center at the focus. Newton was silent about this work. He never forgave Hooke, but in later life Newton did acknowledge that Hooke's correcting the spiral turned him to investigate the elliptical orbits [[124], pp. 195–201].

Then, fortuitously, early in November of 1680 a very bright comet appeared in the skies over Europe and vanished into the sun at the end of the month. Two weeks later another immense comet appeared moving away from the sun. The Royal astronomer, John Flamsteed, believed these were one comet that had simply reversed its direction as it neared the sun. Comets were considered to be foreign bodies not related to the solar system and not governed by the laws of the solar system. The comet would not then be attracted to the sun. So Newton did not subject the cometary orbit to the analysis he had used on planets. Nor did he accept Flamsteed's theory [[122], pp. 155–156], [[124], pp. 202–204].

Then in early autumn of 1682 yet another even brighter comet appeared. This comet was moving away from the sun.

Newton had been thinking about comets since the winter of 1681. In his book Observations on the Comet (1681) the Italian astronomer Giovanni Cassini proposed that the comet of 1680 – 81 had been the same as observed by Tycho Brahe in 1577. This influenced Newton's thoughts. But probably more influential was Hooke's book *Cometa* of 1678. There Hooke dealt with the great comet of 1577 and included a statement of the Law of Inverse Squares as well as the effect of the sun on comet tails [50]. Newton tried a number of trajectories deciding finally on an elliptical trajectory with the sun as a focus and the inverse square gravitational attractive force as also acting on the comet [[124], pp. 203–205].

But there remained a difficulty. What was the agent of the gravitational force? Even though Newton had rejected much of Descartes' mechanical concept of nature, Descartes' mechanical philosophy was still the prevalent basis of thought in the 1680s. Newton had been thinking in terms of 'action at a distance', which required the *aether* as a medium for transmission. Then in the 1680s Newton began to think less in terms of aether and more in terms of the alchemical concept of 'active principle'. This was a radical change in thought. Newton's mentor at Cambridge, Henry More, had written of a 'Spirit of Nature', which acted in some way upon matter and may create and sustain life. White believed this was the concept to which Newton was turning [[124], pp. 206–207]. Max Jammer indicates the possible influence of the 17th century mystic Jacob Böhme on Newton's thought, which would trace Newton's concept of gravitational force back to Neoplatonic and Gnostic traditions. But he also says that there is no convincing evidence that Newton read a single work by Böhme [[55], pp. 134, 143].

1.7.4 Principia

The writing of the *Philosophiae Naturalis Principia Mathematica* grew out of a coffee house discussion in January of 1684 among Hooke, who by now would rate as an enemy of Newton's, Sir Christopher Wren, and Edmond Halley. Halley had asked if the action that keeps the planets in motion around the sun could decrease with the square of the distance. Hooke boasted that he had proven this some years before, but had told no one. Wren was skeptical and said he would give Hooke or Halley two month's time to produce the proof. Halley admitted that he had been unable to prove this and the proof was not forthcoming from Hooke in the time allotted. Then in August Halley decided to approach Newton in Cambridge [[55], p. 11], [[124], pp. 190–192], [[122], pp. 159].

Halley showed up unannounced at Newton's door. According to Newton's recollection it was a rather long time into the conversation when Halley asked what he thought the orbit of a planet would be supposing the force of attraction towards the sun to be the reciprocal to the square of the distance from the sun. Newton replied immediately that it would be an ellipse. Halley was amazed and asked how Newton knew this. He had calculated it, Newton said. Halley then asked if he could see the calculation at which point Newton looked through his papers but was unable to locate it. He then said he would redo the calculation and send it [[122], pp. 160] [[124], Chap. 9].

It seems that the search through the papers was something of a stalling tactic. Newton had produced this proof as a result of his discussions with Hooke. And it exists among Newton's papers. But he had become very cautious about releasing his work. Not giving the paper to Halley had also been wise. He found that he could not so easily reproduce the proof and then discovered an error in the previous calculation.

In November Halley received a nine page document entitled *De Motu Corporum in Gyrum* (On the Motion of Revolving Bodies), hand delivered by the mathematician Edward Paget. The paper contained much more than was promised. Newton demonstrated that an ellipse requires an inverse square force centered on one focus. He also showed that an inverse square force results in an orbit which is a conic section, and an ellipse for velocities below a certain limit. Then beginning from general postulates he derived Kepler's second and third laws and obtained the motion of a projectile in a resistive medium.

Halley recognized that this was nothing short of a revolution and was back in Cambridge without delay to confer with Newton. This began what would completely absorb Newton from August of 1684 until Spring of 1686. Westfall says that Newton had been grasped and that Newton was powerless in the grip of this undertaking. Halley had no need to coax anything from Newton [[122], pp. 160–162].

When Newton began this work that would produce the *Philosophiae Naturalis Principia Mathematica*, which is commonly known simply as the *Principia*, facing him were widely accepted concepts regarding matter and motion. Most of these were metaphysical. And Newton would finally move in a different direction, that of an *experimental and mathematical philosophy*. We can distill these down to two primary concepts, even while we confess that this is a simplification. These are the understanding of mass and of force.

1.7.5 Newtonian Concepts

Newton's concepts of mass, force, space and time as well as his belief in "Rules of Reasoning in Philosophy" are important in any of our attempts to understand what Newton actually did. These are seldom considered in studies at the beginning level in university, and sometimes completely neglected in advanced courses. Newton's laws are often simply served up to students as accomplished facts into which the student is asked to insert the modern understanding of all of these concepts. In the spirit of honesty we provide brief discussions of these concepts and Newton's development of them.

Mass. Kepler's work in astronomy was critical to Newton's understanding of the forces responsible for planetary motion. But behind this lay a fundamental concept that was also Keplerian and was instrumental in bringing Newton to an understanding of mass. When Kepler discovered that planetary motion was elliptical the previous belief that planetary motion was perpetual, because it was circular and hence natural, was no longer tenable. Kepler considered forces to result from motory intelligences or pure Form in a Neoplatonic sense. And his concept of mass was that of pure matter, which was contrary then to force. Neoplatonic tradition held that matter was an impediment to the realization of Form. Matter causes a body to remain in place [[56], p. 53].

With the concept that matter resists motion Kepler was moving from metaphysical speculation to physical reasoning. If celestial bodies had no inertia then no force would be necessary for their motion. The smallest force would cause them to move at infinite velocities. But the periods of motion of celestial bodies are different, which Kepler claimed is clear indication that they have inertia. (55)

Inertia accounts for the inability of matter to transport itself. But inertia is also the resistance of matter to motion by the action of a force. This resistance to motion increases with the *quantity of matter* present in the volume of the body. (56)

Max Jammer considers that with these ideas Kepler *conceptualized mass*, which is a first step in the development of the concept of mass. (59)

Descartes rejected this concept of inertia as resisting motion. He did, however, accept the that inertial mass affected momentum. For Descartes the quantity of matter was volume. This concept of matter as extension made Descartes' ideas contrary to the Catholic claim of Transubstantiation of the bread and wine in the Eucharist and led to the condemnation of his works in November of 1663. (61)

In his treatise *On centrifugal force*, Christiaan Huygens had demonstrated that for bodies rotating in equal circles at equal velocities the ratio of the centrifugal forces are equal to the ratio of the *solid quantities* of the bodies. In modern terms solid quantity is inertial mass and Huygens' result is easily derived. (63)

Force. Newton's concept of force was not simple. His interest in force was primarily in relation to gravitation where his work was related to the motion of celestial bodies. *De Motu* (*De Motu Corporum in Gyrum*) was a treatment of the motion of celestial bodies [[122], p. 116].

According to Westfall the crux of Newton's dynamics was in the relation between *inherent* force and *impressed* force. Inherent force was "the inherent, innate, and essential force of a body." And the impressed force was "the force brought to bear or impressed upon a body." Newton was trying to clarify the concepts of dynamics and to understand the meaning of force [[55], p. 166]. At this time in history force was not understood as it was even in the 18th century. Force was a cumulative property of a moving body [[55], p. 131].

Along with Newton's struggle came the meaning of space and absolute motion. Descartes had claimed the velocity was relative. Descartes' moving body had no definite velocity or definite line of motion, which Newton thought was absurd. Newton was beginning to think of inherent force as the distinguishing characteristic of motion. This was leading him toward a concept of absolute space. These ideas he was developing in revisions of *De Motu* [[122], p. 166].

Newton eventually removed inherent force from his First Law, where it had been the 'power by which it (a body) preserves its state of resting or of moving in a right line' and introduced force of inertia (*vis inertiae*). Inherent force was no longer the cause of motion. And impressed force became action only, no longer remaining in the body when the action has ended. And he had adopted the Keplerian concept of inertia [[122], p. 167].

A consistent statement of the Second Law and a definition of quantity of matter (mass) followed. The impressed force did not result simply in a change in motion, but a change in the *quantity of motion* we now call *momentum*. This quantity of motion is proportional to the quantity of matter, which is proportional to its weight. This defined the mass of a body as a quantity calculable from density and the space occupied by the body, i.e. the volume. 'A body twice as dense in double the space is quadruple in quantity (mass)'. Without this definition of mass the Second Law could not have been consistently formulated [[122], p. 168].

Previously Newton had linked inherent and impressed force by a parallelogram law referring only to a single body [[122], p. 130]. This was no longer tenable. The final statement must involve only the impressed force. And the link must combine the body considered to be acted upon by the force and the body considered to be the cause of the force. This is Newton's Third Law. It is not a trivial addition, but is the line of genius connecting previously obscure concepts [[122], p. 168].

Newton credited his First Law

> Law I: Every body continues in its state of rest, or of uniform motion in a right line, unless it is compelled to change that state by force impressed upon it.

and his Second Law

> Law II: The rate of change of momentum is proportional to the motive force impressed; and is made in the direction of the right line in which that force is impressed.

to Galileo and Huygens [[55], p. 123]. These two laws may be interpreted as a qualitative and a quantitative definition of force. But for Newton the Second Law is not a definition, nor is it simply a method for measuring force. Newton considered force to be an á priori, although, as we have just seen, not a simple concept. The Second Law summarizes a property of force. (124) Jammer contends that the Second Law is a free creation of the human mind. (127)

Ernst Mach claims that the Third Law is Newton's "most important achievement with respect to the principles" of mechanics [[71], p. 243]. This was, however, a principle understood by Kepler, although he never formulated this as such [[55], p. 127].

> Law III: To every action there is always opposed an equal reaction; or, the mutual actions of two bodies upon each other are always equal and directed to contrary parts.

The first two laws deal with the response of a single body to a force and the third law defines the force of interaction between bodies. These form the basis of the new approach to mechanics that Newton introduced.

Space and Time. Newton had considered both space and time and defined his understanding of each rather clearly. According to Newton [[91], pp. 228, 231].

> Absolute space, in its own nature, without relation to anything external, remains always similar and immovable. Relative space is some movable dimension or measure of absolute spaces…
>
> and, similarly
>
> Absolute true, and mathematical time, of itself, and from its own nature, flows equably without relation to anything external, and by another name is called duration: relative, apparent, and common time, is some sensible and external … measure of duration by the means of motion …

Newton also assumed space to be Euclidean, i.e. completely describable in terms of the geometry of Euclid. For Newton's laws of mechanics to be universally true it was, or at least it seemed to be, necessary that space was absolute and immovable. Newton also chose the methods of Euclidean geometry for the mathematical proofs in the *Principia*. This is curious, makes reading tedious, and proved, according to David Park, disastrous to English physics in the post Newtonian period. The *Principia*, since the publication of the first edition in July of 1687, has remained the most important document in the history of science. But when the second edition was published in

1713 the methods of Euclidian geometry were already obsolete, being supplanted by the development of the calculus on the European continent [[91], p. 210].

Newton's introduction of absolute space was unfortunate. As we pointed out, this seemed necessary for the First Law. But the fact that uniform motion cannot be detected was known to Nicole Oresme (ca. 1320–1382 and Nicolaus Cusanus (1401–1464) and had been used by Copernicus, although Copernicus cited Virgil [[91], pp. 92, 93, 145]. Cusanus had also argued that an infinite universe can have no center. Newton's concept of absolute space was then flawed from the beginning.

The Newtonian concept of absolute time was finally dismantled in 1905 by Albert Einstein [[24], pp. 37–65]. And space and time will become a continuum at the hands of Hermann Minkowski (75–91). For most of what we will do in our study of classical mechanics, however, we may simply ignore Newton's belief regarding time.

Philosophy. In Book III of the *Principia*, subtitled *De mundi systemate* (On the system of the world), in the second (1713) and in the third editions (1726) Newton included a section entitled "Rules of Reasoning in Philosophy." In the second edition these are [[91], pp. 214–215]

> Rule I: We are to admit no more causes of natural things than such as are both true and sufficient to explain their appearances.
> Rule II: Therefore to the same natural effects we must, as far as possible, assign the same causes.
> Rule III: The qualities of bodies, which admit neither intensification nor remission of degrees, and which are found to belong to all bodies within the reach of our experiments, are to be esteemed the universal qualities of all bodies whatsoever.
> Rule IV: In experimental philosophy we are to look upon propositions inferred by general induction from phenomena as accurately or very nearly true, notwithstanding any contrary hypothesis that may be imagined till such time as other phenomena occur, by which they may either be made more accurate, or liable to exceptions.

These are not fool proof rules that will guarantee success in Philosophical thought. Newton was a strikingly original thinker who certainly would not have followed a set of rules as he developed his ideas. We saw this above as we tried to follow his deciphering of the inherent and impressed forces. Rather we can consider these as reflections on his thought process. And we can see places in which these were exhibited.

Rule I indicates a faith in the simplicity of nature. And Rule II is almost a rephrasing of Rule I. We should curb our own natural tendency to generate explanations and hypotheses. Appearances indicates observation. We should not pretend to know more than observation yields. Here Newton is distinguishing his method from that of people who invent new explanations for whatever appears to be a new effect. Rule III pertains to universal gravitation. Newton realized that the cometary orbit was elliptical with the sun at the focus, and that, therefore, the gravitational attraction to the sun included comets as well as planets. Rule III extends that to all bodies. Rule IV is like the banner of experimental physics. The laws are induced from and describe the observed phenomena. Hypotheses that may be formed have no bearing on the laws. The laws are only made more accurate by other observations.

1.8 Eighteenth Century

Newton's geometrical methods were used until the 1740s, at which time the methods of Pierre Louis Maupertuis, Johann (Jean, John) Bernoulli,[11] Euler and Lagrange began to be employed. These new methods used the analytical language of the differential calculus and introduced variational concepts.

1.8.1 Maupertuis

In 1740 Maupertuis proposed a new principle which he called the "law of rest". This was to provide the conditions for equilibrium of a collection of n bodies with masses $M_1, M_2, \ldots, M_n$ under the action of a number of central forces, i.e. forces magnitude varies as a power of the distance from the center of the force. He considered that each of the forces acting on a particular body had a separate center and chose the forces to be all of the same character such that if the ith body is located at distances $p_i, q_i, \ldots, w_i$ from the respective centers of the forces then the magnitude of each central force acting on the ith particle is proportional to $p_i^m, q_i^m, \ldots, w_i^m$ where m is integer. He designated the components of the forces (per unit mass) acting on the ith particle along the directions of $p_i, q_i, \ldots, w_i$ as $P_i, Q_i, \ldots, W_i$, and the constants of proportionality for the forces as $\mathcal{P}_i, \mathcal{Q}_i, \ldots \mathcal{W}_i$. Then the forces (per unit mass) acting on the ith particle are related to the distances from the force centers as

$$\begin{aligned} P_i &= \mathcal{P}_i p_i^m \\ Q_i &= \mathcal{Q}_i q_i^m \\ &\vdots \\ W_i &= \mathcal{W}_i w_i^m. \end{aligned} \tag{1.1}$$

Maupertuis' law of rest claimed that the quantity

$$\sum_{i=1}^{n} M_i \left(\mathcal{P}_i p_i^{m+1} + \mathcal{Q}_i q_i^{m+1} + \cdots + \mathcal{W}_i w_i^{m+1}\right) \tag{1.2}$$

is an *extremum*. That is

$$\sum_{i=1}^{n} M_i \left(P_i \mathrm{d}p_i + Q_i \mathrm{d}q_i + \cdots + W_i \mathrm{d}w_i\right) = 0. \tag{1.3}$$

---

[11] Johann was the original name given by his parents. Jean or John appear sometimes, depending upon whether the author is French or English. Johann Bernoulli was born and died in Basel, Switzerland.

We recognize each term $M_i\left(\mathcal{P}_i p_i^{m+1} + \cdots\right)$ in (1.2) as the (negative) potential energy of the mass M_i.Therefore, (1.2) is the total (negative) *potential energy* of the system of particles. Maupertuis' claim is then that the law of rest requires that the total potential energy is an extremum for the system of masses.

From (1.1) in (1.2) we see that Maupertuis' law of rest requires that

$$\sum_{i=1}^{n} M_i\left(\int P_i \mathrm{d}p_i + \int Q_i\, \mathrm{d}q_i + \cdots + \int W_i \mathrm{d}w_i\right) = \min / \max \qquad (1.4)$$

for a system in equilibrium. For the central forces specified in (1.1) this is a general formulation of the law of rest.

Maupertuis does not mention, and he may not have been aware, that Johann Bernoulli had already obtained the formula (1.3). Bernoulli called (1.3) the *principle of virtual velocities* and termed the variations $\mathrm{d}p_i, \mathrm{d}q_i, \ldots, \mathrm{d}w_i$ the *virtual velocities.*

Bernoulli's point was that if the system is at rest there can only be fictitious or *virtual* motion of the masses consistent with the applied forces. This virtual motion is in the differential displacements of the masses $\mathrm{d}p_i, \mathrm{d}q_i, \ldots, \mathrm{d}w_i$. If we think of this virtual motion as taking place in the differential time interval $\mathrm{d}t$ then the virtual velocities associated with this virtual motion are $\mathrm{d}p_i/\mathrm{d}t, \cdots, \mathrm{d}w_i/\mathrm{d}t$. Bernoulli's virtual velocities differ from these by only an arbitrary constant factor $\mathrm{d}t$.

In formulating his law of rest in terms of an extremum, Maupertuis introduced something new into the science of mechanics. This new principle proved to be very fruitful and remains the basis for much of our modern understanding of physics. However, Maupertuis was more interested in connecting the equilibrium of systems with the concept that nature acts in as economical a fashion as possible. In a subsequent memoir of 1744, which dealt with the refraction of light, Maupertuis was able to show that Fermat's law for diffraction follows from minimizing the product of the distances traveled s_k by light in various media k with the speeds of light in those media v_k. That is

$$\sum_{k} s_k v_k \qquad (1.5)$$

is a minimum for light traveling on a path passing through a set of refractive media. The quantity in (1.5) Maupertuis called the *quantity of action*. This success, which may be considered as minor, gave Maupertuis the temerity to propose a rather general principle. He wrote that "Nature, in making its effects, always acts by the simplest means possible[12]" [90] [cf. [22], pp. 267–269].

[12]Maupertuis' metaphysical enthusiasm was based on a limited principle and seems to have blinded him to difficulties. Voltaire turned his terrible vindictive wit on Maupertuis in a parody that nearly destroyed his reputation [[91], p. 252]. And Herbert Goldstein deals with Maupertuis in less than generous terms [[34], footnote p. 368].

1.8.2 Euler

Euler approached the problem of identifying an extremum somewhat differently. In the appendices to his *Methodus inveniendi* (1744) Euler sought an integral the extremum of which provided the correct law of motion. He began with a proposition that the trajectory followed by a body of mass M is that which minimizes the integral

$$\int M v \mathrm{d}s \tag{1.6}$$

where $\mathrm{d}s$ is the differential along the path followed by the body and v is the velocity of the body [90], [[128], pp. 24–25, [22], pp. 273–274].

In his *Mechanica* (1736) [cited in [90]] Euler had studied motion in a plane (x, y) with $\mathrm{d}s = \sqrt{\mathrm{d}x^2 + \mathrm{d}y^2}$ and had shown that the force on a body can be decomposed into two orthogonal components showing that the results were the same as those obtained from Newton's methods. But the application was only to a single body. This was, therefore, not a general principle.

His results were, however, analogous to those of Maupertuis, as he pointed out in a letter to Maupertuis (1745). Euler believed there were mathematical difficulties that must yet be resolved. Maupertuis, however, thought that Euler's work was verification of the general philosophical principle he had proposed in 1744. He now proposed that this was the principle of the least quantity of action [90].

Based on philosophical and theological conviction, Maupertuis believed that a general truth, beyond what the science itself revealed, was emerging.

But Euler worried about the details. His concerns led him to a major investigation between 1748 and 1751 of a more general formulation of the action as an integral. In this investigation Euler worked in the context of the extremum principle of Maupertuis. He asked if

$$\int M v \mathrm{d}s = \min / \max \tag{1.7}$$

is a particular case of a more general condition.

In a study of a hanging string acted upon by a central (gravitational) force and of fluids, which he considered to be collections of particles, Euler identified a term he considered an *analytical invariant*. We presently identify this term, as we did above, as the (negative) potential energy of the system of masses or elements of the string. As Marco Panza notes, the mathematical importance of this term preceded the physical notion of potential energy [90] [96], [cf. [128], p. 46]. He had not yet, however, been able to formulate a general variational principle from which all of mechanics could be obtained.

If, with Euler, we consider the central forces acting on the mass particle j to be λ_{j} and the virtual displacements to be $\mathrm{d}\lambda_{\mathrm{j}}$ then Maupertuis' principle claims that $\sum_{\mathrm{j}} \int \lambda_{\mathrm{j}} \mathrm{d}\lambda_{\mathrm{j}}$ is an extremum. This must be true for all elements $\mathrm{d}s$ of the hanging string. Then the integral of this term over the length of the hanging string must be an extremum. If, from Newton's Second Law, we write $\sum_{\mathrm{j}} \lambda_{\mathrm{j}} d\lambda_{\mathrm{j}} = F\mathrm{d}s = Mv\mathrm{d}v$, where F is the net total force on the body along $\mathrm{d}s$, then the integral of $\sum_{\mathrm{j}} \int \lambda_{\mathrm{j}} \mathrm{d}\lambda_{\mathrm{j}}$ along the length of the string is $\Omega = (1/2)\, Mv^2 + K'$, where K' is a constant. And then the integral of Ωdt is

$$\int \Omega \mathrm{d}t = \frac{1}{2} \int Mv\mathrm{d}s + K't \tag{1.8}$$

and the variational principles $\int \Omega \mathrm{d}t = \min / \max$ and $\int Mv\mathrm{d}s = \min / \max$ are equivalent. This Euler termed the conservation of *vis viva*, which is the 18th century term for (twice) the *kinetic energy*. With our identification of Ω as the negative of the potential energy, we would identify this as simply *conservation of energy* [90].

We may then ask if there was a possible difference in the positions of Euler and of Maupertuis.

Euler was concerned about formulating a general mathematical principle, which would provide a method by which we could identify the actual dynamical path followed by a system if the initial and final conditions of the system were known. To locate this path Euler considered the problem of finding the form of the function $y\,(x)$ that would produce an extreme value of the general integral

$$\int_{\mathrm{a}}^{\mathrm{b}} F\left(y, y', x\right) dx \tag{1.9}$$

with $y' = \mathrm{d}y/\mathrm{d}x$. The values of y at the end points $y\,(a)$ and $y\,(b)$ were fixed.

To solve the problem using the methods of ordinary calculus Euler wrote the integral (1.9) as a sum over differences in the variable x. He could then set the partial derivatives of the resultant sum with respect to y equal to zero. The result was a general equation for the function $y\,(x)$. This strategy provided a general differential equation for $y\,(x)$

$$\frac{\partial F}{\partial y} - \frac{\mathrm{d}}{\mathrm{d}x}\left(\frac{\partial F}{\partial y'}\right) = 0 \tag{1.10}$$

for x on the closed interval $[a, b]$.

The steps bringing Euler to this result required careful treatment of the infinitesimal intervals in x and the corresponding values of $y\,(x)$. Cornelius Lanczos treats this in detail [[65], pp. 51–53]. There are difficulties at the fixed end points resulting from the requirement that the values of $y\,(x)$ are fixed there. And we cannot completely sweep our logical difficulties away in the limit.

The actual path of the system was that for which Newton's Laws held. And Euler had shown that Newton's Laws were recovered from the condition that the first order variation in a specifiic integral vanishes. But if only the first order variation vanishes we have no way of deciding whether the integral has attained a minimum or a maximum value at the extremum. Euler, however, published the principle of least action as an exact dynamical theorem in 1744 [[128], p. 24].

Euler's objective was to establish a mathematical principle. Maupertuis was also bound by the mathematics. But his thoughts went beyond the mathematics to the metaphysics. Euler's did as well when he claimed that the action was least. Except for Mauprtuis' exuberance, the difference seems to be very slight, if there at all.

At this time Joseph-Louis Lagrange was teaching himself mathematics.

1.8.3 Lagrange

The Person. Lagrange was one of 11 children, only two of whom survived to adulthood. He was born in Turin, the capital of the kingdom of Sardinia-Piedmont (now Italy) since 1720, and baptized as Giuseppe Lodovico Lagrangia. Lagrange had French ancestry on his father's side. His paternal great grandfather had been a French cavalry captain and Lagrange was attracted to the French background in his family. As a youth, he signed his name as LaGrange or Lagrange rather than Lagrangia [96].

In 1754, at the age of eighteen, Lagrange had the temerity to send Euler a letter containing results he would later publish in Italian [90]. These results were not at all remarkable and after publication of the paper Lagrange found them already present in a communication between Johann Bernoulli and Gottfried Leibniz. The possibility that he would be considered a plagiarist disturbed him greatly and he increased his efforts to produce good mathematics.

Lagrange then undertook a study of the tautochrone or isochrone curve. A mass released from rest at any point on the tautochrone, and sliding without friction under the force of gravity, will take the same time to reach the lowest point on the curve. The curve had been found using geometrical methods by Christiaan Huygens in 1659. By the end of 1754 Lagrange had some reasonable results related to what Euler would later call the calculus of variations. With well-founded confidence Lagrange again wrote to Euler in the summer of 1755. Euler quickly wrote back indicating that he was impressed with the results [96].

In the Autumn of 1755, at nineteen, Lagrange was appointed professor of mathematics at the Royal Artillery School in Turin [90].

Then in 1756 Lagrange sent Euler some work that generalized the results Euler had obtained earlier. Euler spoke about this to Maupertuis, who was then the president of the Berlin Academy (*Preußische Akademie der Wissenschaften*), and together they agreed that Lagrange should be offered a position in Berlin. Lagrange's strong interest in the principle of least action certainly influenced this. But Lagrange was not interested in prestige. He simply wanted to pursue mathematics. So he politely, and shyly, refused the position. Lagrange was, however, elected to the Berlin Academy in September of 1756, on Euler's recommendation.

In 1757 Lagrange was a founding member of a scientific society in Turin, which included a group of young scientists living there. The primary objective of the society was the publication of the journal *Mélanges de Turin*.[13] And Lagrange contributed to the first three volumes of the journal.

Jean-Baptiste d'Alembert, who was familiar with Lagrange's work and friendly with Friedrich II of Prussia, arranged for Lagrange to be again offered a position at the Berlin Academy in 1765. Lagrange again refused, but this time with the qualification, "It seems to me that Berlin would not be at all suitable for me while M. Euler is there."

With the knowledge that Euler would be leaving the Berlin Academy in 1766 for St. Petersburg, d'Alembert wrote to Lagrange urging him to take the Berlin position. This time Lagrange accepted and succeeded Euler as Director of Mathematics at the Berlin Academy in 1766. He was thirty years old.

Lagrange remained in Berlin until 1787 when he left to become a member of the *Académie des Sciences* in Paris.

During his last years in Berlin, Lagrange wrote the masterpiece *Mécanique analytique*, which was published in Paris in 1788. This was really a summary of what had been done in the science of mechanics since Newton's synthesis in 1687. The methods Lagrange presented were radically different from those of Newton and in the preface he wrote

> One will not find figures in this work. The methods that I expound require neither constructions, nor geometrical or mechanical arguments, but only algebraic operations, subject to a regular and uniform course.

Lagrange died in Paris on 10 April, 1813. He had survived the French Revolution, while many had not [96].

The Contributions. Lagrange entered the discussion in 1761 with two memoirs, both of which were published in volume 2 of the journal *Mélanges de Turin*.

In the first of these he provided a formulation of the Euler principle in terms of a new $\delta-$formalism [62]. In this new formalism Lagrange considered virtual variations in the function $y(x)$ appearing in (1.9). If the integral has an extreme value for a specific function $y(x) = \eta(x)$, then Lagrange could consider the values of the integral when $y(x)$ takes on the form $y(x) = \eta(x) + \delta y(x)$ as powers in the variation $\delta y(x)$ leaving the terms $\delta y(x)$ inside the integral. This approach avoided the end point problems inherent in Euler's method. If the integral has an extremum at $y(x) = \eta(x)$ then the result of Lagrange's approach required that the first order variation in the integral vanished. Lagrange's approach now forms the basis of modern variational calculus [cf. [31], [65], [11], [104], [36], [67], [70]].

In the second memoir he proposed that "all of the problems of dynamics" *(toutes les questions de dynamique)* can be treated by a variational principle, which was a generalization of the Euler principle (1.7) over a collection of particles.

The mathematical approach of Lagrange required reduction to a collection of independent coordinates, now termed *generalized coordinates*, which required Lagrange's method of *undetermined multipliers* to incorporate the constraints.

[13] *Mélanges de Philosophie et de Mathématiques de la Soc. Roy. de Turin.*

These two memoirs, we must recognize, are among the most important in theoretical physics. They carry us from the geometrical methods of Newton to the true Analytical Mechanics of Euler and Lagrange [90] [[128], pp. 30–31].

Lagrange produced a more complete formulation in 1788. This was the *Méchanique analytique*. Here he relied on the ideas of Bernoulli and of d'Alembert. In his *Traité de dynamique* d'Alembert used the virtual velocity concept of Bernoulli to reduce dynamical situations to equivalent problems in statics. Lagrange simplified this. He wrote finally

$$\begin{aligned}\delta W = \sum_{\mathrm{i}=1}^{\mathrm{n}} M_{\mathrm{i}} \Big[& F_{\mathrm{x,i}} \delta x_{\mathrm{i}} + F_{\mathrm{y,i}} \delta y_{\mathrm{i}} + F_{\mathrm{z,i}} \delta z_{\mathrm{i}} \\ & - \frac{\mathrm{d}^2 x_{\mathrm{i}}}{\mathrm{d}t^2} \delta x_{\mathrm{i}} - \frac{\mathrm{d}^2 y_{\mathrm{i}}}{\mathrm{d}t^2} \delta y_{\mathrm{i}} - \frac{\mathrm{d}^2 z_{\mathrm{i}}}{\mathrm{d}t^2} \delta z_{\mathrm{i}} \Big] \\ = 0, \end{aligned} \tag{1.11}$$

where $F_{\mathrm{x,i}}$, $F_{\mathrm{y,i}}$, and $F_{\mathrm{z,i}}$ are the (x, y, z) components of the applied forces (per unit mass) acting on the ith body of mass M_{i}. This is the modern form of *d'Alembert's principle* [cf. [36], p. 22; [21], p. 90].

Lagrange identified the generalized coordinates as φ_{j} and, using his $\delta-$formalism, Lagrange obtained

$$\frac{\mathrm{d}}{\mathrm{d}t} \frac{\delta T}{\delta \left(\mathrm{d}\varphi_{\mathrm{j}} / \mathrm{d}t \right)} - \frac{\delta T}{\delta \varphi_{\mathrm{j}}} - \frac{\delta U}{\delta \varphi_{\mathrm{i}}} = 0 \tag{1.12}$$

from (1.11). In (1.12) T was Lagrange's notation for half the *vis viva*. In modern terminology this is the kinetic energy, which we still designate as T. And U is the 18th century force function

$$F_{\mathrm{x,i}} = \frac{\delta U}{\delta x_{\mathrm{i}}}, \tag{1.13}$$

which was introduced by Lagrange [see [40]]. This is the negative of the modern potential energy. The equations (1.12), which, since they are based on (1.11), are equivalent to Newton's Laws, are Lagrange's equations for the motion of a mechanical system [90]. We will eventually call these the Euler–Lagrange Equations. In 1788, then, Lagrange showed that Newtonian (rational) mechanics results from a *variational principle*. Crucial in this development is Lagrange's method of undetermined multipliers. This is a method by which the constraints are introduced into the variational integral, which removes any algebraic difficulties and, in integral form, incorporates even constraints that can only be written in differential form [[22], pp. 342–344]. Lagrange, however, made no metaphysical claims. His work was purely mathematical.

1.9 Hamilton and Jacobi

1.9.1 Hamilton's Goal

William Rowan Hamilton[14] began with a study of optics. He wished to bring to optics the same beauty, power, and harmony that Lagrange had brought to mechanics [[22], p. 390].

Geometrical optics could be based on either the wave or corpuscular picture of light.[15] Hamilton thought of the wave and corpuscular pictures to be explanatory devices for a single science he called mathematical optics. In this vein Hamilton considered the concept, or law, of least action to rank among the greatest theorems in physics. He was, however, skeptical regarding its pretensions to cosmological necessity based on economy. For Hamilton it was possible to speak of the extremum property of the action. But it was not possible to extend this to a claim that the divine design of the universe is based on economy [[22], p. 391].

Hamilton's work in optics resulted in the definition of an action he termed V, in which the refractive index v_{i} for a medium i was inversely proportional to the speed of light in the medium u_{i} with the proportionality factor $f(x)$ a function of the frequency of the light x. Specifically $v_{\mathrm{i}} f(x) = 1/u_{\mathrm{i}}$. The extremum condition $\delta V = 0$ resulted in equations similar to those of Lagrange [[22], p. 393].

Hamilton's work in mechanics began in 1833. A manuscript entitled *The Problem of Three Bodies by my Characteristic Function* is contained in his Notebook 29. Hamilton's characteristic function in this manuscript is $\int_0^t 2T\mathrm{d}t$, which he identifies as the accumulated *living force* (*vis-viva* - twice the kinetic energy) in the system between the times 0 and t. This manuscript contained the basis for what is now called the *Hamilton–Jacobi method*.

This method, which we can claim is the most general and elegant expression of Analytical Mechanics, formulates the science in terms of a single function. Rather than concentrating his attention on the consequences of a variational principle, Hamilton set for himself the goal of seeking that function which has an extreme value in Analytical Mechanics. This is consistent with his words at the beginning of this chapter.

Hamilton's ideas are set down in two papers, or essays as Hamilton calls them, published in the *Philosophical Transactions of the Royal Society* (London) in 1834 and 1835 [39], [40]. In the first of these papers he refers to Lagrange's work as a "kind of scientific poem".

---

[14] Sir William Rowan Hamilton (1805–1865) was an Irish mathematician and physicist.

[15] An experimental comparison of the speed of light in water and in air, which is considered to have ended the classical (Newtonian) corpuscular theory of light, was conducted by Léon Foucault in 1850 and Hippolyte Fizeau in 1851. Hamilton's principal work in mechanics predated these experiments [[126], p. 136].

1.9.2 *The First Essay*

In the first Essay [39] Hamilton considered a collection of point particles with masses m_i located at the points (x_i, y_i, z_i) and moving in accordance with Newton's Laws. He used Lagrange's notation from the *Méchanique analytique*. The forces came from a force function U (see (1.13)). And Hamilton designated half the *vis-viva* as T. For particles moving according to Newton's Laws, then, $dT = dU$, which integrates to

$$T = U + \mathcal{H}, \tag{1.14}$$

where $\mathcal{H}$ is a constant. At the initial condition

$$T_0 = U_0 + \mathcal{H}. \tag{1.15}$$

If the initial conditions are slightly perturbed there will be perturbations in all three quantities, including the constant term $\mathcal{H}$.

Hamilton then defined the "accumulated of living force" as

$$V = \int_0^t 2T\,dt, \tag{1.16}$$

which is a function of the initial coordinates $(a_i, b_i, c_i, \ldots)$ and the final coordinates $(x_i, y_i, z_i, \ldots)$ at the time t for each particle i. From the variation of this V he obtained the partial derivatives of V as

$$\frac{\delta V}{\delta x_1} = m_1 x_1^{'} \cdots \frac{\delta V}{\delta z_n} = m_n z_n^{'}, \tag{1.17}$$

$$\frac{\delta V}{\delta a_1} = m_1 a_1^{'} \cdots \frac{\delta V}{\delta c_n} = m_n c_n^{'}, \tag{1.18}$$

and

$$\frac{\delta V}{\delta \mathcal{H}} = t. \tag{1.19}$$

Here we use the 18th century notation for the partial derivative.

The importance of the function V was then apparent. If V could be found as a function of coordinates and the constant $\mathcal{H}$, then any problem in dynamics would have been solved. The initial momenta and those at a later time could then be found as solutions to algebraic equations, which were obtained by partial differentiation of V. Hamilton therefore called V the *Characteristic Function* of motion of the system. He called the variation of V the *law of varying action*.

To obtain a method for finding V Hamilton returned to (1.14) and (1.15) and used (1.17) and (1.18). He obtained two partial differential equations

$$\sum \frac{1}{2m_i}\left[\left(\frac{\delta V}{\delta x_i}\right)^2 + \left(\frac{\delta V}{\delta y_i}\right)^2 + \left(\frac{\delta V}{\delta z_i}\right)^2\right] = U + \mathcal{H} \tag{1.20}$$

and

$$\sum \frac{1}{2m_i}\left[\left(\frac{\delta V}{\delta a_i}\right)^2 + \left(\frac{\delta V}{\delta b_i}\right)^2 + \left(\frac{\delta V}{\delta c_i}\right)^2\right] = U_0 + \mathcal{H}. \tag{1.21}$$

Both of these partial differential equations must be satisfied by the characteristic function. According to Hamilton these equations are the principal means of discovering the form of V, and are of essential importance in the theory.

In the final two sections of this essay Hamilton produced a transformation from the characteristic function V to a function of the coordinates, the initial values of the coordinates, and the time. The transformation, which was a *Legendre transformation*,[16] resulted in the function

$$S = V - t\mathcal{H}, \tag{1.22}$$

since $\delta V/\delta\mathcal{H} = t$. With (1.16), (1.14), and (1.21) and recalling that $\mathcal{H}$ is a constant, the function S is then

$$S = \int_0^t (T + U)\,\mathrm{d}t. \tag{1.23}$$

From the variation of S Hamilton obtained

$$\frac{\delta S}{\delta t} = -\mathcal{H},\ \frac{\delta S}{\delta x_i} = m_i x_i',\ \ldots,\ \text{and}\ \frac{\delta S}{\delta a_i} = -m_i a_i',\ \ldots \tag{1.24}$$

Finally Hamilton produced two partial differential equations for S.

$$\frac{\delta S}{\delta t} + \sum \frac{1}{2m_i}\left[\left(\frac{\delta S}{\delta x_i}\right)^2 + \left(\frac{\delta S}{\delta y_i}\right)^2 + \left(\frac{\delta S}{\delta z_i}\right)^2\right] = U \tag{1.25}$$

and

$$\frac{\delta S}{\delta t} + \sum \frac{1}{2m_i}\left[\left(\frac{\delta S}{\delta a_i}\right)^2 + \left(\frac{\delta S}{\delta b_i}\right)^2 + \left(\frac{\delta S}{\delta c_i}\right)^2\right] = U_0. \tag{1.26}$$

[16]The Legendre transformation, which replaces a variable with the derivative with respect to that variable preserving the information content, was published by A.M. Legendre in 1787 [29], [74], [66].

This is the final elegant formulation that was Hamilton's goal. He had found a single function containing the description of the mechanical motion of a system of interacting particles with constraints.

1.9.3 The Second Essay

In the second essay [40] Hamilton fixed his attention on the function S obtained in the Legendre transformation (1.22), which he called the *Principal Function*. He began the essay, however, with a derivation of the second order partial differential equations of Lagrange directly from Newton's Second Law without the use of a least action principle, on which Lagrange had based his derivation. He presented the equations in terms of generalized coordinates η_{j} and velocities $\eta_{\mathrm{j}}^{\prime}$. He defined a new generalized coordinate

$$\bar{\omega}_{\mathrm{i}} = \frac{\delta T}{\delta \eta_{\mathrm{i}}^{\prime}} \tag{1.27}$$

and Legendre transformed his function $\mathcal{H}$ from a function of the generalized coordinates and velocities to one dependent on generalized coordinates η_{i} and $\bar{\omega}_{\mathrm{i}}$.

$$\mathcal{H} = F\left(\bar{\omega}_1 \ldots, \eta_1 \ldots\right) - U\left(\eta_1 \ldots\right). \tag{1.28}$$

From his form of the Lagrange Equations (Euler–Lagrange Equations)

$$\frac{\mathrm{d}}{\mathrm{d}t}\frac{\delta T}{\delta \eta_{\mathrm{i}}^{\prime}} = \frac{\delta\left(T + U\right)}{\delta \eta_{\mathrm{j}}}, \tag{1.29}$$

he obtained

$$\frac{\mathrm{d}}{\mathrm{d}t}\eta_{\mathrm{i}} = \frac{\delta \mathcal{H}}{\delta \bar{\omega}_{\mathrm{i}}} \text{ and } \frac{\mathrm{d}\bar{\omega}_{\mathrm{i}}}{\mathrm{d}t} = -\frac{\delta \mathcal{H}}{\delta \eta_{\mathrm{j}}}. \tag{1.30}$$

In modern terminology the Eq. (1.30) are the *canonical equations*.

At this point Hamilton returned to the formulation in terms of his Principal Function S and showed that the variation in S was

$$\delta S = \sum \left(\bar{\omega}_{\mathrm{i}} \delta \eta_{\mathrm{i}} - p_{\mathrm{i}} \delta e_{\mathrm{i}}\right), \tag{1.31}$$

in which $\bar{\omega}_{\mathrm{i}}$ and η_{i} are the values of the momenta and coordinates at the time t and p_{i} and e_{i} are their values at the time $t = 0$. From (1.31) he then identified

$$\bar{\omega}_{\mathrm{i}} = \frac{\delta S}{\delta \eta_{\mathrm{i}}} \text{ and } p_{\mathrm{i}} = -\frac{\delta S}{\delta e_{\mathrm{i}}}. \tag{1.32}$$

And the partial differential equations for the Principal Function are those in (1.25) and (1.26) then became

$$\frac{\delta S}{\delta t}+\mathcal{H}\left(\frac{\delta S}{\delta\eta_1},\ldots,\eta_1,\ldots\right)=0 \tag{1.33}$$

and

$$\frac{\delta S}{\delta t}+\mathcal{H}\left(\frac{\delta S}{\delta e_1},\ldots,e_1,\ldots\right)=0, \tag{1.34}$$

and once S is known as a function of coordinates and time the canonical momenta can be found from algebraic equations.

The remainder of the essay was then devoted to applications of his theory.

1.10 Jacobi

Carl Gustav Jacobi (1804–1851) was a mathematician, a great teacher, and a prolific writer. "Such were Jacobi's forceful personality and sweeping enthusiasm that none of his gifted students could escape his spell: they were drawn into his sphere of thought, and soon represented a 'school'" [[110], cited in [95]].

Jacobi was born into a Jewish family. He, however, converted to Christianity around 1825, presumably motivated, at least in part, by the requirements for a university teaching position in Prussia.

Jacobi's interest in theoretical mechanics was as a mathematician rather than as a physicist. This is particularly evident in the titles of the two important papers we will consider here. These papers were published in the *Crelle Journal für die reine und angewandte Mathematik.*[17] The first paper was entitled *Zur Theorie der Variations-Rechnungen und der Differential-Gleichungen*[18] [52] and the second was *Über die Reduction der Integration partiellen Differentialgleichungen erster Ordnung zwischen irgend einer Zahl Variabeln auf die Integration eines einzigen Systems ge-wöhnlicher Differentialgleichungen*[19] [53]. Because of the date of publication we will designate these as Jacobi 1837 a, b. Both of these papers, however, dealt explicitly with Hamilton's theory. At the time they were written (received 29 November and 9 December 1836) Jacobi, it seems, was convinced of the validity of an axiomatic approach to theoretical physics, which had been the position of Lagrange. Jacobi's

[17] Crelle Journal for the pure and applied Mathematics. The present title is the Journal für die reine und angewandte Mathematik (Journal for the pure and applied Mathematics), but it is commonly called Crelle's Journal or simply Crelle. The journal has been in continuous publication since 1826.

[18] On the Theory of Variational Calculus and Differential Equations.

[19] On the Reduction of the Integration of Partial Differential Equations of Frst Order in an Arbitrary Number of Variables to the Integration of a single System of Ordinary Differential Equations.

position changed, however. In his lectures on Analytical Mechanics in Berlin (1847–1848) he was critical of Lagrange's philosophical position. This was apparently motivated by a changing evaluation of the role of mathematics in the empirical sciences at that time[20] [95].

1.10.1 Jacobi 1837a

Jacobi began the first paper (Jacobi 1837a) indicating that it had fallen on him to fill a rather large and important hole in the variational calculus. The vanishing of the first variation only guarantees an extremum. It says nothing of whether that extremum is a minimum or a maximum, which must be decided on the sign of the second variation. The first part of the paper deals with this question for variational problems dependent on either the first or the second derivative of the function sought. Then Jacobi began an analysis of Hamilton's theory.

Jacobi pointed out that for his theory Hamilton required the principle of *vis viva*, which follows if the force function is time independent. This limitation, Jacobi showed, is easily removed, resulting in a formulation for which the principle of *vis viva* no longer holds, but the principle of stationary (least) action remains valid.

And Jacobi marveled at the fact that Hamilton required the solution of two partial differential equations for the Principal Function S, i.e. (1.25) and (1.26) or, equivalently, (1.33) and (1.34). It is easy to show, he claimed, that it is sufficient to solve only one of these equations.

Jacobi's most extensive criticism, however, which takes much of the remainder of the paper, was his contention that "Little seems to be gained through this reduction to a partial differential equation of first order ..." Solving the single partial differential equation to which the dynamical problem has been reduced is much more difficult, he pointed out, than solving the usual system of ordinary differential equations. It is also, he continued, a rather important discovery in the theory of partial differential equations that a partial differential equation can often be reduced to a single system of ordinary differential equations, which do not contain the original function itself. This is certainly the case in Analytical Mechanics, where the ordinary differential equations had already been obtained by Lagrange.

It is difficult to escape the impression that these two papers of Jacobi are a pair with the first serving as in introduction to the second, in which Jacobi presented the details that had been outlined in the first. The two papers also appear in the same volume of Crelle's Journal, almost sequentially. But Jacobi did not claim this connection.

[20] This may be compared to David Hilbert's belief that the proper way to develop any scientific subject rigorously required an axiomatic approach, which he specifically called for in his address to the International Congress of Mathematicians in 1900 [cf. [129]].

1.10.2 Jacobi 1837b

In the second paper (Jacobi 1837b) Jacobi primarily developed and considered a theorem. The theorem deals with a system of n unconstrained point particles (with masses m_i) which obey the equations of motion

$$m_i\frac{d^2x_i}{dt^2}=\frac{\partial U}{\partial x_i};\ m_i\frac{d^2y_i}{dt^2}=\frac{\partial U}{\partial y_i};\ \text{and } m_i\frac{d^2z_i}{dt^2}=\frac{\partial U}{\partial z_i} \tag{1.35}$$

for each i, where the force function U is a function of the $3n$ coordinates $x_1, y_1, z_1, \ldots, x_n, y_n, z_n$ and the time t. The corresponding Principal Function S, which is a complete solution to the equation

$$U=\frac{\partial S}{\partial t}+\frac{1}{2}\sum\frac{1}{m_i}\left[\left(\frac{\partial S}{\partial x_i}\right)^2+\left(\frac{\partial S}{\partial y_i}\right)^2+\left(\frac{\partial S}{\partial z_i}\right)^2\right], \tag{1.36}$$

and, apart from an arbitrary additive constant, contains $3n$ arbitrary constants

$$a_1, a_2, \ldots, a_{3n}, \tag{1.37}$$

which are linked to another $3n$ arbitrary constants as

$$\frac{\partial S}{\partial a_1}=\beta_1,\ \frac{\partial S}{\partial a_2}=\beta_2, \ldots, \frac{\partial S}{\partial a_{3n}}=\beta_{3n}. \tag{1.38}$$

Jacobi's theorem, as presented here reproduces the basic results of Hamilton's theory, with some modification. There is no reference to the principle of living force, which had formed a cornerstone of much of the preceding work in Analytical Mechanics, because this principle no longer holds if the force function depends also on the time. And the Principal Function S, in Jacobi's theorem, must only satisfy a single partial differential equation. There is also a somewhat deeper issue that seems to have been ignored by Jacobi. There is no reference to the principle of least action. The principle of least action results in Newton's Second Law, or equivalently Lagrange's Equations (1.35). But the path through the variational calculus to the laws of motion is not a necessity. As Hamilton had shown, the Euler–Lagrange Equations could be obtained without the principle of least action.

In his development of the theorem, Jacobi's discussion was based on differential calculus. In his preceding paper Jacobi had considered the mathematical question of min / max when the first variation of an integral vanishes, and rather clearly indicated why we should base our discussions on ordinary differential equations, rather than partial differential equations.

We may then consider that Jacobi's theorem expands and formalizes Hamilton's theory. Because of Jacobi's theorem we must also associate Jacobi's name with the idea of presenting mechanics in terms of a single function. This is now, and will remain, the *Hamilton–Jacobi Theory*.

Jacobi's path to his theorem is insightful. He began by casting the equations of motion in the form

$$m_i \frac{d^2 x_i}{dt^2} = \frac{\partial U}{\partial x_i} + \lambda \frac{\partial F}{\partial x_i} + \lambda_1 \frac{\partial F_1}{\partial x_i} + \cdots , \tag{1.39}$$

which are Lagrange's equations with the constraints $F = 0$, $F_1 = 0, \ldots$ Using insightful differential relations he showed that Hamilton's definition of the Principal Function [see (1.23)]

$$S = \int_0^t \left\{ U + \frac{1}{2} \sum m_i \left[\left(x_i'\right)^2 + \left(y_i'\right)^2 + \left(z_i'\right)^2 \right] \right\} dt \tag{1.40}$$

results in

$$\begin{aligned} \frac{\partial S}{\partial x_i} &= m_i x_i' & \frac{\partial S}{\partial a_i} &= -m_i a_i' \\ \frac{\partial S}{\partial y_i} &= m_i y_i' & \frac{\partial S}{\partial b_i} &= m_i b_i' \\ \frac{\partial S}{\partial z_i} &= m_i z_i' & \frac{\partial S}{\partial c_i} &= m_i c_i', \end{aligned} \tag{1.41}$$

where $m_i a_i'$, $m_i b_i'$, and $m_i c_i'$ are the initial values of the momenta, which are the constants designated as $-\beta$ in (1.38). Then, considering S to be a (general) function of (t, x, y, z, a, b, c) and using (1.41), Jacobi obtained (1.36).

Jacobi followed the introduction of his theorem with a fairly pointed critique of Hamilton's presentation of his (beautiful) theory. He confessed that he did not know why Hamilton required the construction of a function S, which is a function of the $6n + 1$ variables (x, y, z, a, b, c, t), that must simultaneously satisfy the two partial differential equations (1.25) and (1.26), since he (Jacobi) had shown that it is completely sufficient that the function S satisfies only the partial differential equation (1.36) and contains $3n$ arbitrary constants, to be found from initial conditions. He then pointed out that this resulted in Hamilton's placing his beautiful discovery in a "false light".

Jacobi's concern was not to attack Hamilton's ideas, even though he demonstrated that the approach thorough ordinary differential equations is preferable. Rather he wanted Hamilton's theory to stand on its own, without requiring the separate understanding of the proof Hamilton presented. Hamilton's neglect of the general rules that Lagrange presented in his lectures on function theory for the integration of non-linear partial differential equations of fist order, Jacobi supposed, meant that certain important results in Analytical Mechanics escaped Hamilton. He also considered that Hamilton's limitation to force functions that are time independent was too great a limitation, particularly since this limitation is easily removed. Yourgrau and Mandelstam have a literal translation of a portion of this critique [[128], p. 58].

Jacobi moved these ideas rather quickly into his teaching. His published *Lectures on Dynamics*, presented at the University of Königsberg in the winter semester of 1842–43 [52], contain Hamilton's ideas (Lecture 8) and consequences of adding time dependence to the force function (Lecture 9). Lecture 9 is particularly interesting in that there he considered explicitly the dependence of the function $\mathcal{H} = T - U$ on generalized coordinates and momenta, which he designated as q_{i} and p_{i}. There he wrote as the canonical equations, which Hamilton first wrote as (1.30), in the modern form

$$\frac{\mathrm{d}q_{\mathrm{i}}}{\mathrm{d}t} = \frac{\partial \mathcal{H}}{\partial p_{\mathrm{i}}} \text{ and } \frac{\mathrm{d}p_{\mathrm{i}}}{\mathrm{d}t} = -\frac{\partial \mathcal{H}}{\partial q_{\mathrm{i}}}, \tag{1.42}$$

and he considered the case for which a force function cannot be written. Then, in place of $\partial U/\partial q_{\mathrm{i}}$, we have

$$Q_{\mathrm{i}} = \sum_{\mathrm{j}} \left(X_{\mathrm{j}} \frac{\partial x_{\mathrm{j}}}{\partial q_{\mathrm{i}}} + Y_{\mathrm{j}} \frac{\partial y_{\mathrm{j}}}{\partial q_{\mathrm{i}}} + Z_{\mathrm{j}} \frac{\partial z_{\mathrm{j}}}{\partial q_{\mathrm{i}}} \right),$$

where (X, Y, Z) are the components of the nonconservative force. The canonical equations are then

$$\frac{\mathrm{d}q_{\mathrm{i}}}{\mathrm{d}t} = \frac{\partial T}{\partial p_{\mathrm{i}}} \text{ and } \frac{\mathrm{d}p_{\mathrm{i}}}{\mathrm{d}t} = -\frac{\partial T}{\partial q_{\mathrm{i}}} + Q_{\mathrm{i}}, \tag{1.43}$$

At that time Jacobi gave these lectures he was in failing health [95].

1.11 Summary

In this chapter we have presented a brief history of Analytical Mechanics tracing the development of human thought regarding the motion we see around us. We naturally began with the Presocratic thinkers, who did not accept the simple explanation of occurrences based on the whims of the gods. Still, however, explanations were based on the personal convictions or even prejudices of individuals. These were remarkable and even resembled aspects of modern scientific thought.

And we saw that political and religious history affected intellectual history. This came as no surprise. It led us, however, to an appreciation of Muslim thought and an understanding of the influence of the development of Islam on the simultaneous development of science in the Muslim world. The separation of the western and eastern Roman Empires and the supposedly insignificant closing of the academy in Athens by Justinian I had dramatic consequences on the form in which Greek thought appeared as it entered western Europe through Muslim Spain.

The Roman Catholic Church, first with Augustine, anticipated that science would provide supporting evidence for theology. This has not always proven easy, but scientific and mathematical thinkers of the medieval period, such as Robert Grosseteste and Jean Buridan, were churchmen.

The university system that came with the intellectual revolution in western thought changed the structure politically in western Europe. Even in the 9th century Charlemagne recognized the importance of education and coaxed Alcuin of York to his empire to design an educational system. Nicolaus Copernicus of Poland in the Kingdom of Prussia was educated at the universities at Krakow and at Bologna in Italy. And Johannes Kepler benefitted from the belief held by the Dukes of Württemberg in the importance of education. Science attracts thinkers and the university system that western Europe was developing provided centers for scientific thought and discussion.

We saw a dramatic change in scientific thought with both Copernicus and Kepler. Copernicus abandoned the philosophical position that placed the earth at the center of the cosmos and Kepler made astrophysics of astronomy, smashing the Platonic spheres. The importance of the steps taken by both Copernicus and Kepler cannot be overstated. Although all corners of western civilization were not changed by their discoveries, western intellectual history turned at this point to a reliance on hard data. The next step (Newton) would synthesize the ideas of Kepler, among others, into a theory.

We did not deal directly with the political upheavals of the Thirty Years' War and the English Civil War. But these were important in the subsequent founding of scientific societies in which discussions could be resolved by logic and experiment, rather armed conflict. In our detailed study of Newton we saw conflict, of course. But the conflict between Newton and Robert Hooke remained intellectual.

The publication of Newton's Principia changed the scientific landscape. Newton remained, however, a deeply religious Puritan. And he thought seriously of himself as a theologian [see [124] Chap. 14]. His use of Euclidian geometry as the basis of his proofs, or demonstrations, placed physics at a disadvantage until Maupertuis, Euler, and Lagrange made calculus the language of Analytical Mechanics. The use of calculus was not without controversy [cf. [5]], and we may speculate that Newton rejected the use of his *fluxions* in the *Principia* because of the questions surrounding the new theory that did not plague Euclid's geometry. Euclidian geometry was pure and absolute.

The Analytical Mechanics of Maupertuis, Euler, and Lagrange radically changed our understanding of physical science. It had become, in fact, the experimental and mathematical science that Newton envisioned. With Analytical Mechanics we could actually solve interesting physical problems many which of were of interest to engineers.

The final step in the development of Analytical Mechanics came with the contributions of Hamilton and Jacobi. Hamilton showed that the solution to the motion of any system of interacting particles was contained in a single Principal Function that satisfied a pair of partial differential equations. The Principal Function itself had no physical meaning. But the physical quantities of interest could be found from algebraic equations obtained from the Principal Function. Jacobi's contribution was to greatly simplify Hamilton's theory, while retaining the central idea. It is never easy to solve any interesting problem in physics. But the Hamilton–Jacobi approach comes very close and has provided the path to the study of complex and chaotic systems.

The reader familiar with the basis of the quantum theory will recognize the Principle Function as a possible analogue for the wave function. The wave Function also has no physical meaning. But from it all quantities can be obtained. In Sect. 5.9.2 we reproduce Schrödinger's development of his wave equation from the Hamilton–Jacobi Equation.

In this outline of the history of Analytical Mechanics we have apparently neglected to mention the great step provided by Albert Einstein and Hermann Minkowski. We reserve this our final chapter because any appropriate study of special relativity is tied to a deeper understanding of Analytical Mechanics, which we will then have.

1.12 Questions

1.1. We claimed that we must begin the history of our subject in ancient Greece. Particularly we went back to the Presocratics. These people we claimed were physicists because they were more interested in how the world functioned than in how it should function. In a brief essay speak to the concepts of modern and classical physics that find roots in these Presocratics.

1.2. Cicero wrote that Socrates called down philosophy from the skies. With Socrates the truths that the philosophers, the physicists before him, had sought found their importance with people and with human structures on earth. Philosophy became also the study of ethics and of morals. But Socrates wrote nothing. What we know of Socrates comes to us from Plato, his student. What characterized Plato's thought?

1.3. Aristotle Has had a greater influence on philosophical and scientific thought than any other individual. He was Plato's student. But he disagreed with his master about the structure of reality. How would you characterize Aristotle's approach to understanding the world?

1.4. Greek was the language of the Eastern Roman Empire, which did not so quickly suffer the same deterioration that eventually destroyed the Western Roman Empire in 476 CE (Common Era). The academy in Athens was closed by the Eastern Roman emperor Justinian I in 529 CE. But this alone did not stifle communication between the intellectuals of the East and those of the West. It did, however, move the intellectual center farther East. Reflect on the growth of Muslim (Islamic) science that resulted.

1.5. We concentrated on Muslim developments in mathematics and in astronomy because these were the areas of most interest to us. There we found distinct theological as well as philosophical and scientific interactions that determined the directions of the development of Muslim science. Outline these interactions and the influence they had on the form of Aristotle and Ptolemy that eventually appeared in western Europe.

1.6. In 711 CE the Muslims took Spain creating *al Andalus* (Muslim Spain), which lasted until 1492. This was not a continuous period of stable borders separating al Andalus from Christian Europe. The invasion of forces from the Holy Roman Empire pushed the border of al Andalus progressively southward. But the people of al Andalus became inseparably mixed during the 700 years of Muslim rule. The Muslims brought with them a culture, an educational system, and manuscripts that had a great effect on western Europe. What was the impact of this on Europe?

1.7. As physicists and Engineers we tend to write off medieval physics as of little interest in part because of the influence of the Catholic Church. But there were serious scientists even among those taking Holy Orders. What were the issues that were considered during the late medieval period?

1.8. The modern scientist will quickly point to the idea of Nicolaus Copernicus and his work as the beginning of what we now call classical physics. This was followed by the revolutionary steps taken by Johannes Kepler. With Kepler, as well as the ideas of others, Newton was able to synthesize the basis of mechanics. Place these ideas in context.

1.9. The 18th century saw a great advance in the methods of mechanics beyond what Newton had done. A real Analytical Mechanics emerged at the hands of Pierre Louis Maupertuis, Leonhard Euler and Joseph-Louis Lagrange. Provide a brief explanation of what these people did.

1.10. William Rowan Hamilton and Carl Gustav Jacobi put together the (almost) last parts of the Analytical Mechanics. We have provided an outline of their papers in the chapter. Very briefly describe what they did.

Chapter 2
Lagrangian Mechanics

> The reader will find no figures in this work. The methods which I set forth do not require either constructions or geometrical or mechanical reasonings, but only algebraic operations, subject to a regular and uniform rule of procedure.
>
> Joseph-Louis Lagrange in the preface to *Mécanique Analytique*.

2.1 Introduction

In the preceding chapter we outlined the concept of least action and introduced the basis of Lagrangian mechanics. We shall now drop the terminology of the 18th century and develop this concept in a form which can easily be applied in the 21st century.

Our approach will be to first base our development on Newton's Laws and d'Alembert's principle as Lagrange originally did. This will produce the Euler–Lagrange Equations for any system that obeys Newton's Laws. We will follow closely the development of Edmund T. Whittaker[1] in *A Treatise on the Analytical Dynamics of Particles and Rigid Bodies* [[125], Chap. II]. We will then show that the Euler–Lagrange Equations result from a variational principle. In this we will introduce, as Lagrange did, the method of Lagrange undetermined multipliers. This will also provide for us a simple method for handling general constraints that otherwise may present insurmountable algebraic difficulties.

The result will be the regular and uniform rule of procedure promised by Lagrange.

---

[1] Edmund Taylor Whittaker (1873–1956) was professor of mathematics at the University of Edinurgh from 1911 to the end of his career.

C.S. Helrich, *Analytical Mechanics*, Undergraduate Lecture Notes in Physics, DOI 10.1007/978-3-319-44491-8_2

2.2 Kinematics

We consider the motion of a system consisting of a collection of point particles with masses m_i. Because we are treating *classical* mechanics, these particles will not be molecules, atoms, or subatomic particles. They are *classical point particles*, which are small enough, compared to the dimensions of the system, to be considered mathematically as geometrical points. Their description is then completely provided by the position vectors $\boldsymbol{r}_i$ locating the points i and the velocities $\boldsymbol{v}_i$ of the moving points. Classical point particles have no rotational energy.

Kinematics is the description of the motion of these classical point particles, which is the mathematical representation of the position and velocity vectors of the point particles. If we were attempting to formulate our problem as a direct translation of Newton's Second Law into vector form we would also require the acceleration. However, as we found in the preceding chapter, the Analytical Mechanics of Euler and Lagrange only requires the positions and velocities of the particles.

We first choose any one of the standard coordinate systems of analytical geometry, which is convenient for the system at hand. For general descriptions we normally simply choose a rectangular Cartesian system. For problems with particular symmetries we naturally choose a system that reflects those symmetries, such as polar or spherical systems. The position and velocity vectors for a point in the system we have chosen is then the kinematic description of each particle.

Each of the coordinate systems of analytical geometry has three orthonormal basis vectors. In the rectangular Cartesian system these basis vectors $\left(\hat{e}_x, \hat{e}_y, \hat{e}_z\right)$ are fixed in space and oriented along the axes (x, y, z). In the cylindrical system the basis vectors are $\left(\hat{e}_r, \hat{e}_\vartheta, \hat{e}_z\right)$. Only the basis vector $\hat{e}_z$ is fixed in space while the orientation of the vectors $\hat{e}_r$ and $\hat{e}_\vartheta$ change with the motion of the particle. In the spherical system the basis vectors are $\left(\hat{e}_\rho, \hat{e}_\vartheta, \hat{e}_\phi\right)$ none of which is fixed in space as the particle moves.

In the remainder of this section we will formulate the position vector locating the classical point particle and obtain the velocity in each of the coordinate systems. We shall use the standard (Newtonian) dot notation for the time derivative of a function $\dot{f} = \mathrm{d}f/\mathrm{d}t$.

2.2.1 Rectangular Cartesian

We have illustrated the rectangular Cartesian system in Fig. 2.1. The vector $\boldsymbol{r}$ is represented in the basis $\left(\hat{e}_x, \hat{e}_y, \hat{e}_z\right)$. And the triad is right handed, i.e.

$$\hat{e}_x \times \hat{e}_y = \hat{e}_z. \tag{2.1}$$

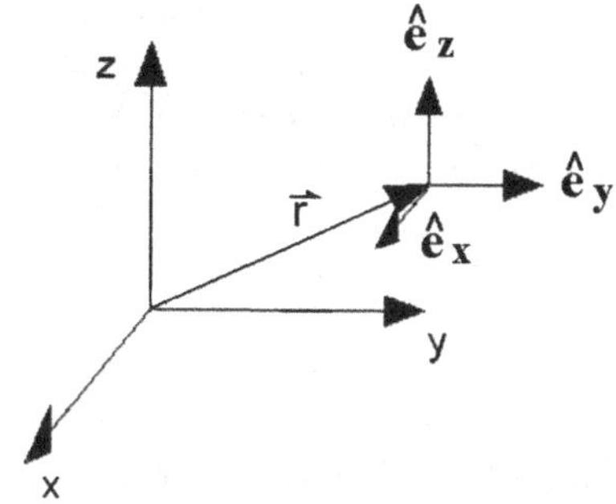

Fig. 2.1 Rectangular Cartesian coordinates

We have drawn the basis vector triad $(\hat{e}_x, \hat{e}_y, \hat{e}_z)$ at the tip of the position vector $\boldsymbol{r}$ for illustrative purposes. The orientation of each of the basis vectors $(\hat{e}_x, \hat{e}_y, \hat{e}_z)$ is fixed in space.

The vector $\boldsymbol{r}$ is

$$\boldsymbol{r} = x\hat{e}_x + y\hat{e}_y + z\hat{e}_z. \tag{2.2}$$

Because only the components (x, y, z) are time dependent the velocity vector $\boldsymbol{v}$ is

$$\boldsymbol{v} = \dot{x}\hat{e}_x + \dot{y}\hat{e}_y + \dot{z}\hat{e}_z. \tag{2.3}$$

2.2.2 Cylindrical

We have illustrated the cylindrical coordinate system in Fig. 2.2. The vector $\boldsymbol{r}$ is represented in the basis $(\hat{e}_r, \hat{e}_\vartheta, \hat{e}_z)$. And the triad $(\hat{e}_r, \hat{e}_\vartheta, \hat{e}_z)$ is right handed so that

$$\hat{e}_r \times \hat{e}_\vartheta = \hat{e}_z. \tag{2.4}$$

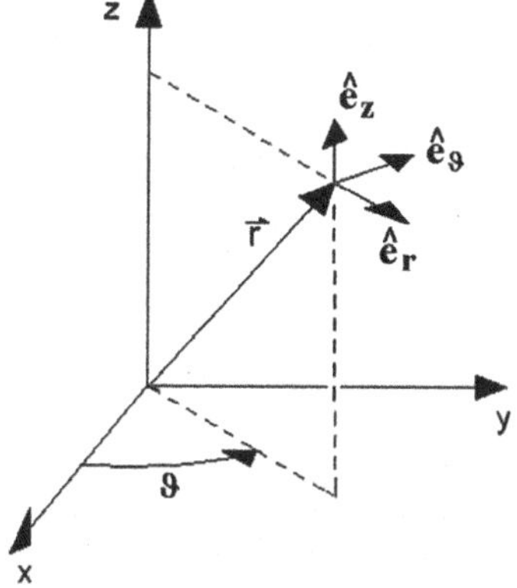

Fig. 2.2 Cylindrical coordinates

We have drawn the vector triad $(\hat{e}_{\mathrm{r}}, \hat{e}_{\vartheta}, \hat{e}_{\mathrm{z}})$ at the tip of the position vector $\boldsymbol{r}$. The basis vector $\hat{e}_{\mathrm{z}}$ remains oriented along the vertical axis and is independent of time. The basis vector $\hat{e}_{\mathrm{r}}$ is in the plane of $\boldsymbol{r}$ and $\hat{e}_{\mathrm{z}}$ and oriented perpendicularly to $\hat{e}_{\mathrm{z}}$. The basis vector $\hat{e}_{\vartheta}$ is perpendicular to $\hat{e}_{\mathrm{r}}$ and $\hat{e}_{\mathrm{z}}$. The basis vectors $\hat{e}_{\mathrm{r}}$ and $\hat{e}_{\vartheta}$ then rotate around an axis parallel to $\hat{e}_{\mathrm{z}}$ as the tip of the position vector $\boldsymbol{r}$ moves.

The position vector $\boldsymbol{r}$ is

$$\boldsymbol{r} = r\hat{e}_{\mathrm{r}} + z\hat{e}_{\mathrm{z}}. \tag{2.5}$$

The differential rotation $\mathrm{d}\vartheta$ of the dyad $(\hat{e}_{\mathrm{r}}, \hat{e}_{\vartheta})$ around $\hat{e}_{\mathrm{z}}$ results in a differential change $\mathrm{d}\hat{e}_{\mathrm{r}} = \mathrm{d}\vartheta\hat{e}_{\vartheta}$ in the basis vector $\hat{e}_{\mathrm{r}}$. The velocity vector $\boldsymbol{v}$ is then

$$\boldsymbol{v} = \dot{r}\hat{e}_{\mathrm{r}} + r\dot{\vartheta}\hat{e}_{\vartheta} + \dot{z}\hat{e}_{\mathrm{z}}. \tag{2.6}$$

2.2.3 Spherical

We have illustrated the spherical coordinate system in Fig. 2.3. The vector $\boldsymbol{r}$ is represented in the basis $(\hat{e}_{\rho}, \hat{e}_{\vartheta}, \hat{e}_{\phi})$. And the triad $(\hat{e}_{\rho}, \hat{e}_{\vartheta}, \hat{e}_{\phi})$ is right handed so that

$$\hat{e}_{\rho} \times \hat{e}_{\phi} = \hat{e}_{\vartheta}. \tag{2.7}$$

The length (magnitude) of the position vector $\boldsymbol{r}$ is ρ.

We have drawn the basis vector triad $(\hat{e}_{\rho}, \hat{e}_{\vartheta}, \hat{e}_{\phi})$ at the tip of the position vector $\boldsymbol{r}$. The basis vector $\hat{e}_{\rho}$ is oriented along the direction of the position vector $\boldsymbol{r}$ and changes in orientation with the position vector. The basis vector $\hat{e}_{\rho}$ is then independent of the length ρ of the position vector $\boldsymbol{r}$ and depends on the *azimuthal*[2] angle ϑ and the *polar* angle ϕ. An infinitesimal rotation $\mathrm{d}\phi$ of the polar angle produces a differential

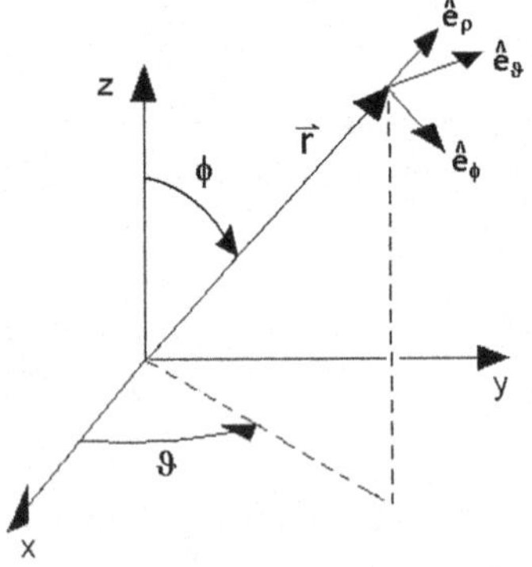

Fig. 2.3 Spherical coordinates

[2] Azimuth comes from the Arabic word as-simt, which means direction, referring to the direction a person faces. In Fig. 2.3 the azimuthal angle ϑ locates the projection of the vestor $\boldsymbol{r}$ on the (x, y) plane with respect to the $x-$axis.

change $\mathrm{d}\hat{e}_\rho = \mathrm{d}\phi\hat{e}_\phi$ in the basis vector $\hat{e}_\rho$, which is analogous to the change $\mathrm{d}\hat{e}_\mathrm{r}$ in the cylindrical basis vector we found above. In the same fashion we deduce the differential change in $\hat{e}_\rho$ resulting from a differential change $\mathrm{d}\vartheta$ in the azimuthal angle as $\mathrm{d}\hat{e}_\rho = \sin\phi \mathrm{d}\vartheta\hat{e}_\vartheta$. The Pfaffian[3] for the basis vector $\hat{e}_\rho$ is then

$$\begin{aligned}\mathrm{d}\hat{e}_\rho &= \left(\frac{\partial \hat{e}_\rho}{\partial \phi}\right)\mathrm{d}\phi + \left(\frac{\partial \hat{e}_\rho}{\partial \vartheta}\right)\mathrm{d}\vartheta \\ &= \mathrm{d}\phi\hat{e}_\phi + \sin\phi \mathrm{d}\vartheta\hat{e}_\vartheta.\end{aligned} \tag{2.8}$$

The position vector $\boldsymbol{r}$ is

$$\boldsymbol{r} = \rho\hat{e}_\rho. \tag{2.9}$$

With (2.8) the velocity vector $\boldsymbol{v}$ is

$$\boldsymbol{v} = \dot{\rho}\hat{e}_\rho + \rho\dot{\vartheta}\sin\phi\hat{e}_\vartheta + \rho\dot{\phi}\hat{e}_\phi. \tag{2.10}$$

2.3 From Newton's Laws

2.3.1 General Formulation

The total force acting on the ith (point) particle we designate as $\boldsymbol{F}_\mathrm{i}$ and consider that this force consists of both forces whose origin is external to the system $\boldsymbol{F}_\mathrm{i}^{\text{ext}}$ and forces arising from interactions among the particles, which we term internal forces $\boldsymbol{F}_\mathrm{i}^{\text{int}}$. That is

$$\boldsymbol{F}_\mathrm{i} = \boldsymbol{F}_\mathrm{i}^{\text{ext}} + \boldsymbol{F}_\mathrm{i}^{\text{int}}.$$

External forces arise from fields, such as gravitational or electromagnetic, and possible contact forces from external bodies constraining the motion.[4] The system of particles we are considering is, therefore, the most general possible in classical terms.

Applying Newton's Second Law to each particle i we obtain

$$m_\mathrm{i}\frac{\mathrm{d}^2}{dt^2}\boldsymbol{r}_\mathrm{i} = \boldsymbol{F}_\mathrm{i}^{\text{ext}} + \boldsymbol{F}_\mathrm{i}^{\text{int}}. \tag{2.11}$$

Written in (rectangular) Cartesian coordinates (x, y, z) (2.11) is a set of three equations,

---

[3] Johann Friedrich Pfaff (1765–1825) was one of Germany's most eminent mathematicians during the 19th century. He is noted for his work on partial differential equations of the first order, which became part of the theory of differential forms. He was also Carl Friedrich Gauss's formal research supervisor.

[4] We realize, of course, that what are considered as contact forces are the result of electromagnetic forces between the atoms making up the particles and those making up the constraining surfaces.

$$m_i \ddot{x}_i = F_{xi}^{ext} + F_{xi}^{int}, \tag{2.12}$$

$$m_i \ddot{y}_i = F_{yi}^{ext} + F_{yi}^{int}, \tag{2.13}$$

and

$$m_i \ddot{z}_i = F_{zi}^{ext} + F_{zi}^{int}. \tag{2.14}$$

We may sum (2.12)–(2.14) over all particles i to obtain the general form of the application of Newton's Laws to our system. In the summation we use Newton's Third Law , which applies to the internal forces of interaction between all pairs of particles. The summation over all of the internal forces is then zero regardless of whether the forces are from interactions between the particles, making up what we see as a single material body, or include interactions between the particles making up separate bodies. Equations (2.12)–(2.14) then become

$$\sum_i m_i \ddot{x}_i = \sum_i F_{xi}, \tag{2.15}$$

$$\sum_i m_i \ddot{y}_i = \sum_i F_{yi}, \tag{2.16}$$

and

$$\sum_i m_i \ddot{z}_i = \sum_i F_{zi}, \tag{2.17}$$

where we drop the superscript *ext* as superfluous. The set of second order differential equations (2.15)–(2.17) constitute a general description of the mechanical behavior of a group of material bodies, provided we require that the particles obey Newton's Laws.

2.3.2 Generalized Coordinates

In most applications we do not need the full set of Cartesian coordinates because of the constraints on the system we are studying. As an example in Fig. 2.4 we consider

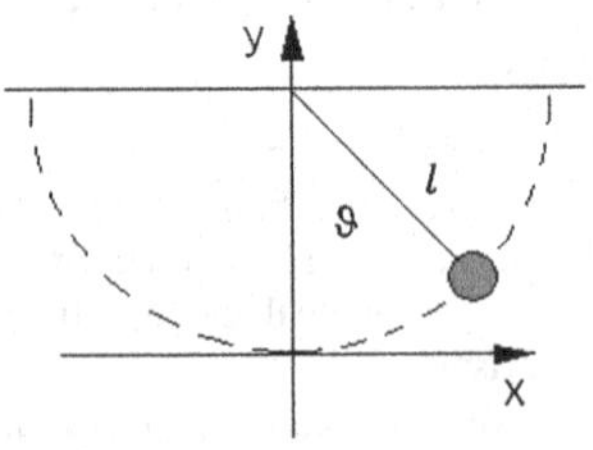

Fig. 2.4 Simple Pendulum with generalized coordinates ℓ and ϑ

a simple pendulum consisting of a mass suspended by a light rod. The motion of the mass is constrained to remain on a circle of radius ℓ. The Cartesian coordinates for this system may be written as

$$x = \ell \sin \vartheta \tag{2.18}$$

and

$$y = \ell (1 - \cos \vartheta) . \tag{2.19}$$

We then have a complete description of the motion in terms of a single variable ϑ. We refer to this single variable, which is completely adequate for the description of the system originally formulated in the two rectangular Cartesian coordinates (x, y), as a *generalized coordinate*. The reduction in coordinates resulted from the single constraint $\ell = \sqrt{x^2 + y^2}$.

Physical constraints on systems generally impose geometric constraints on our representations of those systems. A curve in a plane imposes a relationship between two Cartesian coordinates, which are then no longer independent. We may then reduce the original pair of planar coordinates to a single coordinate. In our example the single independent coordinate was an angle ϑ expressing the relationship between x and y. The situations we encounter in applications will, however, seldom be so simple. For the general case we will require the method of undetermined multipliers introduced by Lagrange.

We may, however, claim that constraints will always impose geometrical relationships between the rectangular coordinates for the particles in our system $(x_\mathrm{i}, y_\mathrm{i}, z_\mathrm{i})$ and the generalized coordinates q_i of the form

$$x_\mathrm{i} = x_\mathrm{i} (q, t) , \tag{2.20}$$

$$y_\mathrm{i} = y_\mathrm{i} (q, t) , \tag{2.21}$$

and

$$z_\mathrm{i} = z_\mathrm{i} (q, t) , \tag{2.22}$$

where we have introduced the shorthand $q = \{q_\mathrm{i}\}$ for the set of generalized coordinates. We must only accept that we may not be capable of writing these relationships in closed form.

From (2.20)–(2.22), Pfaff's[5] differential forms, or Pfaffians,[6] for the rectangular Cartesian coordinates $(x_i, y_i z_i)$ are

$$\begin{aligned} dx_i &= \sum_k \frac{\partial x_i}{\partial q_k} dq_k + \frac{\partial x_i}{\partial t} dt \\ dy_i &= \sum_k \frac{\partial y_i}{\partial q_k} dq_k + \frac{\partial y_i}{\partial t} dt \\ dz_i &= \sum_k \frac{\partial z_i}{\partial q_k} dq_k + \frac{\partial z_i}{\partial t} dt. \end{aligned} \tag{2.23}$$

2.3.3 Virtual Displacement

We introduced the concept of *virtual displacement* of the ith particle $\delta \boldsymbol{r}_i$ in the preceding chapter. This virtual displacement is consistent with the constraints acting on the system and could be an infinitesimal element of the actual path followed by the system. But we hold the time constant, i.e. $dt = 0$. The displacement $\delta \boldsymbol{r}_i$ is then *virtual* in that it does not actually take place. The actual variation would require $dt \neq 0$. From the Pfaffians (2.23), with $dt = 0$, we have

$$\boxed{\delta \boldsymbol{r}_i = \sum_k \left[(\partial x_i / \partial q_k)\, \hat{e}_x + (\partial y_i / \partial q_k)\, \hat{e}_y + (\partial z_i / \partial q_k)\, \hat{e}_z \right] \delta q_k} \tag{2.24}$$

for the virtual displacement.

2.3.4 D'Alembert's Principle

Because (2.11) must hold for every particle at each step along the path followed by the system,

$$m_i \frac{d^2}{dt^2} \boldsymbol{r}_i - \boldsymbol{F}_i = \boldsymbol{0}$$

[5]Johann Friedrich Pfaff (1765–1825) was one of Germany's most eminent mathematicians during the 19th century. He is noted for his work on partial differential equations of the first order, which became part of the theory of differential forms. He was also Carl Friedrich Gauss's formal research supervisor.

[6]Pfaff's differential form for the function $\Psi\ (\xi_1, \ldots, \xi_n)$ is defined as

$$d\Psi = \sum_j^n \left(\frac{\partial \Psi}{\partial \xi_j} \right) d\xi_j \,.$$

at each point and during the next infinitesimal part of the path followed by the particle i. The virtual displacement $\delta \boldsymbol{r}_{\mathrm{i}}$ is a *possible* next infinitesimal part of the path, consistent with the constraints on the system. If $\delta \boldsymbol{r}_{\mathrm{i}}$ were the actual path followed we would have

$$\sum_{\mathrm{i}} \left[m_{\mathrm{i}} \frac{\mathrm{d}^2}{\mathrm{d}t^2} \boldsymbol{r}_{\mathrm{i}} - \vec{F}_{\mathrm{i}} \right] \cdot \delta \boldsymbol{r}_{\mathrm{i}} = 0. \tag{2.25}$$

However, if $\delta \boldsymbol{r}_{\mathrm{i}}$ is a virtual displacement the product in (2.25) may not be identically zero at all points along $\delta \boldsymbol{r}_{\mathrm{i}}$. But the difference between the product in (2.25) and zero would only result in higher order terms in δ's. Therefore, to first order in δ's (2.25), which is *d'Alembert's Principle*, is valid for the virtual displacement $\delta \boldsymbol{r}_{\mathrm{i}}$. [see [65], pp. 88–110]

We note that d'Alembert's Principle is expressed as a scalar equation involving what is termed *virtual work* $\vec{F}_{\mathrm{i}} \cdot \delta \boldsymbol{r}_{\mathrm{i}}$, which is the work that would be done on the mass m_{i} in the virtual displacement $\delta \boldsymbol{r}_{\mathrm{i}}$. This virtual work is equal to a corresponding virtual change in the kinetic energy of the mass m_{i}, which is $m_{\mathrm{i}} \mathrm{d}^2 \boldsymbol{r}_{\mathrm{i}} / \mathrm{d}t^2 \cdot \delta \boldsymbol{r}_{\mathrm{i}}$. This formulation includes, in principle, work by dissipative forces[7] as well. Work and energy then replace forces in a formulation of mechanics based on d'Alembert's Principle.

2.3.5 Euler–Lagrange Equations

In this section we provide the details of the derivation of the Euler–Lagrange Equations from d'Alembert's Principle. We will introduce generalized coordinates q into the virtual displacements, which we wrote above in terms of Cartesian coordinates. Beyond the transition from Cartesian coordinates to the generalized coordinates, this section is primarily a mathematical discussion involving some creative use of partial derivative relations. The passage from Newton's Laws to the Euler–Lagrange Equations without a variational principle is, however, a necessary part of the development.

Using (2.24) d'Alembert's Principle, Eq. (2.25), becomes

$$\begin{aligned} 0 = \sum_{\mathrm{i,k}} & \left[m_{\mathrm{i}} \left(\ddot{x}_{\mathrm{i}} \frac{\partial x_{\mathrm{i}}}{\partial q_{\mathrm{k}}} + \ddot{y}_{\mathrm{i}} \frac{\partial y_{\mathrm{i}}}{\partial q_{\mathrm{k}}} + \ddot{z}_{\mathrm{i}} \frac{\partial z_{\mathrm{i}}}{\partial q_{\mathrm{k}}} \right) \right. \\ & \left. - \left(F_{\mathrm{xi}} \frac{\partial x_{\mathrm{i}}}{\partial q_{\mathrm{k}}} + F_{\mathrm{yi}} \frac{\partial y_{\mathrm{i}}}{\partial q_{\mathrm{k}}} + F_{\mathrm{zi}} \frac{\partial z_{\mathrm{i}}}{\partial q_{\mathrm{k}}} \right) \right] \delta q_{\mathrm{k}}. \end{aligned} \tag{2.26}$$

[7]Dissipative forces are frictional forces. These ultimately result from contact forces between moving bodies and surfaces or moving bodies and moving fluids, all of which are molecular and electromagnetic in nature. If we claim such detailed knowledge these forces are conservative. We may later insert these in modeled form, if they arise.

Equation (2.26) is of the form

$$0=\sum_{\mathrm{k}} \alpha_{\mathrm{k}} \delta q_{\mathrm{k}}. \tag{2.27}$$

Since the generalized coordinates q_{k} are independent of one another, the δq_{k} are arbitrary. Therefore (2.27) can only be valid if each α_{k} is independently zero. That is

$$\begin{aligned} & \sum_{\mathrm{i}} m_{\mathrm{i}}\left(\ddot{x}_{\mathrm{i}} \frac{\partial x_{\mathrm{i}}}{\partial q_{\mathrm{k}}}+\ddot{y}_{\mathrm{i}} \frac{\partial y_{\mathrm{i}}}{\partial q_{\mathrm{k}}}+\ddot{z}_{\mathrm{i}} \frac{\partial z_{\mathrm{i}}}{\partial q_{\mathrm{k}}}\right) \\ & \quad=\sum_{\mathrm{i}}\left(F_{\mathrm{xi}} \frac{\partial x_{\mathrm{i}}}{\partial q_{\mathrm{k}}}+F_{\mathrm{yi}} \frac{\partial y_{\mathrm{i}}}{\partial q_{\mathrm{k}}}+F_{\mathrm{zi}} \frac{\partial z_{\mathrm{i}}}{\partial q_{\mathrm{k}}}\right) \end{aligned} \tag{2.28}$$

for each component k. Equation (2.28) must then be valid if the system obeys Newton's Laws.

Because the Cartesian coordinates are functions of the generalized coordinates and the time (see (2.20)–(2.22)) the time derivative of the coordinate x_{i} is

$$\dot{x}_{\mathrm{i}}=\frac{\mathrm{d} x_{\mathrm{i}}}{\mathrm{d} t}=\sum_{\mathrm{k}} \frac{\partial x_{\mathrm{i}}}{\partial q_{\mathrm{k}}} \dot{q}_{\mathrm{k}}+\frac{\partial x_{\mathrm{i}}}{\partial t}. \tag{2.29}$$

If we now take the partial derivative of (2.29) with respect to $\dot{q}_{\mathrm{k}}$ we obtain

$$\frac{\partial \dot{x}_{\mathrm{i}}}{\partial \dot{q}_{\mathrm{k}}}=\frac{\partial x_{\mathrm{i}}}{\partial q_{\mathrm{k}}} \tag{2.30}$$

Equation (2.30) is often called *cancellation of the dots* because it appears as though we have simply cancelled the dots (time derivatives) in $\partial \dot{x}_{\mathrm{i}} / \partial \dot{q}_{\mathrm{k}}$ to obtain $\partial x_{\mathrm{i}} / \partial q_{\mathrm{k}}$. Mathematically (2.30) is a consequence of the fact that the Cartesian coordinate x_{i} depends only on q and the time t and is independent of the velocities $\dot{q}$.

With (2.30) we can write the terms appearing on the left hand side of (2.28) as

$$\ddot{x}_{\mathrm{i}} \frac{\partial x_{\mathrm{i}}}{\partial q_{\mathrm{k}}}=\ddot{x}_{\mathrm{i}} \frac{\partial \dot{x}_{\mathrm{i}}}{\partial \dot{q}_{\mathrm{k}}}. \tag{2.31}$$

Now

$$\ddot{x}_{\mathrm{i}} \frac{\partial \dot{x}_{\mathrm{i}}}{\partial \dot{q}_{\mathrm{k}}}=\frac{\mathrm{d}}{\mathrm{d} t}\left(\dot{x}_{\mathrm{i}} \frac{\partial \dot{x}_{\mathrm{i}}}{\partial \dot{q}_{\mathrm{k}}}\right)-\dot{x}_{\mathrm{i}} \frac{\mathrm{d}}{\mathrm{d} t}\left(\frac{\partial x_{\mathrm{i}}}{\partial q_{\mathrm{k}}}\right), \tag{2.32}$$

using (2.30) in the last term on the right hand side of (2.32). Since $\partial x_{\mathrm{i}} / \partial q_{\mathrm{k}}$ depends on (q, t) as does x_{i}, the time derivative of $\partial x_{\mathrm{i}} / \partial q_{\mathrm{k}}$ has the same form as (2.29). That is

$$\frac{\mathrm{d}}{\mathrm{d} t}\left(\frac{\partial x_{\mathrm{i}}}{\partial q_{\mathrm{k}}}\right)=\sum_{\mathrm{j}} \frac{\partial^2 x_{\mathrm{i}}}{\partial q_{\mathrm{j}} \partial q_{\mathrm{k}}} \dot{q}_{\mathrm{j}}+\frac{\partial^2 x_{\mathrm{i}}}{\partial t \partial q_{\mathrm{k}}}. \tag{2.33}$$

Also from (2.29) we have

$$\frac{\partial \dot{x}_{\mathrm{i}}}{\partial q_{\mathrm{k}}}=\sum_{\mathrm{j}} \frac{\partial^2 x_{\mathrm{i}}}{\partial q_{\mathrm{j}} \partial q_{\mathrm{k}}} \dot{q}_{\mathrm{j}}+\frac{\partial^2 x_{\mathrm{i}}}{\partial t \partial q_{\mathrm{k}}}, \tag{2.34}$$

since the order of partial differentiation is immaterial.

Therefore from (2.33) and (2.34) we see that

$$\frac{\mathrm{d}}{\mathrm{d}t}\left(\frac{\partial x_{\mathrm{i}}}{\partial q_{\mathrm{k}}}\right)=\frac{\partial \dot{x}_{\mathrm{i}}}{\partial q_{\mathrm{k}}}. \tag{2.35}$$

With (2.35) Eq. (2.32) becomes

$$\begin{aligned} \ddot{x}_{\mathrm{i}} \frac{\partial x_{\mathrm{i}}}{\partial q_{\mathrm{k}}} &= \frac{\mathrm{d}}{\mathrm{d}t}\left(\dot{x}_{\mathrm{i}} \frac{\partial \dot{x}_{\mathrm{i}}}{\partial \dot{q}_{\mathrm{k}}}\right)-\dot{x}_{\mathrm{i}} \frac{\partial \dot{x}_{\mathrm{i}}}{\partial q_{\mathrm{k}}} \\ &= \left[\frac{\mathrm{d}}{\mathrm{d}t} \frac{\partial}{\partial \dot{q}_{\mathrm{k}}}-\frac{\partial}{\partial q_{\mathrm{k}}}\right]\left(\frac{1}{2} \dot{x}_{\mathrm{i}}^2\right). \end{aligned} \tag{2.36}$$

Then using (2.36) the left hand side of (2.28) becomes

$$\begin{aligned} &\sum_{\mathrm{i}} m_{\mathrm{i}}\left(\ddot{x}_{\mathrm{i}} \frac{\partial x_{\mathrm{i}}}{\partial q_{\mathrm{k}}}+\ddot{y}_{\mathrm{i}} \frac{\partial y_{\mathrm{i}}}{\partial q_{\mathrm{k}}}+\ddot{z}_{\mathrm{i}} \frac{\partial z_{\mathrm{i}}}{\partial q_{\mathrm{k}}}\right) \\ &=\left[\frac{\mathrm{d}}{\mathrm{d}t} \frac{\partial}{\partial \dot{q}_{\mathrm{k}}}-\frac{\partial}{\partial q_{\mathrm{k}}}\right] \sum_{\mathrm{i}} \frac{1}{2} m_{\mathrm{i}}\left(\dot{x}_{\mathrm{i}}^2+\dot{y}_{\mathrm{i}}^2+\dot{z}_{\mathrm{i}}^2\right) \end{aligned} \tag{2.37}$$

We recognize the term $\sum_{\mathrm{i}}(1/2)\, m_{\mathrm{i}}\left(\dot{x}_{\mathrm{i}}^2+\dot{y}_{\mathrm{i}}^2+\dot{z}_{\mathrm{i}}^2\right)$ as the kinetic energy , which, in keeping with the notation of Lagrange and Hamilton,[8] we shall designate as

$$T=\sum_{\mathrm{i}} \frac{1}{2} m_{\mathrm{i}}\left(\dot{x}_{\mathrm{i}}^2+\dot{y}_{\mathrm{i}}^2+\dot{z}_{\mathrm{i}}^2\right). \tag{2.38}$$

Then (2.28), which is the requirement that the system obeys Newton's Laws, is

$$\left[\frac{\mathrm{d}}{\mathrm{d}t} \frac{\partial}{\partial \dot{q}_{\mathrm{k}}}-\frac{\partial}{\partial q_{\mathrm{k}}}\right] T=\sum_{\mathrm{i}}\left(F_{\mathrm{xi}} \frac{\partial x_{\mathrm{i}}}{\partial q_{\mathrm{k}}}+F_{\mathrm{yi}} \frac{\partial y_{\mathrm{i}}}{\partial q_{\mathrm{k}}}+F_{\mathrm{zi}} \frac{\partial z_{\mathrm{i}}}{\partial q_{\mathrm{k}}}\right). \tag{2.39}$$

We now recall that the forces remaining are those arising from external fields. In 18th century notation these forces equal to the positive gradient of the force function U. In modern notation these forces are equal to the *negative* gradient of a scalar

[8]Denoting kinetic energy as T is standard modern notation. The fact that T is also used for thermodynamic temperature, and that the kinetic energy of an ideal gas is proportional to thermodynamic temperature, is incidental.

potential V, which is a function only of spatial coordinates. That is

$$F_{\mathrm{xi}} = -\frac{\partial V}{\partial x_{\mathrm{i}}},\ F_{\mathrm{yi}} = -\frac{\partial V}{\partial x_{\mathrm{i}}},\ F_{\mathrm{zi}} = -\frac{\partial V}{\partial z_{\mathrm{i}}}.$$

Therefore, the right hand side of (2.39) is

$$\sum_{\mathrm{i}} \left(\frac{\partial V}{\partial x_{\mathrm{i}}} \frac{\partial x_{\mathrm{i}}}{\partial q_{\mathrm{k}}} + \frac{\partial V}{\partial y_{\mathrm{i}}} \frac{\partial y_{\mathrm{i}}}{\partial q_{\mathrm{k}}} + \frac{\partial V}{\partial z_{\mathrm{i}}} \frac{\partial z_{\mathrm{i}}}{\partial q_{\mathrm{k}}} \right) = \frac{\partial V}{\partial q_{\mathrm{k}}}, \tag{2.40}$$

using the chain rule. With (2.40) Eq. (2.39) becomes

$$\left[\frac{\mathrm{d}}{\mathrm{d}t} \frac{\partial}{\partial \dot{q}_{\mathrm{k}}} - \frac{\partial}{\partial q_{\mathrm{k}}} \right] T = -\frac{\partial V}{\partial q_{\mathrm{k}}}. \tag{2.41}$$

Since the potential V depends only on the coordinates and not on the velocities (2.41) may be written as

$$\left[\frac{\partial}{\partial q_{\mathrm{k}}} - \frac{\mathrm{d}}{\mathrm{d}t} \frac{\partial}{\partial \dot{q}_{\mathrm{k}}} \right] (T - V) = 0. \tag{2.42}$$

There is an equation of the form (2.42) for each of the generalized coordinates q_{k}.

The Eqs. (2.42) are the *Euler–Lagrange Equations*. The combination $T - V$ is called the *Lagrangian*

$$L = T - V. \tag{2.43}$$

The Lagrangian is a scalar function of the generalized coordinates q, the time derivatives of the generalized coordinates $\dot{q}$, and possibly the time t. To obtain the Lagrangian we only need the kinetic energies of the interacting bodies and the potential energies of the external fields.

With (2.43) Eq. (2.42) becomes

$$\boxed{\partial L/\partial q_{\mathrm{k}} - \mathrm{d}(\partial L/\partial \dot{q}_{\mathrm{k}})\,/\mathrm{d}t = 0.} \tag{2.44}$$

The set of Eq. (2.44) is the final form of the Euler–Lagrange Equations.[9]

2.4 Variational Calculus

Our derivation of the Euler–Lagrange Equations in the preceding section was based strictly on the differential calculus and an understanding of the elements of linear algebra (see (2.27)). We had no need of any variational principle, even though we

[9] Some authors prefer to write the Euler–Lagrange equations as $\mathrm{d}(\partial L/\partial \dot{q}_{\mathrm{k}})\,/\mathrm{d}t - \partial L/\partial q_{\mathrm{k}} = 0$. Our choice is based on the fact that this is the natural order resulting from the variational principle.

know from the history of Analytical Mechanics that the principle of least action was central to the thinking, particularly at the Berlin Academy (see Sects. (1.8.1)–(1.8.3)). The formulation of the science in terms of a variational principle, and an introduction of the rudiments of variational calculus, are, therefore, vitally important to our understanding of Analytical Mechanics. Beyond the philosophical importance, a formulation in terms of a variational principle is also absolutely necessary for the incorporation of general constraints into mechanical problems.

In this section we will introduce the basis of variational calculus and show that the Euler–Lagrange equations result from a variational principle.

2.4.1 Functionals

A *functional* defines an operation on a class of functions $\{y(x)\}$ that returns a real number for each function $y(x)$ [cf. [31]]. For example such a functional may be the definite integral of a quantity $F\left[x, y(x), y'(x)\right]$ dependent on the function $y(x)$, its derivative $y'(x)$ and the independent variable x over the interval $[a, b]$. In this case the functional $J[y]$ is the number

$$J[y] = \int_{\mathrm{a}}^{\mathrm{b}} \mathrm{d}x\, F\left[x, y(x), y'(x)\right], \tag{2.45}$$

which is dependent on the function $y(x)$ chosen from the class $\{y(x)\}$ and the values a and b chosen as the limits of integration.

Example 2.4.1 As a specific example we consider the area of the surface of rotation $A_{\mathrm{S}}[y]$ for a class of functions $\{y(x)\}$. We choose our class of functions to be single-valued, have continuous first derivatives, and pass through the two points (x_0, y_0) and (x_1, y_1). The differential area of the surface formed by the rotation of the curve $y(x)$ about the $x-$axis is shown in Fig. 2.5. The differential area of the surface of revolution $\mathrm{d}A_{\mathrm{S}}\left[x, y(x), y'(x)\right]$ between the points x and $x + \mathrm{d}x$ on the $x-$axis is equal to the product of the circumference of the circular cross section of the surface $2\pi y(x)$ and the distance along the curve $\mathrm{d}s$ resulting from the differential distance $x \to x + \mathrm{d}x$ along the $x-$axis. above. That is

$$\begin{aligned} \mathrm{d}A_{\mathrm{S}}\left[x, y(x), y'(x)\right] &= 2\pi y(x) \sqrt{\mathrm{d}x^2 + \mathrm{d}y^2} \\ &= \mathrm{d}x\, [2\pi y(x)] \sqrt{1 + (y'(x))^2} \end{aligned}$$

The total area of revolution is then

$$A_{\mathrm{S}}[y] = \int_{x_2}^{x_1} \mathrm{d}x \left[2\pi y(x) \sqrt{1 + (y'(x))^2}\right]. \tag{2.46}$$

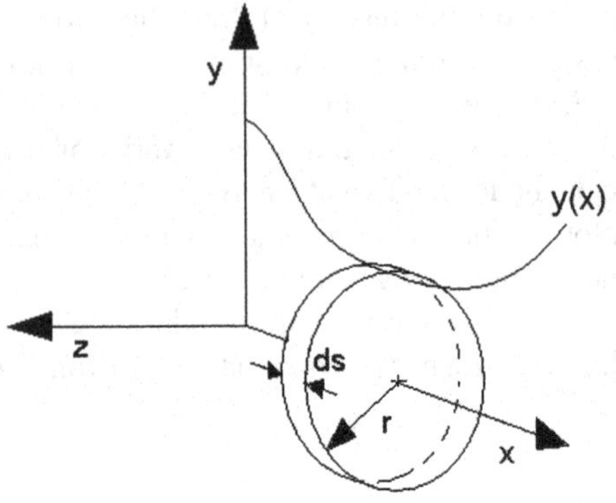

Fig. 2.5 Differential area of the surface of revolution formed by the curve $y = y(x)$

In this example the function $F\left[x, y(x), y'(x)\right]$ is

$$F\left[x, y(x), y'(x)\right] = 2\pi y(x)\sqrt{1 + (y'(x))^2}$$

We may now ask for the function $y(x)$ that results in a maximum or a minimum of the area of rotation in (2.46).

2.4.2 *Extrema of Functionals*

For a function $F\left(x, y, y'\right)$ there may be a specific function $y(x) = \eta(x)$ in the class $\{y(x)\}$ that results in an extreme value of the functional $J[y]$ in (2.45). We then consider all possible functions in the class $\{y(x)\}$ that differ only very slightly (infinitesimally) from $\eta(x)$. That is we write the function $y(x)$, which differs only slightly from $\eta(x)$, as

$$y(x) = \eta(x) + h(x), \tag{2.47}$$

where $h(x)$ is an infinitesimal function of the variable x contained within the class of functions $\{y(x)\}$. We then define $\Delta J[h]$ as the functional

$$\Delta J[h] = J[y] - J[\eta]. \tag{2.48}$$

This functional is an infinitesimal real number for each infinitesimal function $h(x)$ in $\{y(x)\}$. The extremum $J[\eta]$ is a *minimum* if each $h(x)$ increases the value of $J[y]$ and

$$\Delta J[h] = J[y] - J[\eta] \geq 0, \tag{2.49}$$

and is a *maximum* if each $h(x)$ decreases the value of $J[y]$ and

$$\Delta J[h] = J[y] - J[\eta] \leq 0. \tag{2.50}$$

Because $h(x)$ is infinitesimal at every point on the interval in x being considered, the function $F(x, y, y')$ in (2.45) may be written as a generalized Taylor series of the form

$$F(x, y, y') = F(x, \eta, \eta') + h \left.\frac{\partial F}{\partial y}\right|_{y=\eta} + h' \left.\frac{\partial F}{\partial y'}\right|_{y=\eta} + \cdots, \tag{2.51}$$

at each point x on the interval of interest (i.e. $x \in [a, b]$). In (2.51) $\partial F/\partial y|_{y=\eta}$ and $\partial F/\partial y'|_{y=\eta}$ are the partial derivatives $\partial F/\partial y$ and $\partial F/\partial y'$, which are functions of y and y', evaluated in the limit $h(x) \to 0$ and $h'(x) \to 0$. The general functions $y(x)$ and $y'(x)$ are then replaced by the functions $\eta(x)$ and $\eta'(x)$ in the partial derivatives $\partial F/\partial y$ and $\partial F/\partial y'$. Using (2.51) in $\Delta J[h]$,

$$\begin{aligned}\Delta J[h] &= J[y] - J[\eta] \\ &= \int_a^b \mathrm{d}x\, F(x, \eta + h, \eta' + h') - \int_a^b \mathrm{d}x\, F(x, \eta, \eta') \\ &= \int_a^b \mathrm{d}x \left[h(x) \left.\frac{\partial F}{\partial y}\right|_{y=\eta} + h'(x) \left.\frac{\partial F}{\partial y'}\right|_{y=\eta} + \cdots \right].\end{aligned} \tag{2.52}$$

With the definition

Definition 2.4.1 That contribution to $\Delta J[h]$ which is of order n in $h(x)$, $h'(x)$, or products of $h(x)$ and $h'(x)$ is defined as $\delta^{(n)} J[h]$.

Equation (2.52) may be written as

$$\Delta J[h] = \delta J[h] + \delta^{(2)} J[h] + \cdots. \tag{2.53}$$

The necessary condition for an extremum of the functional is that the first order variation vanishes. That is

$$\delta J[h] = 0 \text{ at an extremum} \tag{2.54}$$

This condition alone is called the *weak extremum* because it does not specify whether the extremum is a maximum or a minimum. Whether the extremum is a maximum or a minimum must be determined by investigating the algebraic sign of the variation $\delta^{(2)} J[h]$.

This is the δ–method developed by Lagrange (see Sect. 1.8.3), provided we also require the the values of the class of functions $\{y(x)\}$ are fixed at the end points. We must then require that the infinitesimal functions $h(x)$ vanish at the end points. This restriction was included by Lagrange.

2.4.3 Euler Problem

The Euler problem is to obtain the conditions on $y(x)$ such that the functional (2.45) has a weak extremum, i.e. that $\delta J[y] = 0$. [[31], pp. 14, 15] We shall solve this problem using the δ−method of Lagrange, however, since we realize that there are mathematical difficulties at the end points in Euler's method (see Sect. 1.8.2).

We require that the function $F(x, y, y')$ has continuous first and second (partial) derivatives with respect to all its arguments. And we require that the function $y(x)$ belongs to a class of functions that have continuous first derivatives for $a \leq x \leq b$ (i.e. for $x \in [a, b]$), which satisfies the (fixed) boundary conditions

$$y(a) = A, \qquad y(b) = B.$$

The functions $y(x)$ and $\eta(x)$ in (2.47) are then equal at the end points and, therefore, the function $h(x)$ must vanish at the end points.

From (2.52) a weak extremum requires that

$$\delta J[h] = 0 = \int_{\mathrm{a}}^{\mathrm{b}} \mathrm{d}x \left[h(x) \left. F_{\mathrm{y}} \right|_{\mathrm{y}=\eta} + h'(x) \left. F_{\mathrm{y}'} \right|_{\mathrm{y}=\eta} \right], \tag{2.55}$$

where we have introduced the notation $F_{\mathrm{y}} = \partial F / \partial y$ and $F_{\mathrm{y}'} = \partial F / \partial y'$. Once the form of the function $y(x)$ is specified, $F[x, y(x), y'(x)]$ and its derivatives become functions of x alone. For the extremum specified by (2.55) we have $y(x) = \eta(x)$. Since $(\mathrm{d}/\mathrm{d}x) h(x) F_{\mathrm{y}'}(x) = h'(x) F_{\mathrm{y}'}(x) + h(x) \mathrm{d}F_{\mathrm{y}'}(x) / \mathrm{d}x$ the integral of the second term in (2.55) is

$$\begin{aligned} &\int_{\mathrm{a}}^{\mathrm{b}} \mathrm{d}x h'(x) F_{\mathrm{y}'}(x) \\ &= h(x) F_{\mathrm{y}'}(x) \Big|_{\mathrm{x}=\mathrm{a}}^{\mathrm{x}=\mathrm{b}} - \int_{\mathrm{a}}^{\mathrm{b}} \mathrm{d}x h(x) \frac{\mathrm{d}}{\mathrm{d}x} F_{\mathrm{y}'}(x) \end{aligned} \tag{2.56}$$

Because $h(x)$ vanishes at the end points $x = a, b$, (2.56) becomes

$$\int_{\mathrm{a}}^{\mathrm{b}} \mathrm{d}x h'(x) F_{\mathrm{y}'}(x) = -\int_{\mathrm{a}}^{\mathrm{b}} \mathrm{d}x h(x) \frac{\mathrm{d}}{\mathrm{d}x} F_{\mathrm{y}'}(x) \tag{2.57}$$

and (2.55) is

$$0 = \int_{\mathrm{a}}^{\mathrm{b}} \mathrm{d}x \left[F_{\mathrm{y}} - \frac{\mathrm{d}}{\mathrm{d}x} F_{\mathrm{y}'} \right] h(x) \tag{2.58}$$

for any arbitrary function $h(x)$ in the class of functions $\{y(x)\}$, which satisfies the boundary conditions $h(a) = h(b) = 0$.

Because $h(x)$ is arbitrary, (2.58) holds if and only if

$$F_{\mathrm{y}} - \frac{\mathrm{d}}{\mathrm{d}x} F_{\mathrm{y}'} = 0.$$

We then have the condition for the weak extremum, which we can state as a theorem. [[31], p. 15]

Theorem 2.4.1 *Let $J[y]$ be a functional of the form*

$$J[y] = \int_a^b dx\, F\left(x, y, y'\right),$$

defined on the set of functions $y(x)$, which have continuous first derivatives in $[a, b]$ and satisfy the boundary conditions $y(a) = A$ and $y(b) = B$. Then a necessary condition for $J[y]$ to have an extremum for a given function $y(x)$ is that $y(x)$ satisfies Euler's Equation

$$F_y - \frac{d}{dx} F_{y'} = 0. \tag{2.59}$$

Proof The proof is given in the preceding development.

We define the $\delta-$variation as Lagrange did.

Definition 2.4.2 Let $J[y]$ be a functional of the form

$$J[y] = \int_{\mathrm{a}}^{\mathrm{b}} \mathrm{d}x\, F\left(x, y, y'\right),$$

defined on the set of functions $y(x)$, which have continuous first derivatives in $[a, b]$ and satisfy the boundary conditions $y(a) = A,\ y(b) = B$. Then the **$\delta-$variation** of the functional $J[y]$, indicated by $\delta J[y]$, is that for which the variations $h(x)$ in the function $y(x)$ vanish at the end points a and b.

Example 2.4.2 As an example we consider the function

$$F\left(x, y, y'\right) = \frac{1}{2}\left(y'\right)^2 + 4xy, \tag{2.60}$$

and we choose the $x-$ interval to be $[a, b] = [0, 1]$ with the boundary conditions $y(0) = 0$ and $y(1) = 1$. Then (2.59) becomes

$$F_{\text{y}} - \frac{\text{d}}{\text{d}x} F_{\text{y}'} = 4x - \frac{\text{d}^2}{\text{d}x^2} y = 0. \tag{2.61}$$

The solution to (2.61) is

$$y(x) = \frac{2}{3}x^3 + \frac{1}{3}x. \tag{2.62}$$

And the value of the functional at the weak extremum is

$$\begin{aligned} &J[y] \\ &= \int_0^1 \text{d}x \left[\frac{1}{2} (y')^2 + 4xy \right] \\ &= \int_0^1 \text{d}x \left[\frac{1}{2} \left(2x^2 + \frac{1}{3} \right)^2 + 4x \left(\frac{2}{3}x^3 + \frac{1}{3}x \right) \right] \\ &= 1.6556\ldots \end{aligned} \tag{2.63}$$

Since we have only the condition for a weak extremum, we do not know whether (2.62) results in a maximum or a minimum of the functional.

This Theorem 2.4.1 can easily be generalized to a functional defined on a set of m functions. We define

$$J[y_1 \cdots y_{\text{m}}] = \int_{\text{a}}^{\text{b}} \text{d}x \, F\left(x, y_1 \cdots y_{\text{m}}, y_1' \cdots y_{\text{m}}'\right)$$

with

$$y_{\text{k}}(a) = A_{\text{k}}, \quad y_{\text{k}}(b) = B_{\text{k}},$$

fixed for each $k = 1, \ldots, m$. Following the same steps as before, we have for the first variation

$$\begin{aligned} \delta J &= \int_{\text{a}}^{\text{b}} \text{d}x \sum_{\text{k=1}}^{\text{m}} \left[h_{\text{k}}(x) \, F_{\text{y}_\text{k}} \Big|_{y_\text{k}=\eta_\text{k}} + h_{\text{k}}(x) \, F_{\text{y}'_\text{k}} \Big|_{y_\text{k}=\eta_\text{k}} \right] \\ &= \int_{\text{a}}^{\text{b}} \text{d}x \sum_{\text{k=1}}^{\text{m}} h_{\text{k}} \left[F_{\text{y}_\text{k}} - \frac{\text{d}}{\text{d}x} F_{\text{y}'_\text{k}} \right]. \end{aligned}$$

Because each of the functions $h_{\text{k}}(x)$ is arbitrary and independent of the others, δJ vanishes if and only if each bracketed term [ ] vanishes independently. That is

$$\frac{\partial}{\partial y_{\text{k}}} F - \frac{\text{d}}{\text{d}x} \frac{\partial}{\partial y_{\text{k}}'} F = 0 \tag{2.64}$$

for each k.

2.4.4 Hamilton's Principle

If we choose the time t as the independent variable, the generalized coordinates $q_k(t)$ as the functions of interest, and F as the Lagrangian $L = T - V$, we recognize the set of equations (2.64) as the Euler–Lagrange Equations of Analytical Mechanics (2.44). Specifically we have arrived at a set of equations that are completely equivalent to Newton's Second Law from a variational principle. Therefore, if we define the functional

$$\boxed{S[q] = \int_{t_1}^{t_2} dt L(q, \dot{q}, t),} \tag{2.65}$$

the condition for a weak extremum, using Lagrange's δ (see the Definition 2.4.2),

$$\delta S = \delta \int_{t_1}^{t_2} dt L(q, \dot{q}, t) = 0, \tag{2.66}$$

is that

$$\left[\frac{\partial}{\partial q} - \frac{d}{dt}\frac{\partial}{\partial \dot{q}}\right] L(q, \dot{q}, t) = 0 \tag{2.67}$$

for a single generalized coordinate $q(t)$.

If we consider a set of generalized coordinates $q = \{q_k\}$ the condition that first order variation of the functional

$$S[q] = S[\{q_k\}] = \int_{t_1}^{t_2} dt L(\{q_k\}, \{\dot{q}_k\}, t), \tag{2.68}$$

vanishes, i.e. that the functional $S[\{q_k\}]$ has an extremum, is that the set of equations

$$\left[\frac{\partial}{\partial q_k} - \frac{d}{dt}\frac{\partial}{\partial \dot{q}_k}\right] L(\{q_k\}, \{\dot{q}_k\}, t) = 0 \tag{2.69}$$

is satisfied.

We then have a general theorem of Analytical Mechanics, which is *Hamilton's Principle*.

Theorem 2.4.2 *Let $S[q]$ be the functional*

$$S[q] = \int_{t_1}^{t_2} dt L(q, \dot{q}, t),$$

where

$$L(q, \dot{q}, t) = T - V$$

is the Lagrangian function defined on the set of generalized coordinates $q = \{q_k\}$, *which have continuous first time derivatives* $\dot{q} = \{\dot{q}_k\}$ *on the interval* $[t_1, t_2]$, *and fixed values at the end points* t_1 *and* t_2. *Then a necessary condition for* $S[q]$ *to have an extremum for a given set of generalized coordinates* $q = \{q_k\}$ *is that each generalized coordinate* $q_k(t)$ *satisfies the* Euler–Lagrange Equation

$$\frac{\partial}{\partial q_k} L - \frac{d}{dt} \frac{\partial}{\partial \dot{q}_k} L = 0. \tag{2.70}$$

The system then satisfies Newton's Laws.

We note that here the variation $h(t)$ vanishes at the end points and that this is, therefore, the variation $\delta S[q]$ (see the Definition 2.4.2).

Example 2.4.3 As an example we choose the simple pendulum, which we have drawn in Fig. 2.6. In terms of the generalized coordinate ϑ the kinetic energy is

$$T = \frac{1}{2} m\ell^2 \dot{\vartheta}^2$$

and the potential energy is

$$V = mg\ell (1 - \cos\vartheta),$$

where the reference is $V = 0$ when $\vartheta = 0$. Then the Lagrangian is

$$L = T - V = \frac{1}{2} m\ell^2 \dot{\vartheta}^2 + mg\ell (\cos\vartheta - 1)$$

and

$$S[q] = \int_{t_1}^{t_2} \mathrm{d}t \left[\frac{1}{2} m\ell^2 \dot{\vartheta}^2 + mg\ell (\cos\vartheta - 1) \right].$$

The Euler–Lagrange equation is

$$\frac{\mathrm{d}}{\mathrm{d}t} \left(m\ell^2 \dot{\vartheta} \right) + mg\ell \sin\vartheta = 0$$

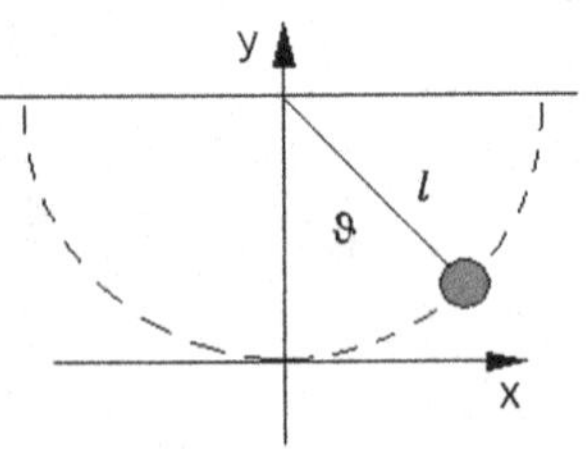

Fig. 2.6 Simple Pendulum with generalized coordinates ℓ and ϑ

or

$$\ddot{\vartheta} = -\left(\frac{g}{\ell}\right)\sin\vartheta.$$

This is a nonlinear equation for $\vartheta(t)$. We may linearize it to obtain the (familiar) equation for the simple pendulum with angular frequency $\omega = \sqrt{g/\ell}$.

2.5 Constraints

The reduction to generalized coordinates depends on the form of the constraints and may not be simple. Indeed in the general case it may be algebraically impossible to obtain a closed form Lagrangian in terms of generalized coordinates.

In this section we shall develop a systematic method for incorporating constraints of any kind into Hamilton's Principle. This will also provide the forces of constraint automatically in the course of the solution of the problem at hand.

We begin our discussion of constraints with a situation for which the constraint can only be formulated in differential terms: the rolling constraint.

2.5.1 Rolling

If a disk of radius R rolls upright and without slipping along a coordinate x the differential angle of rotation of the disk $\mathrm{d}\vartheta$, produces a displacement along the plane, $\mathrm{d}x$, given by

$$\mathrm{d}x = R\mathrm{d}\vartheta$$

or

$$0 = \mathrm{d}x - R\mathrm{d}\vartheta. \tag{2.71}$$

This differential relationship is all the description of rolling gives us. If the rolling is along a straight line we can integrate (2.71) to obtain

$$g_{\mathrm{roll}}(x,\vartheta) = 0 = x - R\vartheta + \text{constant}.$$

Then (2.71) implies the existence of a functional relationship between x and ϑ of the form $g_{\mathrm{roll}}(x,\vartheta) = 0$. And (2.71) is the Pfaffian differential of $g_{\mathrm{roll}}(x,\vartheta)$, i.e.

$$dg_{\mathrm{roll}} = 0 = \left(\frac{\partial g_{\mathrm{roll}}}{\partial x}\right)dx + \left(\frac{\partial g_{\mathrm{roll}}}{\partial \vartheta}\right)d\vartheta. \tag{2.72}$$

Comparing (2.72) with (2.71) we have

$$\left(\frac{\partial g_{\mathrm{roll}}}{\partial x}\right) = 1 \text{ and } \left(\frac{\partial g_{\mathrm{roll}}}{\partial \vartheta}\right) = -R. \tag{2.73}$$

But the integration cannot be performed for the general case of rolling on a surface because the path along which the object is rolling must be obtained from a solution of the equations of motion. And that solution involves the rolling condition as a constraint. We must, therefore, accept that the general rolling condition, or constraint, can only be formulated in differential form.

There is, however, a specific functional relationship between the path followed by the body on a surface and the angle through which the body has rotated, provided there is no slipping. This functional relationship

$$g_{\mathrm{roll}} = g_{\mathrm{roll}}\left(q\right),$$

where q is the set of generalized coordinates, only becomes known through solution of the set of dynamical (Euler–Lagrange) equations. Therefore, even though we are only able to write the rolling constraint initially in differential form, we realize that in general this differential form is an exact differential of a function, which we have designated here as g_{roll}.

In complete generality we may then write the rolling constraint as a Pfaffian in N generalized coordinates as

$$\mathrm{d}g_{\mathrm{roll}} = 0 = \sum_{\mathrm{i}}^{\mathrm{N}} \left(\frac{\partial g_{\mathrm{roll}}}{\partial q_{\mathrm{i}}}\right) \mathrm{d}q_{\mathrm{i}}\,. \tag{2.74}$$

And we will always be able to obtain an algebraic formulation of the terms $(\partial g_{\mathrm{roll}}/\partial q_{\mathrm{i}})$, even though we cannot write an expression for g_{roll}.

A Pfaffian is the exact differential of a function. That is the Pfaffian of the function g_{roll} defines the function g_{roll} by specifying the rule for constructing it from infinitesimals. The fact that we cannot write g_{roll} in closed algebraic form is because the interaction among the coordinates in the dynamical system is generally complicated. But this fact in no way denies the existence of the function.[10]

[10] This situation is common in thermodynamics. There we have a Pfaffian for each of the potentials. But we cannot write down a potential for any but the simplest ideal substance because of the complexity of the interdependence of the thermodynamic properties for real substances.

2.5.2 Holonomic and Nonholonomic Constraints

What we have said here regarding the rolling constraint holds for any constraint for which we can only write a differential expression. Such constraints are called *Nonholonomic*.

Constraints for which we can write a general algebraic expression of the form

$$g_{\mathrm{k}}(q) = 0 \tag{2.75}$$

are called *holonomic*. For N generalized coordinates the differential of a holonomic constraint results in the Pfaffian

$$\mathrm{d}g_{\mathrm{k}} = 0 = \sum_{\mathrm{i}}^{\mathrm{N}} \left(\frac{\partial g_{\mathrm{k}}}{\partial q_{\mathrm{i}}} \right) \mathrm{d}q_{\mathrm{i}} \tag{2.76}$$

The Eqs. (2.74) and (2.76) are both Pfaffians and identical in form. If we can show that only the differential of the constraint is of interest to us in our formulation then we can ignore the difference between holonomic and nonholonomic constraints.

In formulating any problem we always choose the coordinates that are the most logical. This results in a set of coordinates we may consider initially to be generalized coordinates. Each additional algebraic equation of constraint, whether holonomic or nonholonomic, reduces the number of necessary coordinates by one. This is true whether or not we can incorporate the constraint algebraically into the Lagrangian.

Some authors have placed various emphases on the distinction between holonomic and nonholonomic constraints. Whittaker's discussion is quite detailed. His example is, however, for the motion of a sphere on a horizontal surface. He points out that for the slipping motion of a sphere on the surface the constraint is holonomic, while for the non-slipping motion on the surface the rolling constraint is nonholonomic [[125], p. 34]. Louis Hand and Janet Finch also have a detailed description of constraints, subdividing the definitions depending on whether or not time is a variable. Their example of a nonholonomic constraint is based on the rolling of a disk on a rough surface [[41], p. 12, 36]. Cornelius Lanczos, who also uses rolling as an example of a nonholonomic constraint points out that the original definition comes from Heinrich Hertz [[65], p. 25].

The distinction between holonomic and nonholonomic constraints is real and interesting. However, in our formulation in terms of *Lagrange Undetermined Multipliers* here the difference has no immediate practical consequences.

2.5.3 Lagrange Undetermined Multipliers

We consider that there are n constraints. In principle there are then n algebraic functions $g_{\mathrm{k}}(q) = 0$, as we pointed out at the end of Sect. 2.5.1, although we can actually write these only for the holonomic constraints.

For arbitrary functions of the time $\lambda_{\mathrm{k}}(t)$ $(k = 1, \ldots, n)$ we, therefore, also have n equations

$$\lambda_{\mathrm{k}}(t)\, g_{\mathrm{k}}(q) = 0. \tag{2.77}$$

The integrals of the products (2.77) over any arbitrary time interval must also vanish. Then

$$\int_{t_1}^{t_2} \mathrm{d}t \lambda_{\mathrm{k}}(t)\, g_{\mathrm{k}}(q) = 0. \tag{2.78}$$

Adding these integrals to the functional $S[q]$ in (2.65) does not change the value of $S[q]$. So Hamilton's Principle requires that the $\delta-$variation (see Definition 2.4.2 in Sect. 2.4.3 of)

$$S[q] = \int_{t_1}^{t_2} \mathrm{d}t \left[L + \sum_{k=1}^{n} \lambda_{\mathrm{k}}(t)\, g_{\mathrm{k}}(q) \right] \tag{2.79}$$

must vanish. In performing the variation we must consider also variations in the (arbitrary) functions $\lambda_{\mathrm{k}}(t)$. The result is

$$\begin{aligned}\delta S = 0 = \int_{t_1}^{t_2} \mathrm{d}t & \left\{ \sum_{j}^{N} \delta q_{\mathrm{j}} \left[\frac{\partial L}{\partial q_{\mathrm{j}}} - \frac{\mathrm{d}}{\mathrm{d}t}\left(\frac{\partial L}{\partial \dot{q}_{\mathrm{j}}}\right) \right.\right. \\ & \left.\left. + \sum_{k=1}^{n} \lambda_{\mathrm{k}}(t) \left(\frac{\partial g_{\mathrm{k}}}{\partial q_{\mathrm{j}}}\right) \right] + \sum_{k=1}^{n} \delta\lambda_{\mathrm{k}} g_{\mathrm{k}}(q) \right\}.\end{aligned} \tag{2.80}$$

The integral in (2.80), over any arbitrary time interval, vanishes if and only if the integrand vanishes. That is

$$\begin{aligned}0 = \sum_{j}^{N} \delta q_{\mathrm{j}} & \left[\frac{\partial L}{\partial q_{\mathrm{j}}} - \frac{\mathrm{d}}{\mathrm{d}t}\left(\frac{\partial L}{\partial \dot{q}_{\mathrm{j}}}\right) \right. \\ & \left. + \sum_{k=1}^{n} \lambda_{\mathrm{k}}(t) \left(\frac{\partial g_{\mathrm{k}}}{\partial q_{\mathrm{j}}}\right) \right] + \sum_{k=1}^{n} \delta\lambda_{\mathrm{k}} g_{\mathrm{k}}(q).\end{aligned} \tag{2.81}$$

Using (2.75) the last term in (2.81) vanishes regardless of variations in the $\lambda_k(t)$. Therefore

$$0 = \sum_j^N \delta q_j \left[\frac{\partial L}{\partial q_j} - \frac{d}{dt}\left(\frac{\partial L}{\partial \dot{q}_j}\right) + \sum_{k=1}^n \lambda_k(t)\left(\frac{\partial g_k}{\partial q_j}\right)\right] \tag{2.82}$$

is the condition resulting from Hamilton's Principle .

If all of the coordinates in the set $q = \{q_j\}_{j=1}^N$ were independent the variations δq_j would be independent and (2.82) would require that each of the square brackets $[\cdots]$ must be zero. But the set of coordinates $q = \{q_j\}_{j=1}^N$ must also satisfy the n constraints (2.75). There are then only $N - n$ independent coordinates. So we must take another approach.

The other approach is to first judiciously choose all of the n arbitrary functions $\lambda_k(t)$ so that for the first n expressions in the brackets $[\cdots]$ in (2.82) vanish. That is we choose the arbitrary functions $\lambda_k(t)$ such that

$$\frac{\partial L}{\partial q_j} - \frac{d}{dt}\left(\frac{\partial L}{\partial \dot{q}_j}\right) + \sum_{k=1}^n \lambda_k(t)\left(\frac{\partial g_k}{\partial q_j}\right) \equiv 0 \text{ for } j = 1 \ldots N. \tag{2.83}$$

Solving for the $\lambda_k(t)$ with $(k = 1, \ldots, n)$ is, in principle, a straightforward problem in linear algebra. Because the Lagrangian $L = L(q, \dot{q}, t)$ is specified, the first term involving L in (2.83) is a known function of $(q, \dot{q}, t)$ for each j, which we shall call Φ_j. We also know the partial derivatives $(\partial g_k / \partial q_j)$ for all, including the nonholonomic, constraints as functions of $q = \{q_j\}_{j=1}^N$. The Eq. (2.83) are then the n linear algebraic equations for functions $\lambda_k(t)$ for $(k = 1, \ldots, n)$,

$$\Phi_j + \sum_{k=1}^n \lambda_k(t)\left(\frac{\partial g_k}{\partial q_j}\right) = 0,$$

which are soluble in a straightforward fashion.[11] The first n equations (2.83) are then satisfied and the n multipliers $\lambda_k(t)$ are known. There are then $N - n$ equations remaining in the set (2.83). And the set of Eq. (2.82) is reduced to

$$0 = \sum_{j=n+1}^N \delta q_j \left[\frac{\partial L}{\partial q_j} - \frac{d}{dt}\left(\frac{\partial L}{\partial \dot{q}_j}\right) + \sum_{k=1}^n \lambda_k(t)\left(\frac{\partial g_k}{\partial q_j}\right)\right], \tag{2.84}$$

in which the $N - n$ remaining coordinates q_j for $j = n + 1, n + 2, \ldots, N$ and their variations δq_j are independent. Therefore (2.84) requires that

---

[11]The solution to this set of equations produces $\lambda_k(q, \dot{q}, t)$, with t appearing in the event that the Hamiltonian involves t explicitly. The final solution, which produces $q_j = q_j(t)$, results in $\lambda_k(t)$.

$$\frac{\partial L}{\partial q_j} - \frac{d}{dt}\left(\frac{\partial L}{\partial \dot{q}_j}\right) + \sum_{k=1}^{n} \lambda_k(t) \left(\frac{\partial g_k}{\partial q_j}\right) = 0 \text{ for } j = n+1 \ldots N. \tag{2.85}$$

We now note that the Eqs. (2.83) and (2.85) are identical. Therefore,

$$\boxed{\partial L/\partial q_j\text{-d}\left(\partial L/\partial \dot{q}_j\right)/\text{d}t + \sum_{k=1}^{n} \lambda_k(t) \left(\partial g_k/\partial q_j\right) = 0 \quad \forall j.} \tag{2.86}$$

And we need pay no attention to the order of solution. This is what we shall always do in practice. Indeed we may often consider the $\lambda_k(t)$ to be of no interest and never obtain them explicitly.

This is the method of *Lagrange Undetermined Multipliers*. We may use this method to incorporate any and all constraints on our system, whether those constraints are holonomic or nonholonomic, and regardless of the algebraic complexity of the constraints.

Example 2.5.1 We again consider the pendulum and incorporate the constraint that the length is constant using Lagrange Undetermined Multipliers. That is we first write the Lagrangian for the mass m moving freely in space then add the constraint that the radial distance r from the pivot point is fixed and equal to ℓ. We have drawn the situation in Fig. 2.7. The kinetic energy in cylindrical coordinates is

$$T = \frac{1}{2} m \left(\dot{r}^2 + r^2 \dot{\vartheta}^2\right).$$

The potential energy, referenced to the pivot point, is

$$V = -mgr \cos \vartheta.$$

Then

$$L = T - V = \frac{1}{2} m \left(\dot{r}^2 + r^2 \dot{\vartheta}^2\right) + mgr \cos \vartheta.$$

The constraint is

$$g = r - \ell = 0.$$

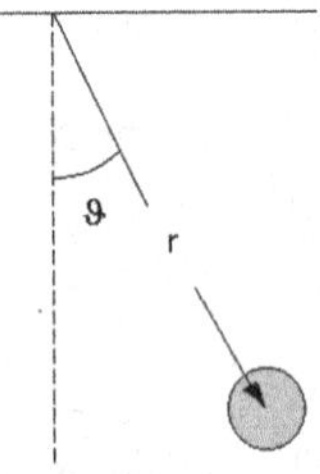

Fig. 2.7 A single generalized coordinate for the simple pendulum is ϑ. Including r and a Lagrange multiplier will allow the calculation of the force in the suspending rod

Then

$$dg = dr = 0.$$

and

$$\left(\frac{\partial g}{\partial r}\right) = 1$$

We have the Euler–Lagrange equations (with multiplier λ) for the two coordinates r and ϑ as

$$\frac{d}{dt}(m\dot{r}) - mr\dot{\vartheta}^2 - mg\ \cos\vartheta + \lambda = 0$$

$$\frac{d}{dt}\left(mr^2\dot{\vartheta}\right) + mgr\ \sin\vartheta = 0.$$

Using the constraint $\ell = r$ we have

$$\lambda = m\ell\dot{\vartheta}^2 + mg\ \cos\vartheta.$$

The second equation is, with $\ell = r$,

$$m\ell^2\ddot{\vartheta} + mg\ell\ \sin\vartheta = 0,$$

or

$$\ddot{\vartheta} = -\frac{g}{\ell}\ \sin\vartheta.$$

The use of Lagrange undetermined multipliers has resulted, then, in the same equation we obtained previously.

2.5.4 Forces of Constraint

We define the *canonical momentum* p_{j} conjugate to the generalized coordinate q_{j} as

$$p_{\mathrm{j}} \equiv \frac{\partial L}{\partial \dot{q}_{\mathrm{j}}}. \tag{2.87}$$

Newton's Second Law equates $\mathrm{d}p_{\mathrm{j}}/\mathrm{d}t$ to the force F_{j} applied in the jth direction. With (2.87) we may then write the Eq. (2.86) as

$$
\begin{aligned}
\frac{\mathrm{d}}{\mathrm{d}t} p_{\mathrm{j}} &= -\frac{\partial V}{\partial q_{\mathrm{j}}} + \sum_{\mathrm{k}=1}^{\mathrm{n}} \lambda_{\mathrm{k}}(t) \left(\frac{\partial g_{\mathrm{k}}}{\partial q_{\mathrm{j}}} \right) \\
&= - \frac{\partial}{\partial q_{\mathrm{j}}} \left[V - \sum_{\mathrm{k}=1}^{\mathrm{n}} \lambda_{\mathrm{k}}(t)\, g_{\mathrm{k}} \right].
\end{aligned} \tag{2.88}
$$

That is

$$
V_{\mathrm{eff}} = V - \sum_{\mathrm{k}}^{\mathrm{n}} \lambda_{\mathrm{k}}(t)\, g_{\mathrm{k}} \tag{2.89}
$$

acts as an *effective potential* in which the system moves. The second part of this effective potential produces the reactive forces on the system due to the constraints. The reactive force in the jth direction arising from the constraints is then

$$
\boxed{F_{\mathrm{j}}^{\mathrm{react}} = \sum_{\mathrm{k}=1}^{\mathrm{n}} \lambda_{\mathrm{k}}(t) \left(\partial g_{\mathrm{k}} / \partial q_{\mathrm{j}} \right)} \tag{2.90}
$$

and the force from the kth constraint alone in the jth direction is

$$
F_{\mathrm{kj}}^{\mathrm{react}} = \lambda_{\mathrm{k}}(t) \left(\frac{\partial g_{\mathrm{k}}}{\partial q_{\mathrm{j}}} \right). \tag{2.91}
$$

In order to find this reactive force we need to find the undetermined multiplier $\lambda_{\mathrm{k}}(t)$. In practice this usually constitutes an additional step in the solution, which is often unnecessary if we only seek the trajectory of the system. The engineer may, however, be very interested in knowing the forces of constraint on a system. These determine the strength of the structure which confines the system. The Lagrangian formulation provides these constraints in a systematic form.

2.6 Cyclic Coordinates

Let us assume that we have been able to obtain a Lagrangian into which all the constraints have been incorporated algebraically. For such a situation the Euler–Lagrange equations are (2.44). The structure of these equations allows us to make some general statements about the behavior of the system.

If the Lagrangian does not depend explicitly on a certain coordinate it is termed *cyclic* in that coordinate. The conjugate momentum corresponding to that coordinate is then a constant of the motion, since from (2.87) and (2.44) we have

$$\frac{\mathrm{d}}{\mathrm{d}t}\left(\frac{\partial L}{\partial \dot{q}_{\mathrm{r}}}\right) = \frac{\mathrm{d}}{\mathrm{d}t} p_{\mathrm{r}} = \frac{\partial L}{\partial q_{\mathrm{r}}} = 0. \tag{2.92}$$

This was first observed by the mathematician Amalie (Emmy) Nöther.[12]

The lack of dependence of the Lagrangian on a particular coordinate indicates a symmetry of the system with respect to that coordinate. For example if there is no dependence in the Lagrangian on the coordinate x the Lagrangian is unchanged by a translation in the $x-$direction and there can, therefore, be no force in the x direction. The canonical momentum p_{x} is then conserved. A symmetry about a particular axis means the angular momentum about that axis is constant. Later we shall see that a symmetry in the time means energy is conserved. The first two of these are intuitive. But the energy-time relationship is usually a surprise.

2.7 Summary

In this chapter we have laid the foundations of Analytical Mechanics. The end result was the Euler–Lagrange Equations.

We began by showing that Newton's Laws result in the Euler–Lagrange equations using d'Alembert's Principle, which is a virtual work principle based mathematically on a scalar equation. So our formulation was based on scalar energy terms, which promises a simplicity not present in a standard vector formulation.

We then turned to the more elegant formulation in terms of the variational calculus and a variational principle we identified as Hamilton's Principle. This formulation also provided us with the only truly general method to incorporate complicated constraints into mechanical problems: the method of Lagrange Undetermined Multipliers.

The only possible disadvantage in our formulation lies in the fact that the equations we have are second order differential equations in the time. Our approach would be simplified if we could find equivalent first order equations. This will be the subject of our next chapter.

---

[12]Amale (Emmy) Nöther (1882–1935) was a German mathematician whose specialty was algebra and ring structures. She was Extraordinary Professor at Göttingen from 1922–1933, then she accepted a professorship at Bryn Mawr College, which she held until her death in 1935.

2.8 Exercises

2.1. Cylindrical coordinates are $\{r, \vartheta, z\}$. The position vector from the origin is $\boldsymbol{r} = r\hat{e}_{\mathrm{r}} + z\hat{e}_{\mathrm{z}}$. Show that the velocity vector is

$$\frac{\mathrm{d}}{\mathrm{d}t}\boldsymbol{r} = \dot{r}\hat{e}_{\mathrm{r}} + r\dot{\vartheta}\hat{e}_{\vartheta} + \dot{z}\hat{e}_{\mathrm{z}}.$$

2.2. Spherical coordinates are $\{\rho, \vartheta, \phi\}$. The position vector is $\overrightarrow{r} = \rho\hat{e}_{\rho}$. Show that the velocity vector is

$$\frac{\mathrm{d}}{\mathrm{d}t}\boldsymbol{r} = \dot{\rho}\hat{e}_{\rho} + \rho\dot{\vartheta}\sin\phi\hat{e}_{\vartheta} + \rho\dot{\phi}\hat{e}_{\phi}.$$

Remark 2.8.1 These exercises should serve to provide a familiarity with the coordinates that we will use extensively. The solution may require some hand drawing and the consideration of small variations.

2.3. In this chapter we introduced Lagrange Undetermined Multipliers to add constraints to a variational principle. The reasoning should work just as well if we wish to find the extremum of an algebraic expression subject to a constraint. For example, if we wish to find the minimum distance from the origin to the straight line

$$y = 3x + 2$$

we seek the minimum distance from the origin to a point (x, y) in the plane and then introduce the constraint that the point lies on the line. The calculation will be easier if we minimize the square of the distance from the origin to the point (x, y), which by Pythagoras' Theorem is

$$f(x, y) = x^2 + y^2.$$

Carry out the calculation to find the point on the line.

Show that the shortest line between the origin and the straight line $y = 3x + 2$ is perpendicular to the line $y = 3x + 2$.

2.4. Show that D'Alembert's Principle results in conservation of mechanical energy $\mathrm{d}(T + V) = 0$ for impressed forces derivable from a scalar potential as $\boldsymbol{F} = -\mathrm{grad}V$.

2.5. Consider the parabola

$$y_1 = -2x_1^2 - 4$$

and the straight line

$$y_2 = 2x_2 + 1$$

First show that the graphs of these two functions never intersect. Having convinced yourself of this, then go on to find the minimum distance between these two functions.

Show also that the minimum distance is a line perpendicular to the given straight line.

2.6. In statistical mechanics we find that the Gibbs expression for the entropy is

$$S = -k_{\mathrm{B}} \sum_{\mathrm{r}} P_{\mathrm{r}} \ln P_{\mathrm{r}}$$

where k_{B} is the Boltzmann constant and P_{r} is the coefficient of probability for the rth state, which is a measure of the density of states in the system phase space.[13] Thermodynamics teaches us that under conditions of constant energy and volume the *entropy of a system will be a maximum.* That is, we have the constraint that

$$\mathcal{E} = \sum_{\mathrm{r}} P_{\mathrm{r}} \mathcal{E}_{\mathrm{r}},$$

which is that the average energy of the system in the ensemble is a constant. We also realize that there is another constraint in the definition of probability. That is

$$1 = \sum_{r} P_{r}$$

Show that the probability that a system of atoms will be in a particular state of total energy $\mathcal{E}_{\mathrm{r}}$ is given by

$$P_{\mathrm{r}} = \exp\left[-1 - \alpha - \beta \mathcal{E}_{\mathrm{r}}\right]$$

where α and β are constants. Do this by maximizing the Gibbs entropy subject to the constraints. The α and β are the Lagrange Undetermined Multipliers.

2.7. Study the following functional

$$J[y] = \int_0^1 \mathrm{d}x \left(y'\right).$$

Determine whether or not it has an extremum. If it does, find that extremum.

2.8. Consider the functional

$$J[y] = \int_0^1 \mathrm{d}x \left(yy'\right).$$

Determine whether or not this functional has an extremum. If so, find that extremum.

[13] The system phase space has an axis for every canonical coordinate and every canonical momentum of every particle (atom/molecule) in the system. This space is called $\Gamma-$space.

2.9. Study the functional

$$J[y] = \int_0^1 \mathrm{d}x\,(xyy').$$

Determine whether or not this functional has an extremum. If it does, find that extremum.

2.10. Find differential equation for the extremum of the functional

$$J[y] = \int_0^1 \mathrm{d}x\left[y^2 + (y')^2 - 2y\sin(x)\right]$$

[answer: $y'' - y = -\sin(x)$]

2.11. Show that the functional of two functions:

$$S[x,y] = \int_{t_1}^{t_2} \mathrm{d}t\Psi\,[t, x, y, \dot{x}, \dot{y}]$$

has an extremum when Euler–Lagrange equations for each of the functions are satisfied.. That is, show that this functional has an extremum when

$$\frac{\partial}{\partial x}\Psi - \frac{\mathrm{d}}{\mathrm{d}t}\frac{\partial}{\partial \dot{x}}\Psi = 0$$

and

$$\frac{\partial}{\partial y}\Psi - \frac{\mathrm{d}}{\mathrm{d}t}\frac{\partial}{\partial \dot{y}}\Psi = 0$$

provided the variations vanish at the end points.

This requires the details of what we indicated in the text.

2.12. Consider that in the functional

$$S[y] = \int_{t_1}^{t_2} \mathrm{d}t\,F(y, \dot{y})$$

the function F does not depend explicitly on the time t. Show that as a consequence

$$F - \dot{y}\frac{\partial F}{\partial \dot{y}} = \text{constant}.$$

2.13. Using the results of the preceding exercise, i.e.

$$F - \dot{y}\frac{\partial F}{\partial \dot{y}} = \text{constant},$$

show that for F as the Lagrangian $F = T - V$ for a single particle of mass m that the total mechanical energy is constant and that

$$\frac{\mathrm{d}}{\mathrm{d}t}\left(F - \dot{y}\frac{\partial F}{\partial \dot{y}}\right) = -\frac{\mathrm{d}}{\mathrm{d}t}\left(T + V\right).$$

2.14. Among all curves joining the points (x_0, y_0) and (x_1, y_1) find that which generates the minimum surface area when rotated around the $x-axis$. Begin with the Pythagorean Theorem that the differential length between two points in the (x, y) −plane is

$$ds = \sqrt{\mathrm{d}x^2 + \mathrm{d}y^2} = \mathrm{d}x\sqrt{1 + (y')^2}.$$

At the point x the distance to the curve $y = y(x)$ is equal to the value of y. So the differential area of the surface of rotation defined by the points x and $x + \mathrm{d}x$ on the x−axis is

$$\mathrm{d}x\left(2\pi y\sqrt{1 + (y')^2}\right).$$

The area of the surface of revolution between x_0 and $x_1$is

$$A_S = 2\pi\int_{x_0}^{x_1}\mathrm{d}x\left(y\sqrt{1 + (y')^2}\right).$$

[Answer: $y = K\cosh\left[(x + C)/K\right]$, where K and C are (integration) constants]

2.15. A particle is released from rest at a point (x_0, y_0) and slides (without friction) down a curve in the (x, y) plane. Since the differential distance down the plane is

$$\mathrm{d}s = \sqrt{\mathrm{d}x^2 + \mathrm{d}y^2} = \mathrm{d}x\sqrt{1 + (y')^2},$$

The speed at which the particle slides is

$$v = \frac{\mathrm{d}s}{\mathrm{d}t} = \sqrt{1 + (y')^2}\frac{\mathrm{d}x}{\mathrm{d}t}.$$

The speed, from energy conservation (no friction) is

$$v = \sqrt{2gy}.$$

Then

$$\mathrm{d}t = \mathrm{d}x\frac{\sqrt{1 + (y')^2}}{\sqrt{2gy}}.$$

What must the curve be, down which the particle slides, so that it reaches the vertical line at $x = b\ (> x_0)$ in the shortest time? We then wish to find the extremum of the functional for the total time

$$T[y] = \int_{x=x_0}^{x=b} dx \frac{\sqrt{1+(y')^2}}{\sqrt{2gy}}$$

This is the brachistochrone problem, which was first posed by Johann Bernoulli in 1696.

[Answer: A cycloid]

2.16. A particle of mass m moves under no forces in the direction x.

Find the Lagrangian and the canonical momentum. Show that the canonical momentum is conserved. Find the energy and show that its total time derivative is zero so that the energy is a constant.

Do this using the Euler–Lagrange equations.

2.17. Consider a particle of mass m in free fall under the influence of gravity. Find the Lagrangian, the canonical momentum, the Euler–Lagrange equation. Show that the energy is constant and integrate the Euler–Lagrange equation.

2.18. Consider a particle of mass m sliding without friction down an inclined plane. We show this in the figure below.

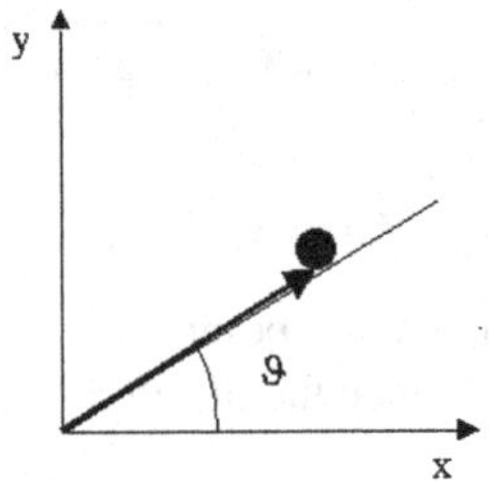

Particle sliding on incline

Find the Lagrangian, the Euler–Lagrange equations, and the energy. Show that the energy is constant and solve the Euler–Lagrange equations. Find the reaction force with the incline using the Lagrange multipliers.

2.19. Consider a mass, m, sliding without friction on the hilly terrain described by the function

$$y = -4x^3 + 5x^2 - x.$$

We have shown the hilly terrain in the figure below.

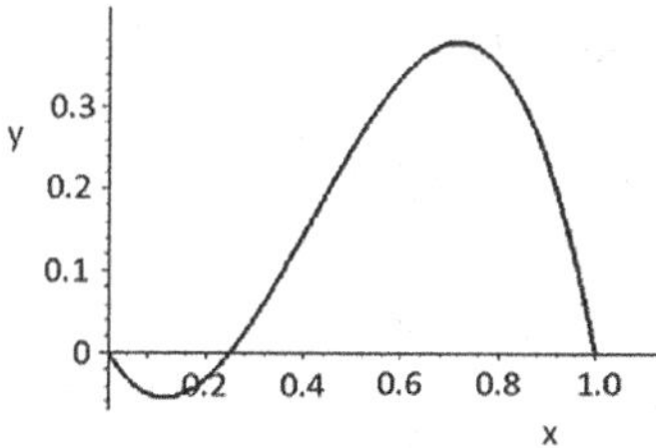

Hilly terrain described by $y = -4x^3 + 5x^2 - x$

Obtain the equations of motion for the particle using Lagrange multipliers.

2.20. In the figure here we have two masses connected by identical springs to one another and to two vertical walls.

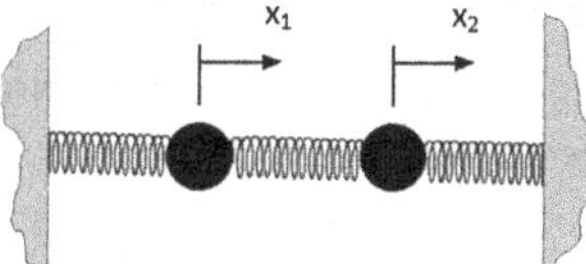

Two masses connected by identical springs to fixed vertical walls.

We neglect gravitational influences.

Study the motion of the system. Find the natural (eigen) frequencies and the corresponding eigenvectors.

2.21. Consider a mass sliding without friction inside a sphere. We have drawn the picture here.

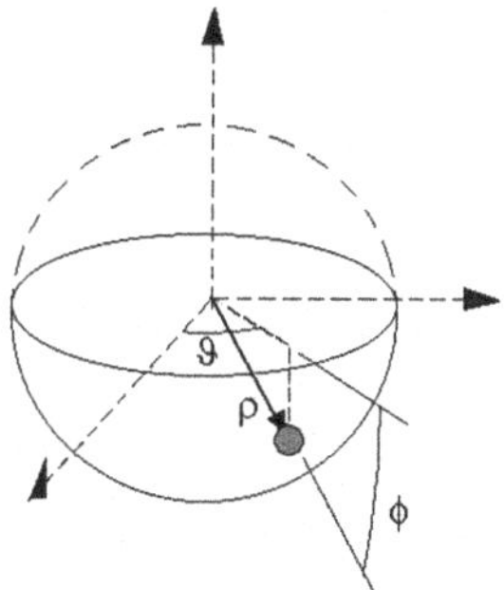

Mass Inside Sphere. Here a mass slides without friction inside a sphere, which we choose to be of glass so the motion can be observed. The coordinates are spherical (ρ, ϑ, ϕ) with the polar angle ϕ measured from zero in the central horizontal plane.

Study the motion by incorporating all constraints directly into the Lagrangian. Find the equilibrium orbit. Study small perturbations around this orbit

2.22. Consider the block, spring and pendulum system shown here.

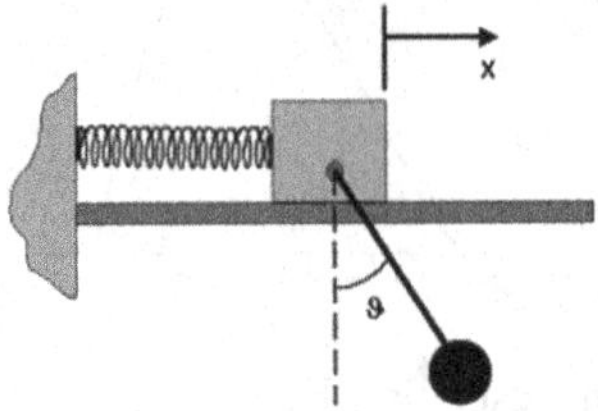

Block, spring, and pendulum

Obtain the Euler–Lagrange equations for this system. Then simplify for equal masses ($M = m$). Consider small vibrations (small x and ϑ). Make the Ansatz that the time dependence is $\exp(i\omega t)$ and find the normal modes of motion.

2.23. In the figure below we have drawn a stationary wire loop with a bead of mass m. The bead is free to move with no friction on the wire.

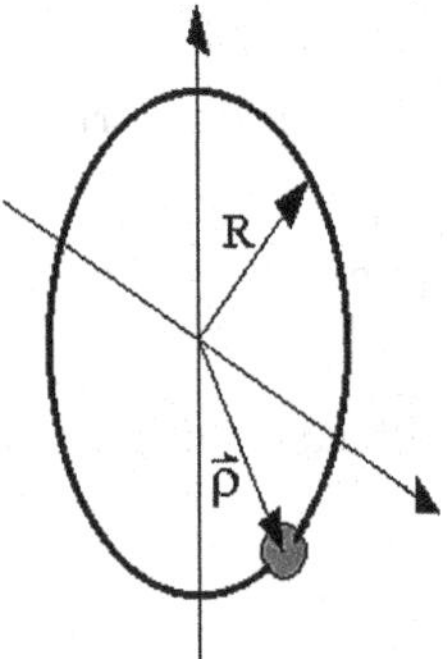

Bead on frictionless, stationary wire loop

Find the Euler–Lagrange equations. Do not attempt a solution.

2.24. Consider now that the loop in the preceding exercise rotates at a constant angular velocity about the vertical axis. That is $\dot{\vartheta} = \Omega =$ constant and $\rho = R =$ constant.

Find the Euler–Lagrange equations. What is the equilibrium location of the bead? Show that the equilibrium is stable, that is small deviations form equilibrium result in sinusoidal oscillations around equilibrium, provided

$$1 + \frac{g}{R\Omega^2} - 2\left(\frac{g}{R\Omega^2}\right)^2 > 0.$$

2.25. A physical pendulum is a uniform rod suspended on an axis constrained to move in one plane about a point other than the center of the rod. In the figure below we have shown a physical pendulum.

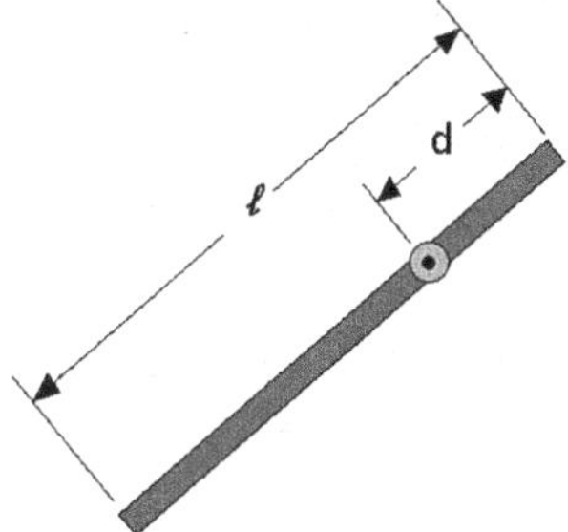

Physical pendulum

Write the Lagrangian for this physical pendulum. Begin by writing the Lagrangian for a differential mass located at a distance x from the axis and then integrate over the rod.

Find the Euler–Lagrange equation for this physical pendulum.

2.26. Consider a uniform rod with linear mass density λ, which is fastened to the floor by a hinge. We have a drawing of the falling rod here.

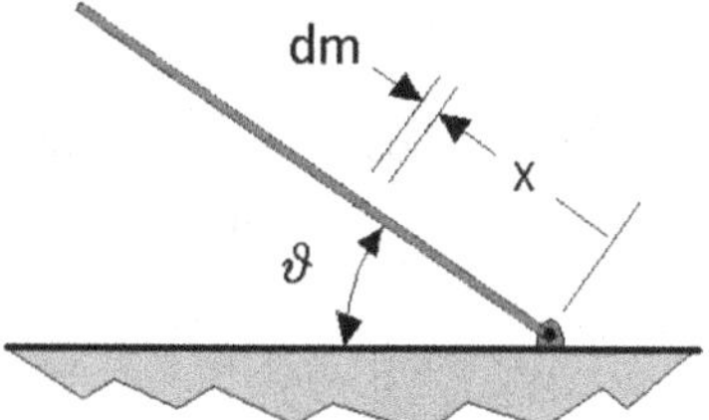

Falling hinged rod

We release the rod from the vertical with a slight nudge so that the angular momentum is initially zero. Obtain the time of fall as an integral. Do not attempt the integration.

[Hint: $\ddot{\vartheta} = (1/2)\, d\dot{\vartheta}^2/\mathrm{d}\vartheta$]

2.27. Now consider the falling rod as in the preceding exercise, except that instead of being hinged the end of the rod is free and the floor is frictionless. We again release the rod at $\vartheta = \pi/2$ with a slight nudge. We have drawn the rod in the figure below.

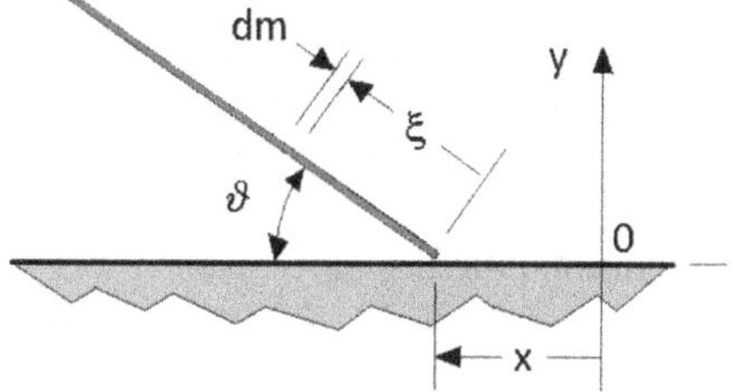

Free rod falling

Find the Euler–Lagrange equations for the falling rod.

2.28. We have drawn a double pendulum in the figure here.

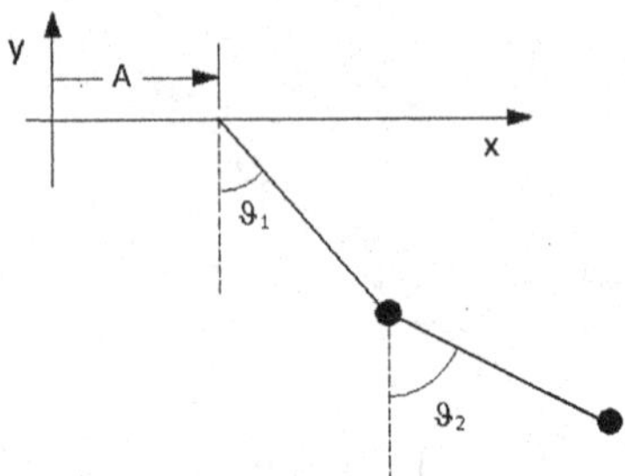

Double pendulum with equal lengths and bobs of equal mass.

Both pendulum lengths are ℓ and the masses of the pendulum bobs are both m. We consider the masses rods connecting the bobs to be zero.

Obtain the Euler–Lagrange equations, linearize these for small angles and find the normal modes of oscillation.

[Answers: for the Euler–Lagrange equations

$$\begin{aligned}&-2m\ell^2\ddot{\vartheta}_1 - m\ell^2\ddot{\vartheta}_2\left(\cos\vartheta_1\cos\vartheta_2 + \sin\vartheta_1\sin\vartheta_2\right)\\&-m\ell^2\dot{\vartheta}_2^2\left(-\cos\vartheta_1\sin\vartheta_2 + \sin\vartheta_1\cos\vartheta_2\right)\\&-2mg\ell\sin\vartheta_1\\&=0\end{aligned}$$

and

$$\begin{aligned}&-m\ell_1^2\ddot{\vartheta}\left(\cos\vartheta_1\cos\vartheta_2 + \sin\vartheta_1\sin\vartheta_2\right) - m\ell_2^2\ddot{\vartheta}\\&-m\ell^2\dot{\vartheta}_1^2\left(-\sin\vartheta_1\cos\vartheta_2 + \cos\vartheta_1\sin\vartheta_2\right)\\&-mg\ell\sin\vartheta_2\\&=0.\end{aligned}$$

For the linearized equations

$$2\ddot{\vartheta}_1 + 2\omega_0^2\vartheta_1 + \ddot{\vartheta}_2 = 0$$

$$\ddot{\vartheta}_1 + \ddot{\vartheta}_2 + \omega_0^2\vartheta_2 = 0,$$

where $\omega_0 = \sqrt{g/\ell}$. For the normal modes

$$\omega = \pm\omega_0\sqrt{2+\sqrt{2}}$$

$$\omega = \pm\omega_0\sqrt{2-\sqrt{2}}$$

the eigenvectors are

$$\text{For } \omega = \omega_0\sqrt{\left(2+\sqrt{2}\right)}: \begin{bmatrix}\Theta_1\\\Theta_2\end{bmatrix} = \frac{2}{\sqrt{6}}\begin{bmatrix}-\frac{1}{2}\sqrt{2}\\1\end{bmatrix}$$

and

$$\text{For } \omega = \omega_0\sqrt{\left(2-\sqrt{2}\right)}: \begin{bmatrix}\Theta_1\\\Theta_2\end{bmatrix} = \frac{2}{\sqrt{6}}\begin{bmatrix}\frac{1}{2}\sqrt{2}\\1\end{bmatrix}.$$

2.29. In the figure below we have a bead moving without friction on a wire. The wire makes an angle α with the vertical and is free to rotate about the vertical axis, also without friction.

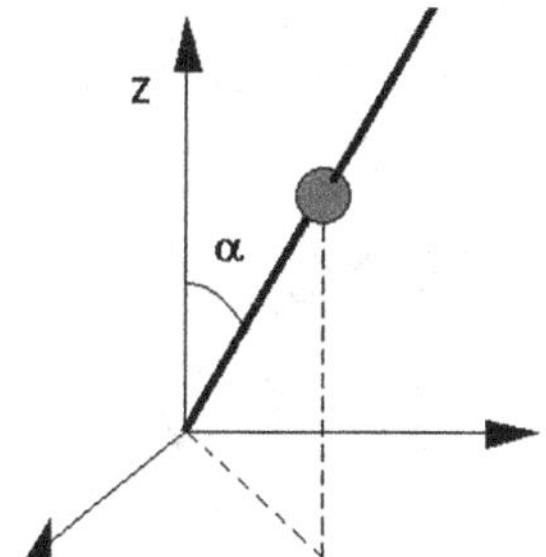

Bead on a frictionless wire

We neglect the mass of the wire.

Show that motion can be described as that of a particle moving in an (effective) potential well

$$V_{\text{eff}} = -\frac{1}{2}\frac{\ell^2}{mr^2} + mg\frac{r}{\tan\alpha}$$

where ℓ is the angular momentum of the bead.

Is there a position of stable equilibrium? This requires consideration of both the radial velocity $\dot{r}$ and the radial acceleration $\ddot{r}$.

Show that the Lagrange Undetermined multiplier is

$$\lambda = -\frac{\ell^2}{mr^3\left(1+\tan^2\alpha\right)} - \frac{mg\tan\alpha}{1+\tan^2\alpha}$$

and that the forces of the wire on the bead are then

$$f_r = -\frac{\ell^2}{mr^3\left(1+\tan^2\alpha\right)} - \frac{mg\tan\alpha}{1+\tan^2\alpha}$$

and

$$f_z = \frac{\ell^2\tan\alpha}{mr^3\left(1+\tan^2\alpha\right)} + \frac{mg\tan^2\alpha}{1+\tan^2\alpha}.$$

Comment on the time dependence of the λ.

2.30. In the preceding exercise the wire was free to rotate about the vertical axis, while the angle remained constant. We now choose to drive the wire at a constant angular velocity Ω about the vertical axis. The angle with the axis will still remain constant at α.

Incorporate the angular constraint and the constant angular velocity using Lagrange undetermined multipliers.

2.31. Consider the situation above with the wire mounted at the origin in a fashion that allows frictionless motion around the vertical and about the pivot point so that the angle to the vertical α varies. Let there be a vertical post erected from the origin. A spring retains the wire so that the pivot angle relative to this vertical post is limited. The spring is mounted at a distance h above the ground on the vertical post by a collar that permits rotation around the post. Consider small vibrations so that the spring remains parallel to the floor. The situation is shown below

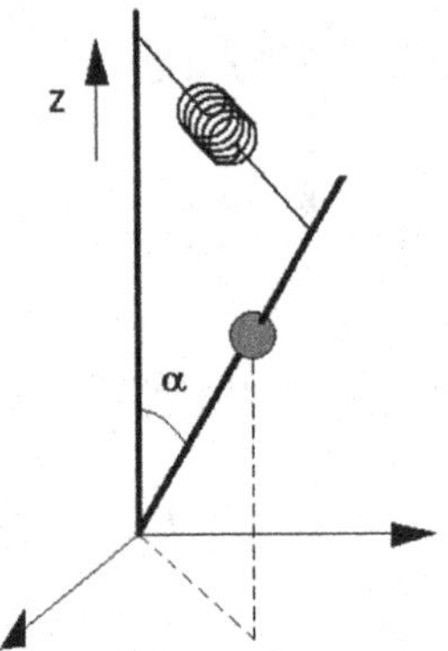

Bead on a frictionless, massless wire with a spring tying the wire to a central pole. The height of the spring is h

Assume that the spring is massless.

Study the motion. Is there an equilibrium orbit for the bead?

2.32. In the figure below we have drawn a cylinder of radius R lying on a laboratory table. Assume that the surface of the cylinder is frictionless. Its axis is parallel to the table top and the ground. The cylinder remains fixed. We then place a small mass m on the uppermost part of the cylinder. If we release the mass and nudge it slightly it will slide without friction on the cylinder.

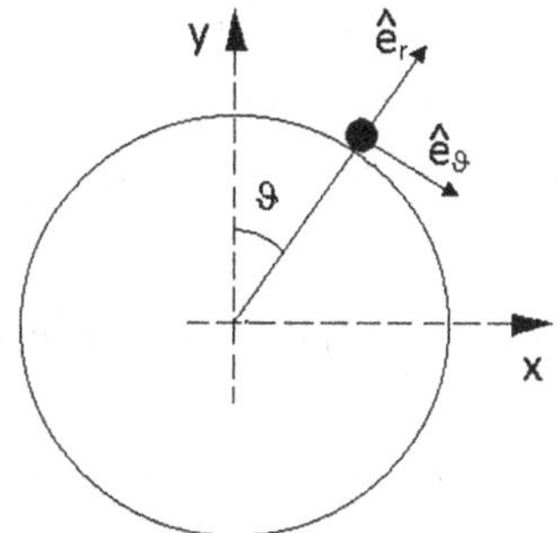

Small mass on a frictionless cylinder

At some point the small mass will fall off the cylinder. Find this point.

2.33. Consider two balls connected by a string of length b. One is suspended through a hole in a table and the other moves on the (frictionless) top of the table. We have drawn the situation in the figure here.

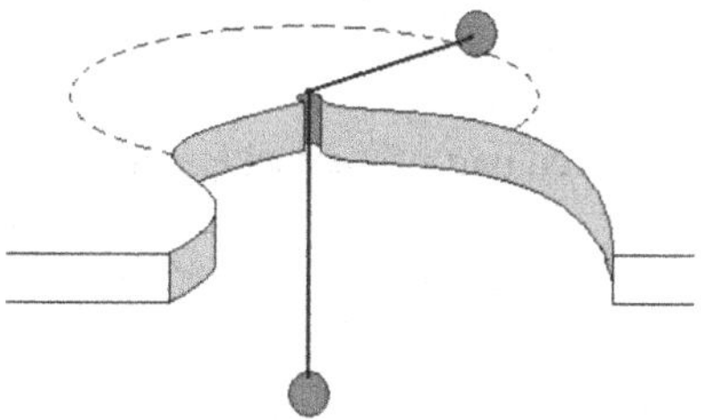

Two balls connected by a string. One is suspended below a frictionless table and the other freely moves on the frictionless table

Investigate the motion. Use Lagrange multipliers. Find an equilibrium point, if there is one. Study the general form of the motion. If you find a point of dynamic equilibrium, consider small oscillations about that point. Determine if the orbit is open or closed.

2.34. In a rocket engine the thermal energy of the burning fuel and oxidant is converted in the nozzle into kinetic energy. This high energy gas is expelled. The momentum carried away by this expelled gas results in an increase in momentum of the rocket. Consider a rocket for which

$$\begin{aligned} m_{\mathrm{r}} &= \text{mass of the rocket excluding fuel} \\ m_{\mathrm{f}} &= \text{mass of the fuel at any instant} \\ m_{\mathrm{e}} &= \text{mass of exhaust gases in the nozzle at any time} \\ \dot{m} &= \text{rate at which fuel is burned.} \end{aligned}$$

Let

$$v = \text{ velocity of the rocket}$$
$$u = \text{ velocity of exhaust gases in space}$$
$$\dot{v} = \text{ acceleration of the rocket.}$$

If we consider that the rocket is in a region of space in which all forces may be neglected, obtain the Euler–Lagrange equation for the rocket. This will be the standard propulsion equation

$$\frac{\mathrm{d}}{\mathrm{d}t} p = M\dot{v} - \dot{m}U = 0,$$

where

$$M = m_{\mathrm{r}} + m_{\mathrm{f}} + m_{\mathrm{e}}.$$

and U is the velocity of the exhaust gas relative to the rocket. Note that the kinetic energy of the rocket, including unburned fuel and the exhaust gases in the nozzle is

$$T = \frac{1}{2} m_{\mathrm{r}} v^2 + \frac{1}{2} m_{\mathrm{f}} v^2 + \frac{1}{2} m_{\mathrm{e}} (v - u)^2 .$$

The exhaust gases are considered part of the rocket until they exit the nozzle.

Chapter 3
Hamiltonian Mechanics

The point is to simplify and to order knowledge. The profession I'm part of has as its whole function the rendering of the physical world understandable and beautiful.

Robert J. Oppenheimer

3.1 Introduction

The Hamiltonian formulation of Analytical Mechanics is completely equivalent to the Lagrangian formulation. It is a transition from a formulation based on a single second order Euler–Lagrange equation for each generalized coordinate to one based on two first order equations, one for each generalized coordinate and one for each conjugate momentum. The Hamiltonian equations are called the canonical equations. As the name suggests, these are the fundamental equations of Analytical Mechanics.

The new perspective considers the generalized coordinates and the conjugate momenta to be equivalent and fundamental. This differs from the Newtonian, and even from the Lagrangian approaches in which the generalized coordinates were considered to be the basis of the description and the momentum was a derived quantity. Because the Hamiltonian formulation provides a logical correspondence with the more fundamental quantum theory, we may accept that the Hamiltonian picture is more natural than that of Lagrange. Nature speaks to us more clearly through the Hamiltonian formulation.

C.S. Helrich, *Analytical Mechanics*, Undergraduate Lecture Notes in Physics, DOI 10.1007/978-3-319-44491-8_3

The absolute equivalence of the two Lagrangian and the Hamiltonian formulations is a result of the fact that the Hamiltonian formulation is obtained from the Lagrangian by a Legendre[1] transformation. This transformation preserves completely the information content. We may recall that this step was used also by Hamilton in his original development.

3.2 Legendre Transformation

In Sect. 1.9.2 we indicated that Hamilton transformed the Lagrangian into his function $\mathcal{H}$ using a Legendre Transformation [see e.g. [44], pp. 47–51]. In this section we shall discuss the important properties of the Legendre Transformation that make it indispensable, and show how to perform a Legendre transformation. Because the visual aspects of the transformation are crucial we base our discussion on an example.

We begin with the Lagrangian for a linear harmonic oscillator with the mass of the oscillator located by the coordinate x. The velocity of the mass is then $\dot{x}$ and the Lagrangian is a function of x and $\dot{x}$, which is a surface above the plane $(x, \dot{x})$. We have plotted this surface in Fig. 3.1. All of the information we have about the oscillator is completely contained in this surface. We now ask for a transformation of this surface which will preserve this information and only this information. Such a transformation will produce a surface that is logically equivalent to the surface in Fig. 3.1, but will introduce a new variable in the place of one of the two variables $(x, \dot{x})$. This can be done if the new variable is obtained directly from the surface and can be shown to reproduce the surface exactly.

We select $\dot{x}$ as the variable to be transformed. We must then replace $\dot{x}$ with a new variable that can be obtained simply from the surface in Fig. 3.1 and can be used to completely construct the surface in Fig. 3.1.

In Fig. 3.1 we constructed the surface $L(x, \dot{x})$ from the value of L at each point in the plane $(x, \dot{x})$. In Fig. 3.2 we have picked a point $\left(x_{\mathrm{j}}, \dot{x}_{\mathrm{j}}\right)$ on the surface $L(x, \dot{x})$ and constructed the plane passing through the point $\left(x_{\mathrm{j}}, \dot{x}_{\mathrm{j}}\right)$ which is perpendicular to the axis of x and parallel to the axis of $\dot{x}$. We have also drawn the intersection of this plane with the surface $L(x, \dot{x})$, which is a curve in the plane located by the point x_{j}. The slope of this curve is $(\partial L(x, \dot{x})/\partial \dot{x})$ holding x constant at the value x_{j}. With this slope we can reconstruct the surface $L(x, \dot{x})$ just as well as we were able using the point values.

To illustrate this we have drawn the curve of Fig. 3.2 with 9 tangent lines in Fig. 3.3. In the left panel of Fig. 3.3 we have drawn the point representation of the curve with small open circles. In the right panel of Fig. 3.3 we have drawn only the 9 tangent lines. The fact that the analytic form of the tangent lines $p = (\partial L(x, \dot{x})/\partial \dot{x})$ provides an infinite number of these lines results in a complete reproduction of the surface in Fig. 3.1 using what is termed line geometry. We have then succeeded

[1] Adrien-Marie Legendre (1752–1833) was a French mathematician who made important contributions to statistics, number theory, abstract algebra and mathematical analysis.

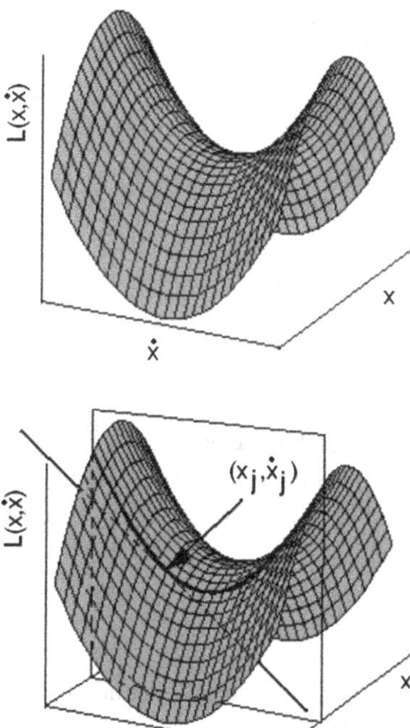

Fig. 3.1 The Lagrangian for a simple harmonic oscillator

Fig. 3.2 Lagrangian for harmonic oscillator with plane parallel to the $\dot{x}$ axis

in finding a transformation that preserves the information content of the original surface in Fig. 3.1 by representing the surface in the new variables (x, p). This is the Legendre transformation .

The transformed function is the Hamiltonian $\mathcal{H}(p, x)$, which we have plotted in Fig. 3.4. We note that the plot of the Hamiltonian in Fig. 3.4 has a distinctly different form from that of the Lagrangian in Fig. 3.1. As we have shown, however, these two functions, and hence the two surfaces, are identical in information content.

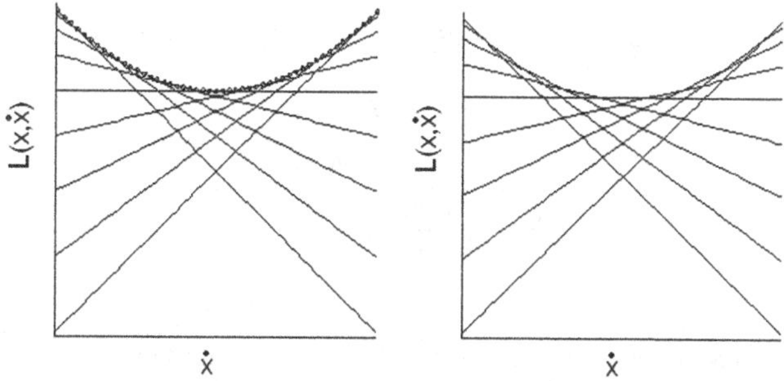

Fig. 3.3 Tangents to Lagrangian for harmonic oscillator $L(x, \dot{x})$. In *left panel* $L(x, \dot{x})$ is plotted with *dots*

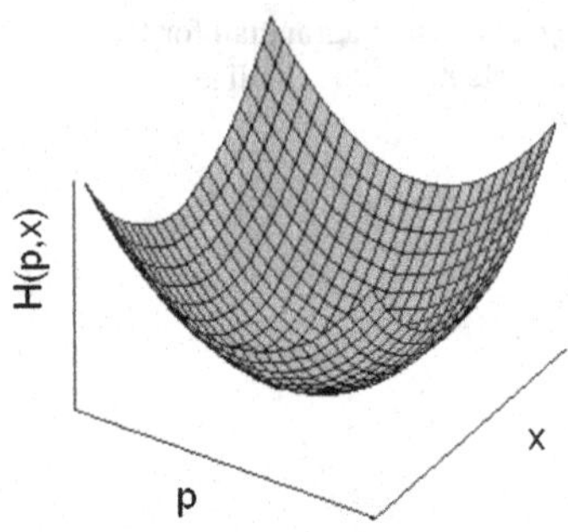

Fig. 3.4 Hamiltonian for the harmonic oscillator obtained from a Legendre Transformation of the Lagrangian $L(x, \dot{x})$ plotted in Fig. 3.1

From what we have shown in terms of an example we can now develop in general. We consider a function of two variables, $\Psi(\xi, \eta)$. We choose to transform out the variable η in favor of the new variable

$$\zeta = \left(\frac{\partial \Psi}{\partial \eta}\right), \tag{3.1}$$

We begin by noting that the dependence of Ψ on (ξ, η) is defined by the Pfaffian differential[2]

$$\mathrm{d}\Psi = \left(\frac{\partial \Psi}{\partial \xi}\right)\mathrm{d}\xi + \left(\frac{\partial \Psi}{\partial \eta}\right)\mathrm{d}\eta. \tag{3.2}$$

In the Legendre Transform to eliminate η in favor of ζ we define[3]

$$\Phi = \eta\zeta - \Psi. \tag{3.3}$$

To determine the variables on which this new function depends we obtain the differential

$$\mathrm{d}\Phi = \eta \mathrm{d}\zeta + \zeta \mathrm{d}\eta - \mathrm{d}\Psi. \tag{3.4}$$

Using (3.3) and (3.2) Eq. (3.4) becomes

$$\begin{aligned}\mathrm{d}\Phi &= \eta \mathrm{d}\zeta + \zeta \mathrm{d}\eta - \left(\frac{\partial \Psi}{\partial \xi}\right)\mathrm{d}\xi - \zeta \mathrm{d}\eta \\ &= \eta \mathrm{d}\zeta - \left(\frac{\partial \Psi}{\partial \xi}\right)\mathrm{d}\xi.\end{aligned}$$

[2]As we indicated in our discussion of rolling, the Pfaffian defines the function by providing a rule for its construction in infinitesimals. We may now interpret that in terms of an infinitesimal line geometry.

[3]The Legendre Transform is normally defined as the negative of (3.3). The form used here will produce a positive Hamiltonian.

That is the function Φ depends on ξ and ζ rather than on ξ and η. We have then succeeded in obtaining the transformation we sought. However, the variable η still appears as a multiplier of $\mathrm{d}\zeta$. We realize that $\partial\psi/\partial\xi$ generally contains η as well. So we must also obtain η algebraically in terms of ζ (and possibly ξ). This step, which is an important part of the Legendre transformation, we must carry out algebraically using (3.1).

3.3 The Hamiltonian

In (2.87) in Sect. 2.5.4 we defined the canonical momentum p_j conjugate to the generalized coordinate q_j as

$$p_\mathrm{j} = \frac{\partial L}{\partial \dot{q}_\mathrm{j}}. \tag{3.5}$$

If we transform out the dependence of the Lagrangian $L(q, \dot{q}, t)$ on $\dot{q} = \{\dot{q}_\mathrm{j}\}$ in favor of a dependence on $p = \{p_\mathrm{j}\}$ using the Legendre Transform we obtain

$$\boxed{\mathcal{H}(p, q, t) = \sum_\mathrm{j} p_\mathrm{j}\dot{q}_\mathrm{j} - L(q, \dot{q}, t).} \tag{3.6}$$

The function $\mathcal{H}$ in (3.6) is the *Hamiltonian.*

Historically the actual point in his development, at which Hamiltonian introduced the Legendre transformation, was to produce his Principle Function S from his Characteristic Function V (see Sect. 1.9.2). In the present approach we transform the functions appearing in the functionals S and V rather than the functionals themselves. This reflects the importance of the Hamiltonian function in the modern form of Analytical Mechanics. We will return to the vision of Hamilton and Jacobi in a later discussion (Chap. 5).

The differential of the Hamiltonian is

$$\begin{aligned}
\mathrm{d}\mathcal{H} &= \sum_\mathrm{j} \left(p_\mathrm{j}\mathrm{d}\dot{q}_\mathrm{j} + \dot{q}_\mathrm{j}\mathrm{d}p_\mathrm{j}\right) - \mathrm{d}L(q, \dot{q}, t) \\
&= \sum_\mathrm{j} \left(p_\mathrm{j}\mathrm{d}\dot{q}_\mathrm{j} + \dot{q}_\mathrm{j}\mathrm{d}p_\mathrm{j}\right) - \sum_\mathrm{j} \left[p_\mathrm{j}\mathrm{d}\dot{q}_\mathrm{j} + \left(\frac{\partial L}{\partial q_\mathrm{j}}\right)\mathrm{d}q_\mathrm{j}\right] - \frac{\partial L}{\partial t}\mathrm{d}t \\
&= \sum_\mathrm{j} \left[-\left(\frac{\partial L}{\partial q_\mathrm{j}}\right)\mathrm{d}q_\mathrm{j} + \dot{q}_\mathrm{j}\mathrm{d}p_\mathrm{j}\right] - \frac{\partial L}{\partial t}\mathrm{d}t.
\end{aligned} \tag{3.7}$$

The Hamiltonian is, therefore, a function of (q, p, t). We also recall that p_j is (normally) simply proportional to $\dot{q}_\mathrm{j}$. Then to eliminate $\dot{q}_\mathrm{j}$ in favor of p_j in the Hamiltonian is rather simple in many cases.

Remark 3.3.1 This elimination of the $\dot{q}_{\mathrm{j}}$ from the Hamiltonian must always be carried out. It is imperative that the Hamiltonian depend only on $p_{\mathrm{j}}, q_{\mathrm{j}}$ and possibly the time, t.

3.4 The Canonical Equations

If we write the Euler–Lagrange Equations (2.69) in the form $\partial L/\partial q_{\mathrm{j}} = \mathrm{d}\left(\partial L/\partial \dot{q}_{\mathrm{j}}\right)/\mathrm{d}t$ and use (3.5), we have

$$\frac{\partial L}{\partial q_{\mathrm{j}}} = \frac{\mathrm{d}}{\mathrm{d}t} p_{\mathrm{j}} = \dot{p}_{\mathrm{j}}.$$

The Pfaffian for the Hamiltonian (3.7) then becomes

$$\mathrm{d}\mathcal{H} = \sum_{\mathrm{j}} \left[-\dot{p}_{\mathrm{j}} \mathrm{d}q_{\mathrm{j}} + \dot{q}_{\mathrm{j}} \mathrm{d}p_{\mathrm{j}} \right] - \frac{\partial L}{\partial t} \mathrm{d}t \tag{3.8}$$

The general form of the Pfaffian of $\mathcal{H}(q, p, t)$ is

$$\mathrm{d}\mathcal{H} = \sum_{\mathrm{j}} \left[\left(\frac{\partial \mathcal{H}}{\partial q_{\mathrm{j}}} \right) \mathrm{d}q_{\mathrm{j}} + \left(\frac{\partial \mathcal{H}}{\partial p_{\mathrm{j}}} \right) \mathrm{d}p_{\mathrm{j}} \right] + \frac{\partial \mathcal{H}}{\partial t} \mathrm{d}t. \tag{3.9}$$

We may then equate the corresponding terms in (3.8) and (3.9) to obtain

$$\boxed{\dot{p}_{\mathrm{j}} = -\partial \mathcal{H}/\partial q_{\mathrm{j}},} \tag{3.10}$$

and

$$\boxed{\dot{q}_{\mathrm{j}} = \partial \mathcal{H}/\partial p_{\mathrm{j}}.} \tag{3.11}$$

But the partial derivative $\partial \mathcal{H}/\partial t$ in (3.9) was taken holding the variables q and p constant while the partial derivative $-\partial L/\partial t$ in (3.8) was taken holding q and $\dot{q}$ constant. Therefore these partial derivatives cannot be equated. We can, however, evaluate $-\partial L/\partial t$ from (3.8) as

$$-\frac{\partial L}{\partial t} = \frac{\mathrm{d}\mathcal{H}}{\mathrm{d}t} + \sum_{\mathrm{j}} \left[\dot{p}_{\mathrm{j}} \frac{\mathrm{d}q_{\mathrm{j}}}{\mathrm{d}t} - \dot{q}_{\mathrm{j}} \frac{\mathrm{d}p_{\mathrm{j}}}{\mathrm{d}t} \right] = \frac{\mathrm{d}\mathcal{H}}{\mathrm{d}t}, \tag{3.12}$$

since $\mathrm{d}q_{\mathrm{j}}/\mathrm{d}t = \dot{q}_{\mathrm{j}}$ and $\mathrm{d}p_{\mathrm{j}}/\mathrm{d}t = \dot{p}_{\mathrm{j}}$. The Hamiltonian is then a constant of the system motion provided the Lagrangian does not depend explicitly on the time t. This is another of Emmy Nöther's observations. As in Sect. 2.6 we require here that the Lagrangian completely represents the system with all constraints incorporated algebraically.

The Eqs. (3.10) and (3.11) are the *canonical equations of Hamilton*, which we first saw in Sect. 1.9.3 as (1.30). These equations are, as their title implies, the fundamental equations of Analytical Mechanics.[4]

The evident advantage of the canonical equations in the solution of a dynamical problem comes from the fact that they are first order differential equations, rather than the second order equations of Euler and Lagrange and of Newton. Specifically the Runge–Kutta[5] method of solution of differential equations, which forms the basis of much numerical work, is applicable to first order equations. We must only recognize that the foundational structure of Analytical Mechanics is based on the Hamiltonian from which the position and the momentum emerge as basic.

If we choose we may obtain the canonical equations directly from Hamilton's Principle based on the Hamiltonian. Hamilton's Principal Function, written in terms of the Hamiltonian, is

$$S = \int_{t_1}^{t_2} \mathrm{d}t \left[\sum_j p_j \dot{q}_j - \mathcal{H} \right] \mathrm{d}t.$$

A δ-variation (see Definition 2.4.2 in Sect. 2.4.3) of this expression leads directly to the canonical equations (see exercises).

3.5 Constraints

We introduce the constraints into the Hamiltonian formulation in the same way we introduced them into the Lagrange formulation. The only difference is that we have $2N$ independent variables in the Hamiltonian formulation: the N generalized coordinates and N conjugate momenta. We consider that there are also n constraint equations of the form $g_k(q) = 0$ (see (2.75) in Sect. 2.5.2), which are functions of the generalized coordinates. There are then $2N - n$ variables which may be varied independently.

We again have integrals (2.78) over the products of $\lambda_k(t)$ and the constraint functions $g_k(q)$, which vanish identically. Hamilton's Principal Function with constraints incorporated is then

$$S = \int_{t_1}^{t_2} \mathrm{d}t \left\{ \left[\sum_j^N p_j \dot{q}_j - \mathcal{H} \right] + \sum_k^n \lambda_k(t)\, g_k(q) \right\}. \tag{3.13}$$

---

[4]Canonical is an adjective derived from canon. It essentially means "standard", "generally accepted" or "part of the backstory".

For over a century mathematicians have used the word canonical to refer to concepts that have a kind of uniqueness or naturalness, and are (up to trivial aspects) "independent of coordinates".

[5]The Runge–Kutta method is a numerical method for solving differential equations published in 1901 as the joint work of the German mathematicians Carl Runge (1856–1927) and Martin Wilhelm Kutta (1867–1944).

In the δ-variation δq_j and δp_j vanish at the end points t_1 and t_2. The δ-variation of (3.13) is then

$$\delta S = \int_{t_1}^{t_2} dt \left\{ \sum_{j}^{N} \left[\dot{q}_j - \frac{\partial \mathcal{H}}{\partial p_j} \right] \delta p_j - \sum_{j}^{N} \left[\dot{p}_j + \frac{\partial \mathcal{H}}{\partial q_j} - \sum_{k}^{n} \lambda_k (t) \frac{\partial g_k}{\partial q_j} (q) \right] \delta q_j \right\} = 0. \tag{3.14}$$

As in the Lagrange formulation, we choose the n functions $\lambda_k(t)$ so that the first n sets of equations

$$\dot{q}_j - \frac{\partial \mathcal{H}}{\partial p_j} = 0 \tag{3.15}$$

and

$$\dot{p}_j + \frac{\partial \mathcal{H}}{\partial q_j} - \sum_{k}^{n} \lambda_k (t) \frac{\partial g_k}{\partial q_j} (q) = 0 \tag{3.16}$$

for $j = 1 \ldots n$ are satisfied. As in the Lagrangian formulation these are linear algebraic equations for the $\lambda_k(t)$ and may be, in principle, be solved for the n functions $\lambda_k(t)$.[6] We are then left with

$$\delta S = \int_{t_1}^{t_2} dt \left\{ \sum_{j=n+1}^{N} \left[\dot{q}_j - \frac{\partial \mathcal{H}}{\partial p_j} \right] \delta p_j - \sum_{j=n+1}^{N} \left[\dot{p}_j + \frac{\partial \mathcal{H}}{\partial q_j} - \sum_{k}^{n} \lambda_k (t) \frac{\partial g_k}{\partial q_j} (q) \right] \delta q_j \right\} = 0. \tag{3.17}$$

In (3.17) all of the $N - n$ variations δp_i and δq_j are independent. Therefore we have $N - n$ sets of equations

$$\boxed{\dot{q}_j - \partial \mathcal{H} / \partial p_j = 0} \tag{3.18}$$

and

$$\boxed{\dot{p}_j + \partial \mathcal{H} / \partial q_j - \sum_{k}^{n} \lambda_k (t) \, \partial g_k (q) / \partial q_j = 0} \tag{3.19}$$

together with the n sets of equations (3.16), which are identical to the $N - n$ equations (3.19). Therefore we finally have N equations (3.18) and N equations (3.19). We also have n constraint equations of the form $g_k(q) = 0$, which are functions of the generalized coordinates. As in the case of the Euler–Lagrange equations we then

[6] As in the Euler–Lagrange case, $\lambda_k(q, p, t)$ is the result of the linear algebra solution.

have a method by which we can obtain the required equations for a compete solution of the system. The mechanism by which the solution is obtained does not differ in each case.

Just as we did in Sect. 2.5.4, and for the same reason, we may identify here the force of the kth constraint affecting the momentum p_{j} as

$$\boxed{\mathrm{F}_{\text{kj}}^{\text{react}} = \lambda_{\text{k}}(t)\, \partial g_{\text{k}} / \partial q_{\text{j}}} \tag{3.20}$$

As an example for the application of the canonical equations we consider the *Kepler Problem*. The Kepler Problem is of considerable historical importance. This was the problem posed by Edmond Halley in the coffee house. It was then the problem that Newton considered and on which he sent Halley a detailed 9 page account. It is, of course, also the problem on which Kepler spent 8 years of his life. In Newton's case, however, the solution was geometrical, and in Kepler's case it was deduced from Tycho Brahe's data and some ingenuity [see [33]]. We shall use the Analytical Mechanics of Euler, Lagrange, and Hamilton. And we will simply study the motion of a particle of mass m in a spherical potential $V(\rho)$, where ρ is the spherical radial coordinate. We can only imagine the delight that Kepler and, perhaps, even Newton would have found in this.[7]

Example 3.5.1 The Kepler Problem The kinetic energy of a particle of mass m in spherical coordinates is

$$T = \frac{1}{2} m \left[\dot{\rho}^2 + \rho^2 \dot{\vartheta}^2 \sin^2 \phi + \rho^2 \dot{\phi}^2 \right].$$

We choose a general form for the potential

$$V(\rho) = -K \rho^{-n}.$$

The force in the radial direction may be either attractive or repulsive depending on the algebraic sign of K.

$$K = \begin{cases} > 0 \text{ for attraction} \\ < 0 \text{ for repulsion} \end{cases}. \tag{3.21}$$

The Lagrangian is then

$$L = \frac{1}{2} m \left[\dot{\rho}^2 + \rho^2 \dot{\vartheta}^2 \sin^2 \phi + \rho^2 \dot{\phi}^2 \right] + K \rho^{-n}. \tag{3.22}$$

[7]Newton's personality must be considered.

Because L is cyclic in the time t the Hamiltonian will be a constant of the motion. The Lagrangian is also cyclic in the coordinate ϑ. Therefore

$$\begin{aligned} p_\vartheta &= \frac{\partial L}{\partial \dot{\vartheta}} \\ &= m\rho^2\dot{\vartheta}\sin^2\phi = \ell = \text{ constant}, \end{aligned} \tag{3.23}$$

which is the (canonical) angular momentum about the vertical axis. The canonical momenta p_ϕ and p_ρ are

$$\begin{aligned} p_\phi &= \frac{\partial L}{\partial \dot{\phi}} \\ &= m\rho^2\dot{\phi} \end{aligned} \tag{3.24}$$

and

$$\begin{aligned} p_\rho &= \frac{\partial L}{\partial \dot{\rho}} \\ &= m\dot{\rho}. \end{aligned} \tag{3.25}$$

We then have algebraic expressions for $\dot{\vartheta}$, $\dot{\phi}$ and $\dot{\rho}$ in terms of the canonical momenta, which we must use to obtain the Hamiltonian as a function of coordinates and momenta.

$$\dot{\vartheta} = \frac{\ell}{m\rho^2\sin^2\phi}, \tag{3.26}$$

$$\dot{\phi} = \frac{p_\phi}{m\rho^2}, \tag{3.27}$$

and

$$\dot{\rho} = \frac{p_\rho}{m}. \tag{3.28}$$

Written in terms of the velocities the Hamiltonian is

$$\mathcal{H} = \frac{1}{2}m\left[\dot{\rho}^2 + \rho^2\dot{\vartheta}^2\sin^2\phi + \rho^2\dot{\phi}^2\right] - K\rho^{-\mathrm{n}}. \tag{3.29}$$

We note the constant Hamiltonian $\mathcal{H}$ is equal to the total energy $\mathcal{E}$ of the particle, which we have shown is constant. With (3.26)–(3.28) the Hamiltonian becomes a function of generalized coordinates and momenta.

$$\mathcal{H} = \frac{p_\rho^2}{2m} + \frac{\ell^2}{2m\rho^2\sin^2\phi} + \frac{p_\phi^2}{2m\rho^2} - K\rho^{-\mathrm{n}}. \tag{3.30}$$

The canonical equations for the momenta p_ρ and p_ϕ are

$$
\begin{aligned}
\dot{p}_\rho &= -\frac{\partial}{\partial \rho}\left(\frac{p_\rho^2}{2m} + \frac{\ell^2}{2m\rho^2 \sin^2\phi} + \frac{p_\phi^2}{2m\rho^2} - K\rho^{-\mathrm{n}}\right) \\
&= \frac{\ell^2}{m\rho^3 \sin^2\phi} + \frac{p_\phi^2}{m\rho^3} - nK\rho^{(\mathrm{n}-1)} \qquad (3.31)
\end{aligned}
$$

and

$$
\begin{aligned}
\dot{p}_\phi &= -\frac{\partial}{\partial \phi}\left(\frac{p_\rho^2}{2m} + \frac{\ell^2}{2m\rho^2 \sin^2\phi} + \frac{p_\phi^2}{2m\rho^2} - K\rho^{-\mathrm{n}}\right) \\
&= \frac{\ell^2}{m\rho^2 \sin^3\phi} \cos\phi. \qquad (3.32)
\end{aligned}
$$

Of course we realize that motion is in the equatorial plane and we normally argue for this from the spherical symmetry of the potential and the constancy of the angular momentum p_ϑ. But this is not in the spirit of Hamiltonian mechanics, which requires that limitation to equatorial motion must be based on a simple mathematical result. From (3.32) we see that $\dot{p}_\phi = 0$ in the equatorial plane ($\phi = \pi/2$). That is in the equatorial plane $p_\phi = p_\phi(\pi/2)$ is a constant.

Using the chain rule we may write (3.32) as

$$
\dot{p}_\phi = \frac{p_\phi}{m\rho^2}\frac{\mathrm{d}p_\phi}{\mathrm{d}\phi} = \frac{\ell^2}{m\rho^2 \sin^3\phi}\cos\phi. \qquad (3.33)
$$

casting the $\dot{p}_\phi$ equation in this form provides us an equation that we may integrate to obtain p_ϕ as a function of ϕ. Carrying out the integration of (3.33) between the equatorial plane ($\phi = \pi/2$) and an arbitrary value of ϕ we have

$$
\begin{aligned}
\frac{1}{2}\left(p_\phi^2 - p_\phi^2(\pi/2)\right) &= \int_{\pi/2}^{\phi} \frac{\ell^2}{\sin^3\phi}\cos\phi \mathrm{d}\phi \\
&= -\frac{\ell^2}{2}\left(\frac{\cos^2\phi}{\sin^2\phi}\right),
\end{aligned}
$$

Which is

$$
p_\phi^2 = p_\phi^2(\pi/2) - \ell^2\left(\frac{\cos^2\phi}{\sin^2\phi}\right). \qquad (3.34)
$$

If we insert (3.34) into the Hamiltonian (3.30) we obtain

$$
\mathcal{H} = \frac{p_\rho^2}{2m} + \frac{\left(p_\phi^2(\pi/2) + \ell^2\right)}{2m\rho^2} - K\rho^{-\mathrm{n}} = \mathcal{E},
$$

which is independent of ϕ. Therefore $\dot{p}_\phi = 0$, the canonical equation for $\dot{p}_\phi$ vanishes identically, and p_ϕ is constant for all values of ϕ. From (3.32) we then have $\cos\phi = 0$ for all time and motion is in the equatorial plane $\phi = \pi/2$. Then $p_\phi(\pi/2) = 0$.[8]

We are then left with the Hamiltonian

$$\mathcal{H} = \frac{p_\rho^2}{2m} + \frac{\ell^2}{2m\rho^2} - K\rho^{-n} \tag{3.35}$$

and with the canonical equations (3.28) and (3.31), which we may now write as

$$\dot{p}_\rho = \frac{\ell^2}{m\rho^3} - \frac{nK}{\rho^{n+1}}. \tag{3.36}$$

A trajectory of the mass, which is the orbit, is obtained from $\rho(\vartheta)$. We obtain a differential equation for $\rho(\vartheta)$ from (3.28) and (3.26) using the chain rule (and setting $\phi = \pi/2$). That is

$$\frac{\mathrm{d}\rho}{\mathrm{d}\vartheta} = \frac{p_\rho}{\ell}\rho^2. \tag{3.37}$$

We may obtain a differential equation for $p_\rho(\vartheta)$ in a similar fashion. Using the chain rule and combining (3.36) and (3.26) we have

$$\frac{\mathrm{d}p_\rho}{\mathrm{d}\vartheta} = \frac{\ell}{\rho} - \frac{mnK}{\ell\rho^{n-1}}. \tag{3.38}$$

The set of Eqs. (3.37) and (3.38) are coupled differential equations in ϑ. So the solution will give us the orbit $\rho(\vartheta)$ and the momentum $p_\rho(\vartheta)$ if we want that. The Eqs. (3.37) and (3.38) are awkward. They may, however, be brought into a nice form by introducing the variable $u = 1/\rho$ instead of ρ. Then

$$\frac{\mathrm{d}u}{\mathrm{d}\vartheta} = -\frac{p_\rho}{\ell} \tag{3.39}$$

and

$$\frac{\mathrm{d}^2u}{\mathrm{d}\vartheta^2} = \frac{mnK}{\ell^2}u^{n-1} - u. \tag{3.40}$$

For the special case of the gravitational (or electrical) field $n = 1$ and (3.40) becomes

$$\frac{\mathrm{d}^2u}{\mathrm{d}\vartheta^2} = -u + \frac{mK}{\ell^2}. \tag{3.41}$$

---

[8]In our later discussion of the action and angle variables we will argue from a slightly different mathematical perspective that the momentum p_ϕ must be identically zero in the Kepler Problem.

The solutions to (3.41) are sinusoids. If we choose our coordinates so that ρ attains a minimum (u attains a maximum) value when $\vartheta = 0$ the solution is

$$\frac{1}{\rho} = A\cos\vartheta + \frac{mK}{\ell^2} \tag{3.42}$$

If we return to the constant value of the Hamiltonian we can find A. With (3.42) and (3.39), the Hamiltonian (3.35) becomes

$$\mathcal{H} = \frac{A^2\ell^2}{2m} - \frac{K^2 m}{2\ell^2}. \tag{3.43}$$

Then

$$A = \sqrt{\frac{2m\mathcal{H}}{\ell^2} + \frac{m^2K^2}{\ell^4}}$$

and the orbit (3.42) becomes

$$\frac{1}{\rho} = \frac{mK}{\ell^2}\left(\sqrt{2\mathcal{H}\frac{\ell^2}{mK^2} + 1}\cos\vartheta + 1\right). \tag{3.44}$$

Equation (3.44), written as

$$\frac{1}{\bar{\rho}} = \pm\left(\varepsilon\cos\vartheta + 1\right), \tag{3.45}$$

with $\bar{\rho} = \rho/\left(\ell^2/mK\right)$ and

$$\varepsilon = \sqrt{\left(2\mathcal{H}\ell^2/mK^2\right) + 1}, \tag{3.46}$$

is the general polar form for the conic sections. The x-axis is the conic axis and the $\pm$ accounts for the algebraic sign of the potential constant K, which enters in the original factor mK/ℓ^2 in (3.44). We have drawn the three possible orbits in Figs. 3.5, 3.6, and 3.7. In each figure the center of the potential is indicated by a small closed circle.

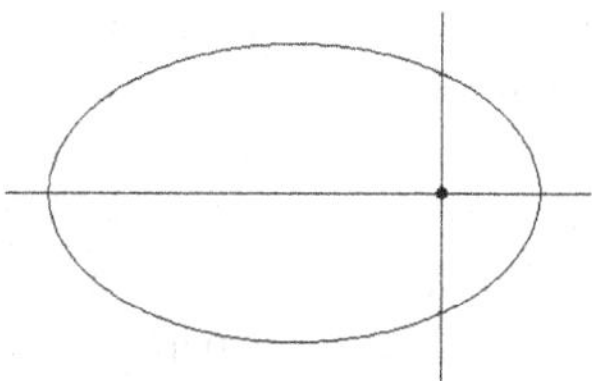

Fig. 3.5 The Kepler (attractive $K > 0$) elliptic orbit. This is the most familiar of the possible Kepler orbits because it is the form of the planetary orbits, although the eccentricity of the planetary orbits is much closer to the circular $\varepsilon = 0$ (see (3.45)) than the $\varepsilon = 0.7$ chosen here

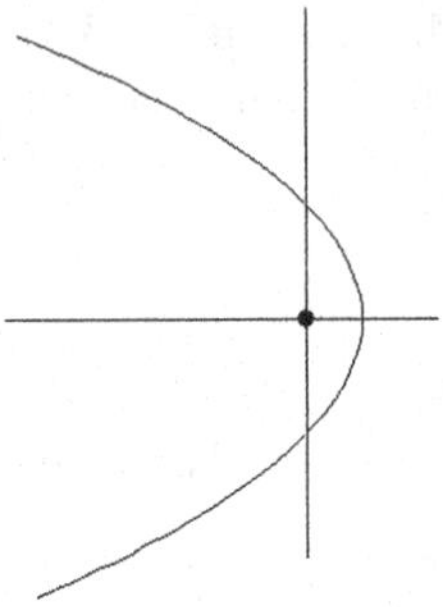

Fig. 3.6 The Kepler (attractive $K > 0$) *parabolic* orbit. For this orbit the eccentricity is $\varepsilon = 1.0$

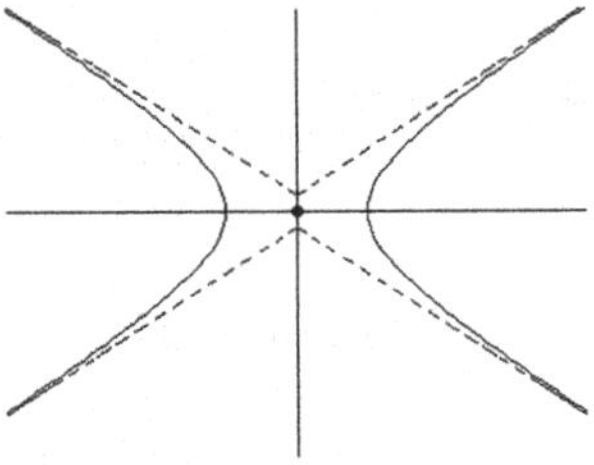

Fig. 3.7 The Kepler (repulsive $K < 0$) *hyperbolic* orbit. This is the orbit for scattering of charged projectiles by charged targets, such as Rutherford Scattering of α-particles by metallic nuclei. For this figure we chose $\varepsilon = 1.5$

We have come quite a distance beyond the point at which Kepler found himself when, in 1605, he finally realized that the orbit of Mars was an ellipse with the sun at a focus [see [119], p. 67]. Kepler had removed astronomy from a philosophical subject to a subject for what would become astrophysics. But he lacked the tools we have used with ease here. He realized that a force from the sun was responsible for the elliptical orbit. But he was limited by the weight of 2,000 years of Aristotelian physics [see [91], pp. 176–78]. What is so clear to us now was a source of astonishing mystery to Kepler.

3.6 Rutherford Scattering

Hans Geiger, the John Harling Fellow, and an undergraduate student, Ernest Marsden, a Hatfield Scholar, conducted a set of experiments in Ernest Rutherford's laboratory at the University of Manchester in England between 1908 and 1910 [30]. The original experiment was to verify Joseph John (J.J.) Thomson's model of the atom, in which there was considerable confidence. The experiment in 1910 permitted a rather precise measurement of the scattering angle for α-particles from various metal foils from which some statements regarding probable scattering angles could be given as functions of atomic masses and film thicknesses. The comment of Rutherford that this was like firing a twelve-inch shell at tissue paper and having it come back and hit you, is legendary in the history of physics. And the analysis leading to the scattering probabilities that Rutherford produced in 1911 is standard fare in any Analytical Mechanics or modern physics course [103] [cf. [117], pp.131–39].

The experimental results of Geiger and Marsden, and particularly those of Geiger in 1910, had convinced Rutherford of the existence of a point-like positive charge in the atom. Geiger had specifically measured the most probable scattering angle for a number of metallic films. The Geiger results, Rutherford first showed, were not consistent with multiple scattering collisions with atoms in which there was a uniform or a saturnian-like charge distribution. Had they been the model of Sir J.J. Thomson, which was consistent with β-particle scattering, would also work for α-particle scattering [116]. He then showed that the experimental results were consistent with what was obtained from single collisions with point-like positive charges [103].

Rutherford's paper did not include any of the analysis of the Kepler Problem, which was also standard fare for any student of physics at the end of the 19th century. He included a figure, which was one half of the orbit in Fig. 3.7, for the evaluation of the collisional cross section. Our Fig. 3.8 is a slight amplification of Rutherford's figure [see [103], Fig. 1] The angle indicated as ϑ in Fig. 3.8 provides the range of the angle ϑ in (3.45). We can find the extreme values of ϑ appearing in Fig. 3.8 by setting $\rho \to \infty$ in (3.44). At these extreme values the limiting magnitude of $\cos\vartheta$ is $1/\varepsilon$. We have shown the scattering angle as Φ in Fig. 3.8.

The scattering angle Φ is dependent on what is known as the impact parameter b, using the notation of James C. Maxwell . The impact parameter would be the distance of closest approach between the projectile and target if there were no interaction force. In Fig. 3.9 we have presented the results of a calculation of the hyperbolic orbit for increasing values of the impact parameter. A similar figure first appeared in Maxwell's paper on the dynamical theory of gases in 1866 [75].

In Fig. 3.10 we have drawn the orbit of Fig. 3.8 oriented such that the projectile approaches the target from the negative polar axial direction. And we have added the sphere of influence around the target atom, which encompasses the region in which the interactive force has a noticeable effect on the projectile. Corresponding to what we have in Fig. 3.9, we can see that the collision is symmetrical around the polar axis and that an impact parameter in the range $b \to b + \mathrm{d}b$ will result in a scattering angle in the range $\Phi \to \Phi + \mathrm{d}\Phi$. We have introduced a capital Φ for the scattering

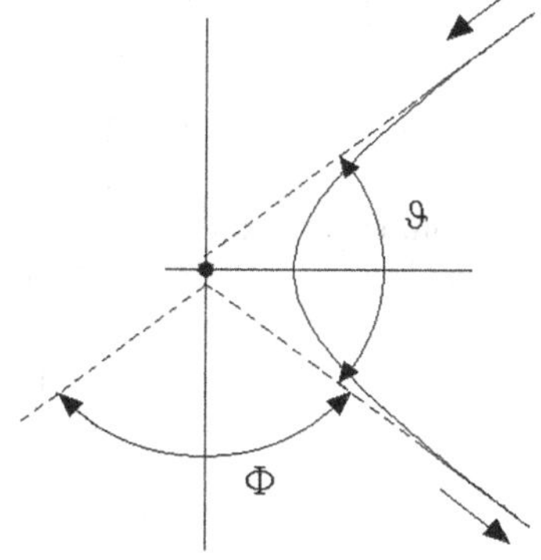

Fig. 3.8 Rutherford's figure for evaluation of the collision cross section. The asymptotic orbit ($\rho \to \infty$) is slightly offset from the target resulting in angular momentum $\ell \neq 0$

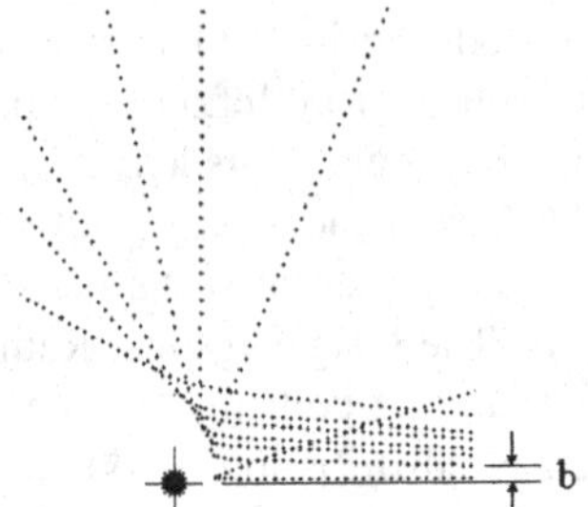

Fig. 3.9 The scattering angle as a function of impact parameter b

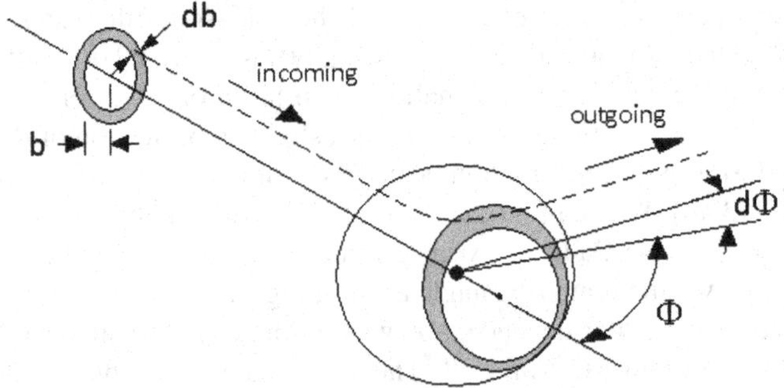

Fig. 3.10 Rutherford Scattering. The axes have been oriented such that the projectile enters the collision along the polar axis. The impact parameter lies within the limits $b \rightarrow b + \mathrm{d}b$ and the projectile is scattered through the polar (scattering angle) $\Phi \rightarrow \Phi + \mathrm{d}\Phi$. The target atom is located at the *filled circle*

angle to avoid confusion with the angles used in the calculation of the orbits above. The differential area of the sphere centered on the origin of the potential (the nucleus of the target atom) is

$$\mathrm{d}S = R^2 \sin\Phi \mathrm{d}\Theta \mathrm{d}\Phi$$

and the shaded area shown on the interaction sphere in Fig. 3.10 is the integral of this over the azimuthal angle Θ, which locates the orbit around the axis, i.e.

$$\int_{\Theta=0}^{2\pi} \mathrm{d}\Theta R^2 \sin\Phi \mathrm{d}\Phi = 2\pi R^2 \sin\Phi \mathrm{d}\Phi,$$

The number of projectiles entering the area $2\pi b\mathrm{d}b$ is related to the *differential scattering cross section* $\sigma(\Phi)$ by

$$2\pi I b\mathrm{d}b = -2\pi I \sigma(\Phi) \sin\Phi \mathrm{d}\Phi, \tag{3.47}$$

where the negative sign accounts for the fact that $\mathrm{d}\Phi/\mathrm{d}b < 0$. Using (3.44) the cosines of the angles $\vartheta_{\pm\infty}$ at the asymptotes in Fig. 3.8 are $\cos\vartheta_{\pm\infty} = 1/\varepsilon$. And from Fig. 3.8 we see that $\pi = \Phi + 2\vartheta_{\pm\infty}$. Then

$$\sin\frac{\Phi}{2} = \frac{1}{\varepsilon}. \tag{3.48}$$

From (3.48) and (3.46) we have

$$\varepsilon^2 - 1 = \left(\frac{\cos\frac{\Phi}{2}}{\sin\frac{\Phi}{2}}\right)^2 = \left(\frac{2\mathcal{H}b}{K}\right)^2. \tag{3.49}$$

From (3.49)

$$b = \pm\frac{K}{2\mathcal{H}}\frac{\cos\frac{\Phi}{2}}{\sin\frac{\Phi}{2}} \tag{3.50}$$

and

$$\frac{\mathrm{d}b}{\mathrm{d}\Phi} = +\frac{K}{4\mathcal{H}}\frac{1}{\sin^2\frac{\Phi}{2}}. \tag{3.51}$$

We chose the plus sign in (3.51) since K and $\mathrm{d}b/\mathrm{d}\Phi < 0$. Combining (3.47) and (3.51) we have

$$\sigma(\Phi) = -\frac{bK}{4\mathcal{H}}\frac{1}{\sin^2\frac{\Phi}{2}}\frac{1}{\sin\Phi}. \tag{3.52}$$

With $\sin\Phi = 2\sin(\Phi/2)\cos(\Phi/2)$ (3.52) becomes

$$\sigma(\Phi) = -\frac{bK}{8\mathcal{H}}\frac{1}{\sin^3\frac{\Phi}{2}}\frac{1}{\cos\frac{\Phi}{2}}. \tag{3.53}$$

and with (3.50) Eq. (3.53) is

$$\sigma(\Phi) = \frac{1}{4}\left(\frac{K}{2\mathcal{H}}\right)^2\frac{1}{\sin^4\frac{\Phi}{2}}. \tag{3.54}$$

Our Eq. (3.54) is essentially equation numbered (5) in Rutherford's paper. He then wrote

> The angular distribution of the α particles scattered from a thin metal sheet affords one of the simplest methods of testing the general correctness of this theory of single scattering. This has been done recently for α rays by Dr. Geiger, who found that the distribution for particles deflected between 30° and 150° from a thin gold-foil was in substantial agreement with the theory.

With those words we may consider that Rutherford established the existence of the nucleus in an atom. At least the work in Rutherford's laboratory was fundamental to the atomic theory that was developing with the dawn of the 20th century. But we must also recognize that the theory of x-ray diffraction being developed in Max von Laue's laboratory was also crucial in formulating a model of the atom. Von Laue's results showed that the number of electrons in an atom was of the order of the atomic number and not the thousands required for the electromagnetic radiation balance in the Thomson atom.

3.7 Summary

In the introduction to this chapter we said that nature speaks to us more clearly through the canonical equations of Hamiltonian than through the equations of Euler and Lagrange. As we may recall from the chapter on history (see Sects. 1.9.2 and 1.9.3) Hamilton was himself not interested in the canonical equations *per se*. He was interested in the general formulation of Analytical Mechanics in terms of the Principal Function, a functional which he called S. He had obtained S from a Legendre transformation of the *vis-viva* or living force (twice the kinetic energy). And for him the canonical equations, which he obtained in the second essay (Sect. 1.9.3), were incidental to the argument. So, although we have followed a procedure that is very logical in modern terms, seeking a simpler set of differential equations than those of Euler and Lagrange, this was not the path followed by Hamilton.

This path would have been more logical to Jacobi, who pointed out that the approach through the differential equations was easier, although he did not wish to detract from Hamilton's ideas (see Sect. 1.10.2). At this point in our study we may then accept the point Jacobi was making, stay with the differential equations, acknowledge the greater simplicity of first order equations, and recognize that we can achieve that by a Legendre transformation of the Lagrangian to produce the Hamiltonian. The canonical equations then result from a δ-variation of the Principal Function S.

The application of Lagrange Undetermined Multipliers followed as nicely in the Hamiltonian formulation as in the formulation of Euler and Lagrange. And the identity of the undetermined multipliers in terms of forces of constraint followed in the same way as in the Euler–Lagrange formulation.

We may then consider that the step from the Euler–Lagrange to the canonical equations is a convenient alternative. The general approach of Hamilton and Jacobi that lies behind what we are doing will, however, become clearer in subsequent sections as we move to more complex modern applications, including the origins of the quantum theory.

3.8 Exercises

Beginning with Exercise 3.10 we will consider electrodynamics. The correct treatment of electrodynamics is in the context of Special Relativity. In these exercises we shall then use the Hamiltonian, which we develop in the chapter on Relativity.

3.1. Written in terms of the Hamiltonian, Hamilton's Principal Function is

$$S = \int_{t_1}^{t_2} \mathrm{d}t \left[\sum_j p_j \dot{q}_j - \mathcal{H} \right] \mathrm{d}t.$$

Hamilton's Principle requires that the δ-variation (a first order variation with fixed end points) must vanish. Show that a δ-variation of this form of S results in the canonical equations of Hamilton.

3.2. In the text we worked out the Kepler Problem as an example. There we transformed the differential equation for ρ into one for $u = 1/\rho$. The solution to this equation was obviously sinusoidal. And in general it would have consisted of both a sine and a cosine. That is

$$u = \frac{1}{\rho} = A \cos \vartheta + B \sin \vartheta + \frac{mK}{\ell^2}.$$

But we claimed that we could orient the axes such that u attained a maximum, i.e. ρ attained a minimum (closest approach) at $\vartheta = 0$ and that then we needed only the cosine term. That is we chose to consider our orbits as symmetrical about the x-axis. This is an arbitrary choice.

Show that this choice does result in dropping the sine term in the above function.

3.3. We consider again the two masses on springs as shown in the figure here.

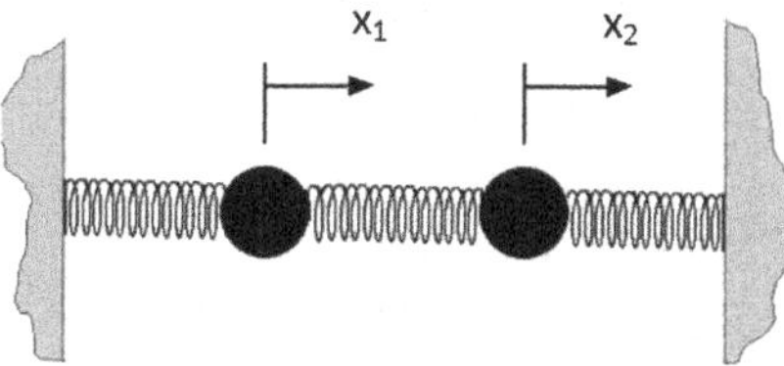

Two equal masses connected by identical springs between walls.

Study this system using the canonical equations of Hamilton. Obtain the normal frequencies as eigenvalues of the canonical equations and obtain the eigenvectors corresponding to those eigenvalues.

3.4. In the figure below we have drawn a pendulum of length ℓ and with bob of mass m located at the point x_0, which is driven horizontally as

$$x_0(t) = A + a\cos\omega t,$$

where A and a are constants.

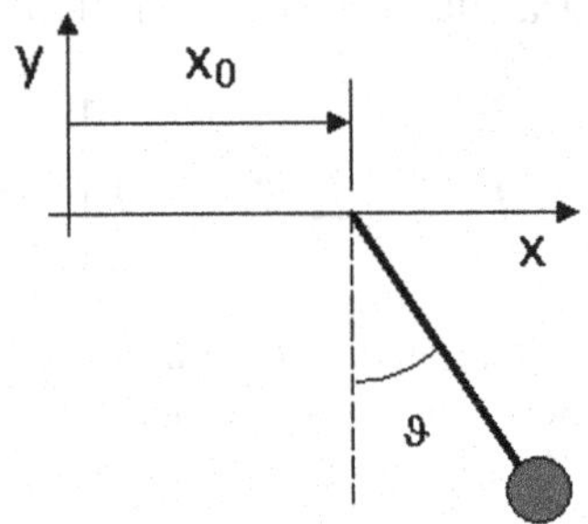

Pendulum with driven mount.

Study the motion of this pendulum. Consider particularly small angles with corresponding small excursions of the mount. Use the canonical equations of Hamilton.

Explain why you cannot incorporate the constraint using a Lagrange multiplier. [Answers: $\vartheta = -a\omega^2\cos\omega t/\left(\ell\omega^2 - g\right)$ and $p_\vartheta = ma\omega\ell g\sin\omega t/\left(\ell\omega^2 - g\right)$]

3.5. Consider the picture below of a bead on a circular, frictionless wire driven at a constant angular velocity ω about the vertical axis as we have shown here.

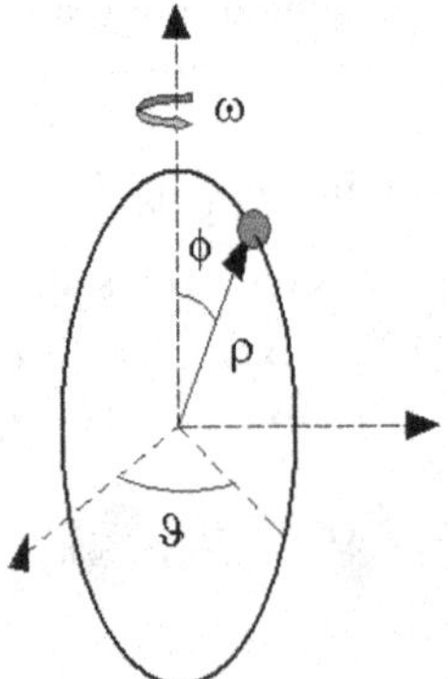

bead on a driven, circular, frictionless wire loop.

In this exercise it is advisable to use spherical coordinates. It is also convenient to take the horizontal plane through the center as the zero of potential.

Investigate the problem. There are certain things to look for. You will want to see if there are any constants of the motion, i.e. first integrals. You will also want to look for any equilibrium points and you will want to consider motion about those points to see if it is stable or not. You may want to investigate the forces of constraint as well.

Investigate the system using both the Euler–Lagrange and the canonical equations.

3.6. Consider again the two identical balls of mass m connected by a string of length b with one ball suspended through a hole in a table while the other moves on the (frictionless) top of the table. We have drawn the situation here.

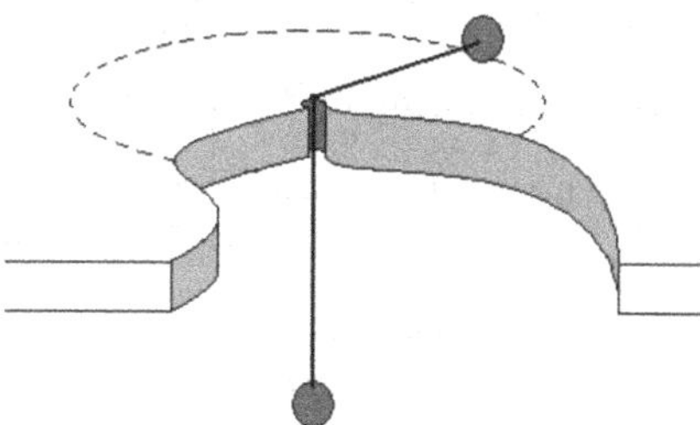

Two identical balls of mass m connected by a string of length b.

Investigate the motion. Now use canonical equations and Lagrange multipliers. Find an equilibrium point, if there is one. Study the general form of the motion. Consider small oscillations about equilibrium. Determine if the orbit is open or closed.

If you have access to appropriate software, plot the effective potential, the orbit of the ball on the table for less than a complete rotation and for a long time, and the phase plot of the ball on the table (p_r versus r).

3.7. In the drawing here we have a mass m sliding without friction on the surface of a sphere. We release the mass an infinitesimal distance from the top of the sphere.

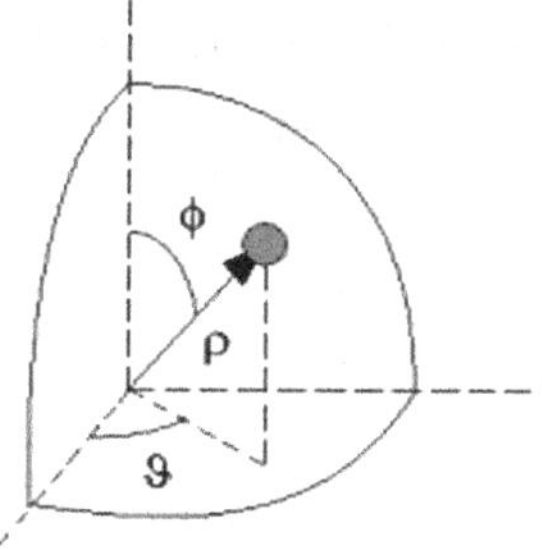

Mass sliding without friction on the surface of a sphere

At some point (some value of ϕ) the mass will leave the surface of the sphere. What is this value of ϕ? Use canonical equations.

3.8. Consider (again) the block, spring and pendulum we have drawn here.

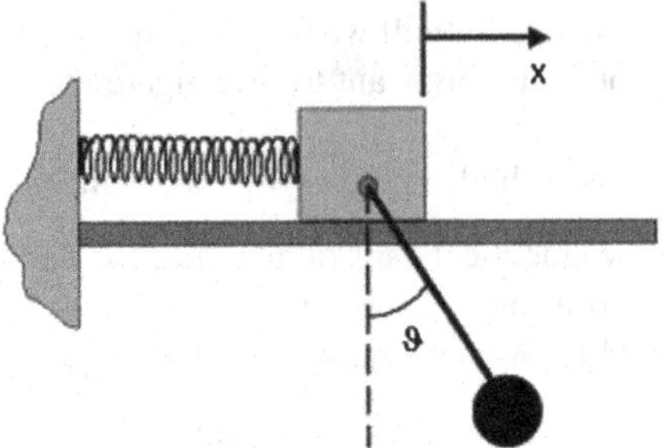

A block, spring and pendulum.

Consider the motion now in terms of the canonical equations. Linearize these for small displacements and find the natural frequencies. Choose the masses of the block and the pendulum bob to be equal.

3.9. In the figure here we have a schematic picture of the double pendulum.

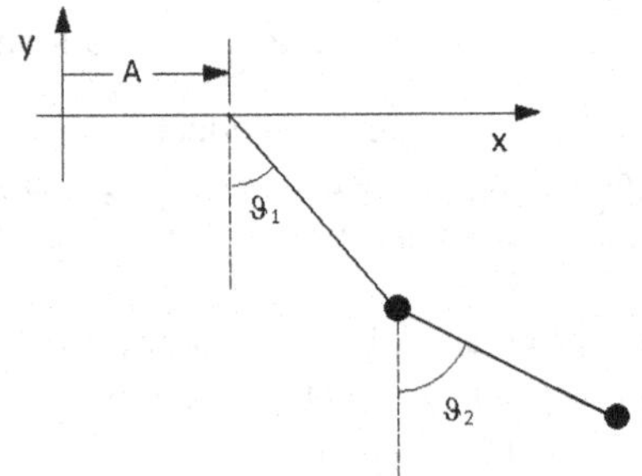

Double pendulum.

Both pendulum lengths are ℓ and both masses are m. Study the motion using the canonical equations. Obtain the modes of natural oscillation for small angles.

3.10. In the static case the magnetic field induction vector $\boldsymbol{B}$ is obtained from the vector potential as

$$\boldsymbol{B} = \text{curl}\boldsymbol{A},$$

provided A satisfies the Coulomb gauge

$$\text{div}\boldsymbol{A} = 0.$$

Show that the magnetic component of the Lorentz electromagnetic force (per unit charge) may be written in terms of the magnetic vector potential $\boldsymbol{A}$ as

$$(\boldsymbol{v} \times \boldsymbol{B})_\mu = \frac{\partial A_\nu}{\partial q_\mu}\dot{q}_\nu - \frac{\partial A_\mu}{\partial q_\nu}\dot{q}_\nu$$

where the velocity of the point charge is $v = \dot{q}_\nu \hat{e}_\nu$. We use the Einstein sum convention in which the repeated Greek indices are summed from 1 to 3 over the indices for the three Cartesian coordinates.

This will require some steps in vector algebra, which are always easier if we use subscript notation for cross products. We use the Levi-Civita density $\varepsilon_{\mu\nu\sigma}$ defined by

$$\varepsilon_{\mu\nu\sigma} = \begin{cases} +1 \text{ if } \mu\nu\sigma \text{ is an even permutation of } 1, 2, 3 \\ -1 \text{ if } \mu\nu\sigma \text{ is an odd permutation of } 1, 2, 3 \\ \ 0 \text{ if any 2 of the indices } \mu, \nu, \sigma \text{ are alike} \end{cases}$$

Then the μth component of $\boldsymbol{v} \times \boldsymbol{B}$ is

$$(\boldsymbol{v} \times \boldsymbol{B})_\mu = \varepsilon_{\mu\nu\gamma} \dot{q}_\nu B_\gamma$$

and the γth component of $\boldsymbol{B} = \text{curl}\boldsymbol{A}$ is

$$B_\gamma = \varepsilon_{\gamma\alpha\beta} \frac{\partial A_\beta}{\partial q_\alpha}.$$

Put these together to obtain $(\boldsymbol{v} \times \boldsymbol{B})_\mu$.

3.11. Show that the canonical equations for a charged particle with charge Q

$$\begin{aligned} \dot{q}_\mu &= \frac{\partial \mathcal{H}}{\partial p_\mu} \\ &= \frac{1}{m} \left(p_\mu - Q A_\mu \right), \end{aligned}$$

and

$$\begin{aligned} \dot{p}_\mu &= -\frac{\partial \mathcal{H}}{\partial q_\mu} \\ &= \frac{Q}{m} \left(p_\nu - Q A_\nu \right) \frac{\partial A_\nu}{\partial q_\mu} - Q \frac{\partial \phi}{\partial q_\mu} \end{aligned}$$

result in the standard form of Newton's Second Law

$$m \ddot{q}_\mu = Q \left(\boldsymbol{v} \times \boldsymbol{B} \right)_\mu - Q E_\mu.$$

3.12. In the text we show that the low energy (nonrelativistic) form the relativistic Hamiltonian for a classical charged point particle with mass m and charge Q is

$$\mathcal{H} = \frac{1}{2m} \sum_\mu \left(p_\mu - Q A_\mu \right)^2 + Q\varphi,$$

where φ is that scalar potential from which we find the electric field, in the static case, as $\boldsymbol{E} = -\mathrm{grad}\varphi$. We have used Q to designate the charge because q is the designation for generalized coordinate. The vector potential $\boldsymbol{A}$ has components A_μ ($\mu = 1, 2, 3$). The magnetic field induction $\boldsymbol{B}$ is obtained from the curl of $\boldsymbol{A}$ as

$$\boldsymbol{B} = \mathrm{curl}\boldsymbol{A},$$

and $\boldsymbol{A}$ is limited by the requirement that

$$\mathrm{div}\boldsymbol{A} = 0,$$

which is the Coulomb gauge for time independent fields.

Consider the motion of a charge Q in a region containing only a magnetic field with induction $\boldsymbol{B} = \hat{e}_z B$. Show that this induction results from

$$\begin{aligned}\boldsymbol{A} &= \frac{B}{2}\left(-y\hat{e}_x + x\hat{e}_y\right),\\ &= -By\hat{e}_x,\\ &= Bx\hat{e}_y,\end{aligned}$$

or

$$\boldsymbol{A} = \frac{1}{2}Br\hat{e}_\vartheta$$

in cylindrical coordinates.

Let the charge be released with non-vanishing velocities in the x and y directions. Show that the orbit of the charge is a circle.

(a) Using $\boldsymbol{A} = -By\hat{e}_x$
(b) Using $\boldsymbol{A} = -\frac{B}{2}y\hat{e}_x + \frac{B}{2}x\hat{e}_y$.

Because we already know that a positive charge moves clockwise in a field with induction $\hat{e}_z B$, we choose initial conditions ($t = 0$) as $x(0) = -R$, $\dot{x}(0) = 0$, $y(0) = 0$, and $\dot{y}(0) = v$.

3.13. Consider the motion of a charge Q in a region containing only a magnetic field. Let the charge be released with non-vanishing velocities in the x, y, and z directions. Show that the charge will "spiral" along the magnetic field lines. Assume that the magnetic field changes slowly in space so that it may always be considered approximately constant. This phenomenon is important in plasma physics and forms the core of some of the ideas proposed for "trapping" charges in a fusion reactor. The radiation from such charge motion also forms the "northern lights".

3.14. Consider the motion of a charged particle in a region of space in which there is a uniform magnetic field with induction $\boldsymbol{B} = \hat{e}_z B$, with the vector potential

$$\boldsymbol{A} = -\hat{e}_x\frac{B}{2}y + \hat{e}_y\frac{B}{2}x$$

and a uniform electric field $\boldsymbol{E} = \hat{e}_{\text{y}} E$. Show that the motion is cycloidal as we have shown here.

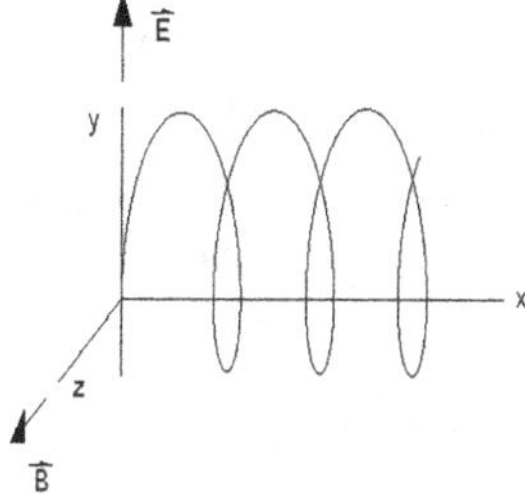

The trajectory of a charged particle moving in a region containing electric and magnetic fields perpendicular to one another.

3.15. If we treat the motion of a charged point particle of mass m and charge Q moving in a constant magnetic field of induction $\boldsymbol{B} = \hat{e}_{\text{z}} B$ using cylindrical coordinates the vector potential is

$$\boldsymbol{A} = \frac{1}{2} B r \hat{e}_{\vartheta}.$$

We note that

$$\text{div}\boldsymbol{A} = \frac{1}{r}\frac{\partial}{\partial \vartheta}\left(\frac{1}{2} B r\right) = 0$$

and with

$$\begin{aligned}\text{curl}\boldsymbol{F} &= \hat{e}_{\text{r}}\left[\frac{1}{r}\frac{\partial F_{\text{z}}}{\partial \vartheta} - \frac{\partial F_{\vartheta}}{\partial z}\right] + \hat{e}_{\vartheta}\left[\frac{\partial F_{\text{r}}}{\partial z} - \frac{\partial F_{\text{z}}}{\partial r}\right] \\ &+ \hat{e}_{\text{z}}\frac{1}{r}\left[\frac{\partial}{\partial r}\left(r F_{\vartheta}\right) - \frac{\partial F_{\text{r}}}{\partial \vartheta}\right]\end{aligned}$$

that

$$\text{curl}\boldsymbol{A} = \hat{e}_{\text{z}}\frac{1}{r}\frac{\partial}{\partial r}\left(\frac{1}{2} B r^2\right) = \hat{e}_{\text{z}}\left(B\right).$$

So the vector potential above satisfies the Coulomb gauge and produces the magnetic field induction we desire. Obtain the canonical equations and the (constant) Hamiltonian for this situation.

In the cylindrical case it will be easiest to simply begin with the low energy (nonrelativistic) approximation to the electromagnetic Lagrangian we developed in our chapter on special relativity. The Lagrangian is the function appearing in the Hamilton's Principal Function. So we return to the Lagrangian for anything other than rectangular Cartesian coordinates. This is

$$L = \frac{1}{2} m \dot{q}_\mu \dot{q}_\mu - Q\varphi + Q A_\mu \dot{q}_\mu$$

where the summation is over the three spatial components.

3.16. In the early work on magnetic confinement of fusion plasmas we considered magnetic bottles to trap the electric charges. Magnetic fields of (almost) any geometry can be produced by arrangements of external electric currents. Magnetic bottles are based on the universal principle that charged particles move on circles with radii that decrease with increasing magnetic induction.

The vector potential

$$\boldsymbol{A} = -\hat{e}_x y \frac{B}{2} \exp(az) + \hat{e}_y x \frac{B}{2} \exp(az),$$

for example, produces the magnetic induction

$$\begin{aligned} B_x &= -x \left(a \frac{B}{2} \right) \exp(az) \\ B_y &= -y \left(a \frac{B}{2} \right) \exp(az) \\ B_z &= B \exp(az), \end{aligned}$$

which, with z-axis vertical, has the form shown here.

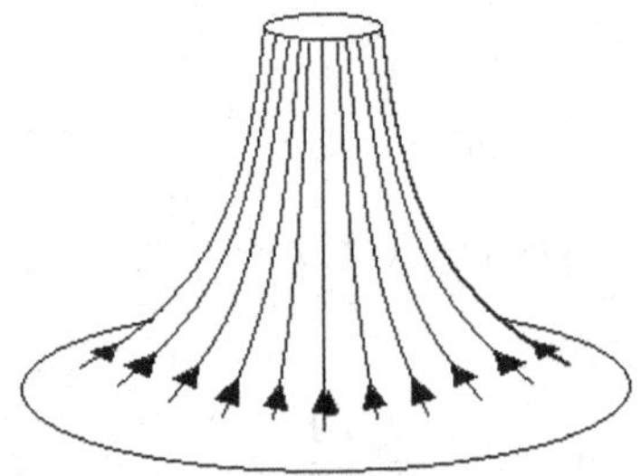

Magnetic field induction $\boldsymbol{B}$ from the vector potential $\boldsymbol{A} = -\hat{e}_x y \frac{B}{2} \exp(az) + \hat{e}_y x \frac{B}{2} \exp(az)$.

Charged particles in this region follow the lines of induction in a corkscrew motion of decreasing radius.

We obtained the particle trajectory from a numerical integration of the canonical equations using a Runge-Kutta algorithm.[9] In the numerical solution we released the charged particle on the x-axis at $x = 1$ with a momentum in the y- and z-directions. The result was the trajectory shown here.

[9]These very important numerical techniques for the solution of first order differential equations were developed around 1900 by the German mathematicians C. Runge and M.W. Kutta.

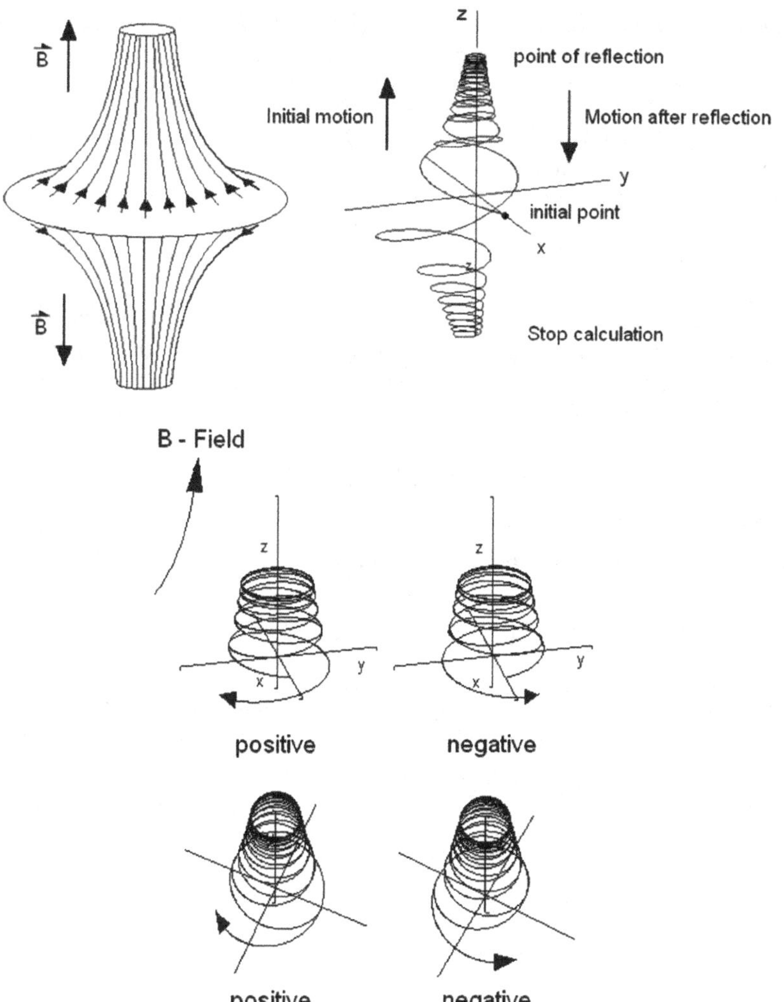

Spiral motion of charges in spatially varying magnetic field.

In this figure we have plotted results for both positive and for negative charges. The charges spiral along the magnetic field lines moving in the positive z-direction until they are deflected and then they spiral out with growing radius along the negative z-direction. The top images are for a small initial momentum and the bottom for a larger initial momentum. The larger momentum makes the spiral of the charge more evident.

The results from a region containing oppositely converging magnetic fields demonstrates the magnetic bottle effect. We show this in the figure.

Show that the vector potential actually results in the magnetic field induction above and that the Coulomb gauge $\operatorname{div}\boldsymbol{A} = 0$ is satisfied by $\boldsymbol{A}$. Then obtain the Hamiltonian and the canonical equations for motion of a charge in this field. Consider planes of constant z to show the decrease in radius with increasing z.

3.17. In the text we discussed the problem of Rutherford scattering. In his analysis Rutherford assumed that only the Coulomb force acted on the α-particle scattered by the nucleus. The potential was then

$$\varphi = \frac{Q_{\mathrm{N}} Q}{4\pi\varepsilon_0} \frac{1}{\rho}$$

We may, depending on the nucleus, also have a nuclear magnetic moment. This will have an affect as well on the moving α-particle. The vector potential at a distance $\boldsymbol{r}$ (using spherical coordinates $|\boldsymbol{r}| = \rho$) from a nucleus with magnetic moment $\boldsymbol{M}$ is [see, e.g. [45], pp. 138–141].

$$\boldsymbol{A}(\boldsymbol{r}) = \frac{\mu_0}{4\pi\rho^2} \boldsymbol{M} \times \boldsymbol{r},$$

which, carrying out the cross product, becomes

$$\boldsymbol{A}(\boldsymbol{r}) = \frac{\mu_0 M}{4\pi} \frac{1}{\rho^3} \sin\phi \hat{e}_\vartheta.$$

We shall simplify our problem by confining motion to the horizontal plane. In spherical coordinates the polar angle is then $\phi = \pi/2$.

Find the Hamiltonian for Rutherford scattering when the nucleus has a magnetic moment. Linearize this for small values of M. Comment on the effect of the nuclear magnetic moment.

Chapter 4
Solid Bodies

> The growth of the use of transformation theory ... is the essence of the new method in theoretical physics. ...[This] symbolic method, however, seems to go more deeply into the nature of things. It enables one to express the physical laws in a neat and concise way, and will probably be increasingly used in the future ...
>
> Paul A.M. Dirac [May, 1930]

4.1 Introduction

For the purpose of describing the motion of a solid body, and later for an analysis of the dynamics of the body, we will consider that the body can be divided into a very large collection of classical point particles. Each of these particles is made up of a (large) number of atoms, but is infinitesimal in size and mass compared to the dimensions and mass of the body. We will make no attempt to consider atomic bonds in our (imagined) construction of the rigid body from these particles. We simply accept that Newton's Third Law must hold between all pairs of classical point particles and that, therefore, all internal forces among them cancel.

Bodies may be rigid, elastic or plastic. In a rigid body the distances between the classical point particles is fixed and not subject to change. In an elastic body these distances change under external forces of compression or tension and return to their previous values if those forces are removed. In a plastic body the distances will not return to their previous values on removal of external forces. In this chapter we will treat only rigid bodies.

The general motion of a solid body will consist of a translation of the body as a whole and a rotation about some axis in the body. We can describe the motion of

C.S. Helrich, *Analytical Mechanics*, Undergraduate Lecture Notes in Physics, DOI 10.1007/978-3-319-44491-8_4

the body as a whole in terms of a vector locating the center of mass (CM) of the body relatively to an external (fixed) coordinate system. We then require a second coordinate system, fixed in the body and with origin at the CM, to describe the motion of the particles relative to the CM. In a rigid body each particle is located at a fixed point in the body coordinate system. The motion of each point particle is then determined by the rotation of the body coordinate system. This rotation may be specified by rotations about each of the three axes of the body coordinate system. There are, therefore, six coordinates required for the description of the motion of each classical point particle.

Our choice of the CM for location of the origin of the rigid body coordinate system is more judicious than it may seem. This choice will result in a separation of the canonical equations into two independent sets. One of these will determine the motion of the CM and the second will determine the rotation of the rigid body about the CM.

For our mathematical work with the rotating body coordinate system we will use the vectors and a *transformation theory* developed by Paul A.M. Dirac for the quantum theory. There is an elegant simplicity resulting from the use of Dirac's transformation theory. For the reader unfamiliar with Dirac's vectors we develop what is needed in the first section of this chapter. The benefit will far outweigh the effort required to learn this approach.

4.2 The Vector Space

4.2.1 Dirac Vectors

For much of what we will do we may consider that Dirac has simply introduced a different notation for vectors. We will, for example, identify the position vector of the ith classical point particle by what Dirac called a *ket vector* $|r_{\mathrm{i}}\rangle$ instead of $\boldsymbol{r}_{\mathrm{i}}$. The brace $|\ldots\rangle$ indicates a vector just as the arrow above a term, or the bold representation of a term, serves in the standard vector notation. Inside the brace we will place the identifier of the vector as in the case of $|r_{\mathrm{i}}\rangle$. This is particularly helpful in quantum mechanics where the vector will (often) be an eigenvector common to a group of operators. The set of quantum numbers of the operators appear then in the brace.

The Dirac vectors $|a\rangle$, $|b\rangle, \ldots$ are elements of a *vector space* $\mathcal{V}$ $(|a\rangle\,, |b\rangle\,, \ldots \in \mathcal{V})$ if they satisfy the postulates of a vector space. These are

1. **closure under addition**: for each $|a\rangle$ and $|b\rangle$ which are elements of $\mathcal{V}$ there is a unique sum $|a\rangle + |b\rangle$ that is a vector in the space. That is[1]

$$\text{For } |a\rangle \in \mathcal{V} \text{ and } |b\rangle \in \mathcal{V} \ \exists \ |a\rangle + |b\rangle = |c\rangle \in \mathcal{V}.$$

2. **addition is associative**:

$$(|a\rangle + |b\rangle) + |c\rangle = |a\rangle + (|b\rangle + |c\rangle)\,.$$

3. **addition is commutative**:

$$|a\rangle + |b\rangle = |b\rangle + |a\rangle\,.$$

4. there exists a *zero vector* $|0\rangle$ in the space $\mathcal{V}$ defined by the requirement that the addition of this zero vector to any vector in the space results in the original vector. That is[2]

$$\exists\,|0\rangle \ni |a\rangle + |0\rangle = |a\rangle\,\forall\,|a\rangle \in \mathcal{V}.$$

5. for each vector $|a\rangle$ in the space $\mathcal{V}$ there exists a *negative* vector, $-\,|a\rangle$ defined by the requirement that the sum of $-\,|a\rangle$ and $|a\rangle$ produces the zero vector. That is

$$\forall\,|a\rangle \in \mathcal{V} \ \exists \ -|a\rangle \ni |a\rangle + (-\,|a\rangle) = |0\rangle\,.$$

6. **closure under multiplication by a scalar**: For every number from the field of complex numbers C (a real number is a complex number with an imaginary part equal to zero) and every vector $|a\rangle$ in the space $\mathcal{V}$ there is a unique vector $C\,|a\rangle$ contained in the space $\mathcal{V}$. Multiplication by a scalar satisfies

$$C\,(|a\rangle + |b\rangle) = C\,|a\rangle + C\,|b\rangle\,.$$

$$(C + D)\,|a\rangle = C\,|a\rangle + D\,|a\rangle\,.$$

$$(CD)\,|a\rangle = C\,(D\,|a\rangle)\,.$$

$$\mathbf{1}\,|a\rangle = |a\rangle\,.$$

In the last expression **1** is the number *one*, known as *unity*.

[1] The symbol $\in$ means "is an element of" and $\exists$ means "there exists." These symbols are commonly used as shorthand in physics just as in mathematics.

[2] The symbol $\ni$ is "such that" and $\forall$ is "for all".

4.2.2 Scalar Product

A scalar product is a product between two vectors which results in a complex number. For the vectors $|a\rangle$ and $|b\rangle$ the scalar product is written as $\langle a\ |b\rangle$. The postulates of the scalar product are:

1.
$$\langle a\ |b\rangle = \langle b\ |a\rangle^*$$

2.
$$\langle c|\left(|a\rangle + |b\rangle\right) = \langle c\ |a\rangle + \langle c\ |b\rangle$$

3.
$$\langle a\ |a\rangle \geq 0$$

4.
$$\langle a\ |a\rangle = 0 \text{ if and only if } |a\rangle \equiv |0\rangle\ .$$

5.
$$\left(\langle a|\, C\right)|b\rangle = C^*\,\langle a\ |b\rangle$$

and
$$\langle a|\left(C\,|b\rangle\right) = C\,\langle a\ |b\rangle$$

The asterisk in postulates 1 and 5 indicates *complex conjugate*. In this chapter we will only use real vectors. So we may ignore the complex conjugate.

The fact that the scalar product takes on the form of a bracket $\langle a\ |b\rangle$ is the reason for the names Dirac gave to the vectors $|a\rangle$ and vector in the dual space $\langle b|$. He called $\langle b|$ a *bra vector* and $|a\rangle$ a *ket vector*. The combination is then a bra-ket or *bracket*.

4.2.3 Representation

General vectors such as $|a\rangle$ and $|b\rangle$ are *abstract vectors*. We can speak of them in general abstract terms and even write abstract equations for them, such as Newton's Second Law $|F\rangle = m\,|a\rangle$. But we cannot conduct concrete mathematical operations on them until we have represented them in terms of a *basis*. Then we are able to deal with the components of the vector, which are numbers, algebraic expressions, or expressions involving derivatives.

We represent vectors in terms of a set of *linearly independent vectors*, which we construct independently of the physical situation we are modeling. The number of such vectors we require is the dimension of the vector space.

Linear Independence. A set of N vectors, $\{|q_n\rangle\}_{\mathrm{n=1}}^{\mathrm{N}}$ in a space $\mathcal{V}$, is *linearly independent* if

$$\sum_{\mathrm{n=1}}^{\mathrm{N}} C_{\mathrm{n}} \left|q_{\mathrm{n}}\right\rangle = 0$$

is satisfied only when each of the coefficients, C_{n}, is identically zero.

If any vector, $|v\rangle \in \mathcal{V}$ can be written as a sum

$$|v\rangle = \sum_{\mathrm{n=1}}^{\mathrm{N}} v_{\mathrm{n}} \left|q_{\mathrm{n}}\right\rangle \tag{4.1}$$

then the set $\{|q_n\rangle\}_{\mathrm{n=1}}^{\mathrm{N}}$ spans the space. A *basis* is any set of linearly independent vectors that spans the space. The smallest number of linearly independent vectors that span the space is the *dimension* of the space. In (4.1) the vector $|v\rangle$ is *represented* in the basis $\{|q_n\rangle\}_{\mathrm{n=1}}^{\mathrm{N}}$.

We may always construct an *orthonormal basis* such that $\langle q_{\mathrm{n}}|\, q_{\mathrm{m}}\rangle = \delta_{\mathrm{nm}}$ where

$$\boxed{\delta_{\mathrm{nm}} = \begin{cases} 1 \text{ if } n = m \\ 0 \text{ if } n \neq m \end{cases}} \tag{4.2}$$

is the *Kronecker delta*. So we will always assume that any basis we use is orthonormal.

Vectors and Matrices. The postulates of addition and scalar multiplication for the vector space are the same as the laws of addition and scalar multiplication for matrices. We may then write vectors in the form of *matrices*.

In a three-dimensional space, for example, the basis vectors can be written as

$$|1\rangle = \begin{bmatrix} 1 \\ 0 \\ 0 \end{bmatrix};\ |2\rangle = \begin{bmatrix} 0 \\ 1 \\ 0 \end{bmatrix};\ \text{and } |3\rangle = \begin{bmatrix} 0 \\ 0 \\ 1 \end{bmatrix}. \tag{4.3}$$

Then the general vector $|a\rangle$ becomes

$$\begin{aligned} |a\rangle &= \begin{bmatrix} a_1 \\ a_2 \\ a3 \end{bmatrix} \\ &= a_1 \begin{bmatrix} 1 \\ 0 \\ 0 \end{bmatrix} + a_2 \begin{bmatrix} 0 \\ 1 \\ 0 \end{bmatrix} + a_3 \begin{bmatrix} 0 \\ 0 \\ 1 \end{bmatrix} \\ &= a_1 |1\rangle + a_2 |2\rangle + a_3 |3\rangle\,. \end{aligned} \tag{4.4}$$

In (4.4) we have represented the vector $|a\rangle$ in the basis $\{|1\rangle, |2\rangle, |3\rangle\}$, which we have chosen then to write in matrix form. We could have chosen the standard vector representation with the unit vectors $(\hat{e}_x, \hat{e}_y, \hat{e}_z)$ instead, which obey a corresponding set of laws.

The matrix form will be simpler for our purposes in treating rigid body dynamics than the standard vector form.

Scalar Product of Matrices. For vectors $|a\rangle$ and $|b\rangle$ represented in the basis $\{|1\rangle, |2\rangle, |3\rangle\}$ and written as column matrices

$$|a\rangle = \begin{bmatrix} a_1 \\ a_2 \\ a3 \end{bmatrix} \text{ and } |b\rangle = \begin{bmatrix} b_1 \\ b_2 \\ b_3 \end{bmatrix}$$

the scalar product in matrix form is

$$\begin{aligned} \langle a\,|b\rangle &= \begin{bmatrix} a_1^* \; a_2^* \; a_3^* \end{bmatrix} \begin{bmatrix} b_1 \\ b_2 \\ b_3 \end{bmatrix} \\ &= a_1^* b_1 + a_2^* b_2 + a_3^* b_3. \end{aligned} \tag{4.5}$$

Consistent with this definition, the *dual* of the column vector $|a\rangle$ is

$$\langle a| = \begin{bmatrix} a_1^* \; a_2^* \; a_3^* \end{bmatrix}. \tag{4.6}$$

That is, the dual of a vector is formed by taking the *Hermitian conjugate* of the column matrix form of the vector.

Projector. The representation of a vector in an arbitrary basis is accomplished by projecting the vector onto the basis using a *projection operator*, or simply *projector*. For an N–dimensional basis $\{|\lambda\rangle\}_{\lambda=1}^{N}$ the projector is defined by

$$\boxed{\mathbf{P} \equiv \sum\nolimits_{\lambda}^{N} |\lambda\rangle \langle\lambda|.} \tag{4.7}$$

The operation of $\mathbf{P}$ on a general vector $|a\rangle$ in the space spanned by the basis $\{|\lambda\rangle\}_{\lambda=1}^{N}$ results in

$$\mathbf{P}\,|a\rangle = \sum_{\lambda}^{N} |\lambda\rangle \langle\lambda|\,a\rangle. \tag{4.8}$$

The quantity $\langle\lambda|\,a\rangle$ is the λth component of the vector $|a\rangle$, which we may write as $\langle\lambda|\,a\rangle = a_\lambda$. Then (4.8) becomes

$$\mathbf{P}\,|a\rangle = \sum_{\lambda}^{N} a_\lambda\,|\lambda\rangle\,. \tag{4.9}$$

When we deal with general vector spaces the question of *completeness* becomes important [see e.g. [21], pp. 36–37; [42], pp. 130–132]. In broad terms completeness is required by any representation of an abstract vector. A vector space is complete provided there is no loss of information when the vector is represented in the basis of the space. This is true if the representation in (4.9) is exact. Simply stated, completeness means $P\,|a\rangle = |a\rangle$, which is true if, and only if, mathematically

$$\boxed{\mathbf{P} = \sum\nolimits_{\lambda}^{N} |\lambda\rangle\,\langle\lambda| = \mathbf{1},} \tag{4.10}$$

the *identity operator*. This is true for the case of the three dimensional spaces that concern us in this chapter, as our example here shows.

Example 4.2.1 We consider the basis (4.3). The components of the projector in this basis, written in matrix form, are

$$|1\rangle\,\langle 1| = \begin{bmatrix} 1\,0\,0 \end{bmatrix}\begin{bmatrix} 1 \\ 0 \\ 0 \end{bmatrix} = \begin{bmatrix} 1\,0\,0 \\ 0\,0\,0 \\ 0\,0\,0 \end{bmatrix}$$

$$|2\rangle\,\langle 2| = \begin{bmatrix} 0\,1\,0 \end{bmatrix}\begin{bmatrix} 0 \\ 1 \\ 0 \end{bmatrix} = \begin{bmatrix} 0\,0\,0 \\ 0\,1\,0 \\ 0\,0\,0 \end{bmatrix}$$

$$|3\rangle\,\langle 3| = \begin{bmatrix} 0\,0\,1 \end{bmatrix}\begin{bmatrix} 0 \\ 0 \\ 1 \end{bmatrix} = \begin{bmatrix} 0\,0\,0 \\ 0\,0\,0 \\ 0\,0\,1 \end{bmatrix}.$$

The projector is then

$$|1\rangle\,\langle 1| + |2\rangle\,\langle 2| + |3\rangle\,\langle 3| = \begin{bmatrix} 1\,0\,0 \\ 0\,1\,0 \\ 0\,0\,1 \end{bmatrix},$$

which is the identity matrix **1**.

If the operator $\boldsymbol{P}$ is an identity then $\mathbf{PP} = \mathbf{1}$. That is

$$\begin{aligned} \mathbf{PP} &= \sum_{\lambda,\nu} |\lambda\rangle\,\langle\lambda|\,\nu\rangle\,\langle\nu| \\ &= \sum_{\lambda} |\lambda\rangle\,\langle\lambda|\,. \end{aligned}$$

This will be true if $\langle\lambda|\,\nu\rangle = \delta_{\lambda\nu}$, which, as we pointed out above, we may always choose.

Unless we specify otherwise we shall implicitly use the standard basis, which we introduced in matrix form in (4.3). In standard vector notation these are the unit Cartesian vectors $(\hat{e}_1, \hat{e}_2, \hat{e}_3)$ or more commonly $(\hat{e}_x, \hat{e}_y, \hat{e}_z)$. Mathematically they are simply three orthonormal vectors.

Operators. In general the action of an operator on a vector produces another vector in the same vector space, which we assume to be complete. If **Q** is a general operator and if $|f\rangle$ is a vector in the vector space, then

$$\mathbf{Q}\,|f\rangle = |g\rangle\,, \tag{4.11}$$

where $|g\rangle$ is a vector in the same space as $|f\rangle$. If the basis vectors in this space are $\{|\mu\rangle\}_{\mu=1}^{N}$ we may represent the vector equation in (4.11) using the operator $\mathbf{P} = \sum_{\mu}^{N} |\mu\rangle\langle\mu|$. To obtain the representation of (4.11) we must represent both of the vectors in the basis. Because the projector is an identity operator we may include projectors at any point within the vector equation (4.11). The result is

$$\sum_{\mu,\nu}^{N} |\mu\rangle\langle\mu|\, \boldsymbol{Q}\, |\nu\rangle\langle\nu|\, f\rangle = \sum_{\mu}^{N} |\mu\rangle\langle\mu|\, g\rangle\,. \tag{4.12}$$

In (4.12) we have both sides of the vector equation (4.11) and both of the vectors individually represented in the basis. We also have the operator **Q** represented in the basis as $\langle\mu|\,\mathbf{Q}\,|\nu\rangle$. The operator, therefore, has two indices when projected onto our basis. If we were to write the vectors in matrix form the operator would be a square matrix.

4.3 Einstein Sum Convention

In his 1916 paper on the foundations of the general theory of relativity, published in *Annalen der Physik*, Einstein noted that the summation symbol $\sum$ appeared "... with respect to the indices which occur twice under a sign of summation, and only with respect to indices which appear twice." He, therefore, concluded that it was possible, without loss of clarity, to omit the summation sign [[24], p. 122]. With this he introduced his convention: If an index occurs twice in a term it will always be summed unless the contrary is expressly stated.

In the paper Einstein was using Greek indices. Therefore, it has become common practice in the physics community to use the *Einstein summation convention* for Greek indices, and not (necessarily) for Latin. Particularly here, as we introduce products of projectors in our development, the Einstein summation convention will be helpful. Therefore, unless we state explicitly that the Einstein summation convention is not used, Greek indices which appear twice in a term will always be summed.

The indices will normally be on basis vectors and components of vectors represented in a basis. In the present picture, there are three spatial dimensions. Therefore, the implied summations will be over the indices 1, 2, 3 unless we note otherwise.

4.4 Kinematics

Kinematics is the description of motion. In our presentation here we will consider the description of motion of a classical point particle referred to a translating and rotating frame. We will then specialize this result to a rigid body constructed of such classical point particles.

4.4.1 Reference Frames

There are two general types of motion of the rigid body. The first is motion of the center of mass (CM) of the rigid body with respect to a stationary set of basis vectors $\{|X_\mu\rangle\}$. We may locate the origin of this set of stationary basis vectors at any fixed point external to the body. The second type of motion is the rotation of the rigid body about the CM. To describe this rotation we define a second set of basis vectors $\{|x_\mu(t)\rangle\}$ that is fixed in the rigid body and rotates with the body. We choose the origin of this second set of basis vectors to be the CM of the body. In our development we will show that the motion of the CM and the rotation about the CM are then dynamically separable.

We begin our description by considering a vector $|r_\text{i}(t)\rangle$ which locates one of the moving classical point particles (designated by the subscript i) that make up the body. We choose the vector $|R(t)\rangle$, represented in the stationary basis $\{|X_\mu\rangle\}$, to locate the CM of the body and the vector $|r_\text{i}'(t)\rangle$, represented in the basis $\{|x_\mu(t)\rangle\}$, to locate the particle relative to the CM. We then write the vector $|r_\text{i}(t)\rangle$ as

$$|r_\text{i}(t)\rangle = |R(t)\rangle + |r_\text{i}'(t)\rangle. \tag{4.13}$$

In Fig. 4.1 we have drawn a picture of the situation

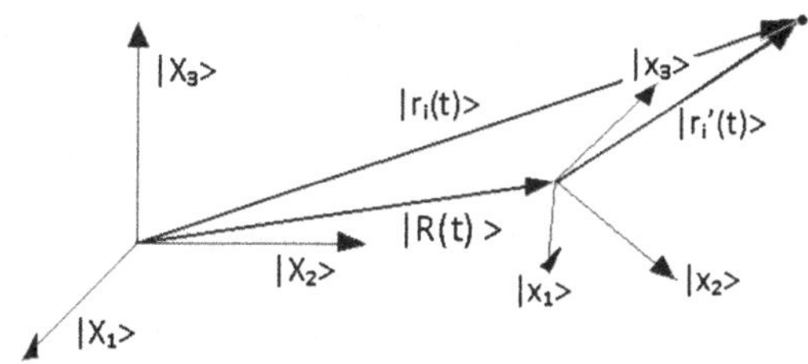

Fig. 4.1 Fixed and rotating coordinate frames. The fixed frame has the basis $\{|X_\mu\rangle\}$ and the rotating frame has the basis $\{|x_\mu(t)\rangle\}$. The vector $|r_\text{i}(t)\rangle$ has been decomposed into vectors in the fixed and rotating systems

Although we have chosen these coordinate systems and bases to provide a convenient description of the motion of a rigid body, until we actually fix our classical point particle located by $\left|r_{\mathrm{i}}'(t)\right\rangle$ in a rigid body, our two bases could be used as well for the description of the motion of a classical point particle in any translating and rotating frame. We could, for example, choose $|R(t)\rangle$ to locate a point on the surface of the rotating earth or a turntable. The basis $\left\{\left|x_{\mu}(t)\right\rangle\right\}$ would then be fixed on the earth's surface or in the turntable and the vector $\left|r_{\mathrm{i}}'(t)\right\rangle$ would describe the motion of a point particle as referred to the earth's surface or the rotating turntable. Locating the particle in a rigid body will require that the vector $\left|r_{\mathrm{i}}'(t)\right\rangle$ is constant in length and orientation in the basis $\left\{\left|x_{\mu}(t)\right\rangle\right\}$. We can relax this limitation for part of our discussion.

Using the Einstein summation convention, the projection operators for these bases are

$$\mathbf{P}_{\mathrm{X}} = \left|X_{\mu}\right\rangle\left\langle X_{\mu}\right| = \mathbf{1} \text{ and } \mathbf{P}_{\mathrm{x}} = \left|x_{\mu}(t)\right\rangle\left\langle x_{\mu}(t)\right| = \mathbf{1}. \tag{4.14}$$

Here we have introduced capital and lower case subscripts to distinguish the projection operators $\mathbf{P}_{\mathrm{X}}$ and $\mathbf{P}_{\mathrm{x}}$, which are both identities. Our representation of the vector $|r_{\mathrm{i}}(t)\rangle$ then becomes

$$\begin{aligned} |r_{\mathrm{i}}(t)\rangle &= \mathbf{P}_{\mathrm{X}}\,|R(t)\rangle + \mathbf{P}_{\mathrm{x}}\left|r_{\mathrm{i}}'(t)\right\rangle \\ &= \left|X_{\mu}\right\rangle\left\langle X_{\mu}\right| R(t)\rangle + \left|x_{\mu}(t)\right\rangle\left\langle x_{\mu}(t)\right| r_{\mathrm{i}}'(t)\rangle, \end{aligned} \tag{4.15}$$

To obtain (4.15) we only applied the projectors to the right hand side of (4.13) leaving the left hand side as an abstract vector. We can do this because the projectors are both identity operators.

The components of the vectors

$$\left\langle X_{\mu}\right| R(t)\rangle \tag{4.16}$$

and

$$\left\langle x_{\mu}(t)\right| r_{\mathrm{i}}'(t)\rangle \tag{4.17}$$

are functions of the time. When we limit our description to the rigid body, which we shall soon do, (4.17) will depend on time only through the rotation of the basis vectors.

We obtain the instantaneous velocity of the point located by $|r_{\mathrm{i}}(t)\rangle$ by differentiating (4.15) with respect to time. This requires differentiation of the components (4.16), the components (4.17), and of the basis vectors $\left\{\left|x_{\mu}(t)\right\rangle\right\}$. That is

$$\begin{aligned}\frac{\mathrm{d}}{\mathrm{d}t}\left|r_{\mathrm{i}}(t)\right\rangle &= \left|X_{\mu}\right\rangle\left(\frac{\mathrm{d}}{\mathrm{d}t}\left\langle X_{\mu}\right|R(t)\right\rangle\Big) \\ &+\left|x_{\mu}(t)\right\rangle\left(\frac{\mathrm{d}}{\mathrm{d}t}\left\langle x_{\mu}(t)\right|r_{\mathrm{i}}^{\prime}\right\rangle\Big) \\ &+\left|x_{\mu}(t)\right\rangle\left(\left\langle x_{\mu}(t)\right|\left.\frac{\mathrm{d}}{\mathrm{d}t}\right]_{\text{basis}}\left|x_{\nu}(t)\right\rangle\right)\left\langle x_{\nu}(t)\right|r_{\mathrm{i}}^{\prime}\rangle . \end{aligned} \tag{4.18}$$

We have included the projector $\left|x_{\mu}(t)\right\rangle\left\langle x_{\mu}(t)\right|$ or $\left|x_{\nu}(t)\right\rangle\left\langle x_{\nu}(t)\right|$ operating on all vector terms on the right hand side of (4.18) to obtain the components of each vector. We can only actually perform the derivative operation on the vector components, which are functions of the time t. In the third term on the right hand side of (4.18) we have included a second projector $\left|x_{\mu}(t)\right\rangle\left\langle x_{\mu}(t)\right|$ to guarantee that the final result is projected onto the basis $\left\{\left|x_{\mu}(t)\right\rangle\right\}$. The term $\left(\left\langle x_{\mu}(t)\right| \mathrm{d}/\mathrm{d}t]_{\text{basis}}\left|x_{\nu}(t)\right\rangle\right)$, as we shall see, is an operator.

On the left hand side of (4.18) we have the abstract form of the rate of change of the position vector $\left|r_{\mathrm{i}}(t)\right\rangle$. The first two terms on the right hand side of (4.18) are representations of vectors with components $\mathrm{d}\left\langle X_{\mu}\right| R(t)\rangle/\mathrm{d}t$ and $\mathrm{d}\left\langle x_{\mu}(t)\right| r_{\mathrm{i}}^{\prime}\rangle/\mathrm{d}t$ in the bases $\left\{\left|X_{\mu}\right\rangle\right\}$ and $\left\{\left|x_{\mu}(t)\right\rangle\right\}$. Differentiating the components $\left\langle X_{\mu}\right| R(t)\rangle$ and $\left\langle x_{\mu}(t)\right| r_{\mathrm{i}}^{\prime}\rangle$ presents us with no mathematical difficulty, since they are simply algebraic functions of the time t. And $\mathrm{d}\left\langle X_{\mu}\right| R(t)\rangle/\mathrm{d}t$ and $\mathrm{d}\left\langle x_{\mu}(t)\right| r_{\mathrm{i}}^{\prime}\rangle/\mathrm{d}t$ are scalar algebraic quantities, which are components of velocities represented in the two bases. The product of each of these terms with the ket vectors $\left|X_{\mu}\right\rangle$ and $\left|x_{\mu}(t)\right\rangle$ (together with the implied summation) means that each of these terms is an instantaneous vector velocity represented in each specific basis.

The first term in (4.18) is the velocity of the CM of the rigid body $\mathrm{d}|R\rangle/\mathrm{d}t$ represented in the stationary basis $\left\{\left|X_{\mu}\right\rangle\right\}$. We shall designate this as $|V\rangle$ and the components V_{μ} as (V_1, V_2, V_3). In matrix form $|V\rangle$ is

$$\frac{\mathrm{d}}{\mathrm{d}t}|R\rangle=|V\rangle=\left[\begin{array}{c} V_1 \\ V_2 \\ V_3 \end{array}\right]. \tag{4.19}$$

The second term in (4.18) is similar to the first. It is the velocity of the point $\left|r_{\mathrm{i}}^{\prime}(t)\right\rangle$ represented in the basis of the moving system $\left\{\left|x_{\mu}(t)\right\rangle\right\}$. This term is the velocity of the point particle as observed by a person who is at rest in the moving system. We shall designate this as $\left|v_{\mathrm{i}}^{\prime}\right\rangle$ and the components $v_{\mu}^{\prime}$ as $\left(v_{\mathrm{i}1}^{\prime}, v_{\mathrm{i}2}^{\prime}, v_{\mathrm{i}3}^{\prime},\right)$. In matrix form $\left|v_{\mathrm{i}}^{\prime}\right\rangle$ is

$$\frac{\mathrm{d}}{\mathrm{d}t}\left|r_{\mathrm{i}}^{\prime}\right\rangle=\left[\begin{array}{c} \dot{r}_{\mathrm{i}1}^{\prime} \\ \dot{r}_{\mathrm{i}2}^{\prime} \\ \dot{r}_{\mathrm{i}3}^{\prime} \end{array}\right]=\left|v_{\mathrm{i}}^{\prime}\right\rangle=\left[\begin{array}{c} v_{\mathrm{i}1}^{\prime} \\ v_{\mathrm{i}2}^{\prime} \\ v_{\mathrm{i}3}^{\prime} \end{array}\right] \tag{4.20}$$

The third term in (4.18) represents something different from the previous two. This term is that part of the velocity vector $\mathrm{d}\left|r_{\mathrm{i}}^{\prime}(t)\right\rangle/\mathrm{d}t$ resulting from a rotation of the basis $\left\{\left|x_{\mu}(t)\right\rangle\right\}$ about an arbitrary axis, while the vector $\left|r_{\mathrm{i}}^{\prime}\right\rangle$ remains fixed in that basis. We have indicated this in part by writing the time derivative operator for this term as $\left.\mathrm{d}/\mathrm{d}t\right]_{\text{basis}}$. Now we must develop a mathematical formulation that will produce this operator.

4.4.2 Rotation of the Basis

We can produce a rotation of the basis $\{|x_\mu(t)\rangle\}$ about an arbitrary axis from a sequence of rotations about the individual basis vectors. If the angles of rotation about the basis vectors are *finite* the result will depend on the order in which the rotations are performed. We can convince ourselves of this by rotating a book through two finite angles about two perpendicular axes defined by the edges of the book. The final orientation of the book depends on the order in which the two rotations are performed. If the angles of rotation about the basis vectors are *infinitesimal*, however, the final orientation of the body is independent of the order of rotation.

The first step in our problem is then to find a mathematical operation that will describe the change of a position vector locating a classical point particle $|r_{\mathrm{i}}'\rangle$ as the basis rotates through an angle about one of the basis vectors. This will be a rotation of the basis alone. The locations of all of the point particles, i.e. all $|r_{\mathrm{i}}'\rangle$, remain unchanged in this operation. We may, therefore, dispense with any reference to a classical point particle and replace the position vector $|r_{\mathrm{i}}'\rangle$ of the point particle in question by a (dummy) constant vector $|a\rangle$. We may then designate this vector as $|a'\rangle$ after rotation, which will simplify our picture of the rotation.

We begin with a rotation ϑ_3' of the basis vectors $\{|x_\mu(t)\rangle\}$ about the basis vector $|x_3(t)\rangle$. We retain the prime notation for the rotation angles to indicate the basis $\{|x_\mu(t)\rangle\}$. In Fig. 4.2 we have illustrated the rotation. We consider that the rotation takes place between the time $t = t_1$ and $t = t_2$. In Fig. 4.2 we designate the bases before and after rotation as $\{|x_\mu(t_1)\rangle\} = \{|x_1\rangle, |x_2\rangle, |x_3\rangle\}$ and

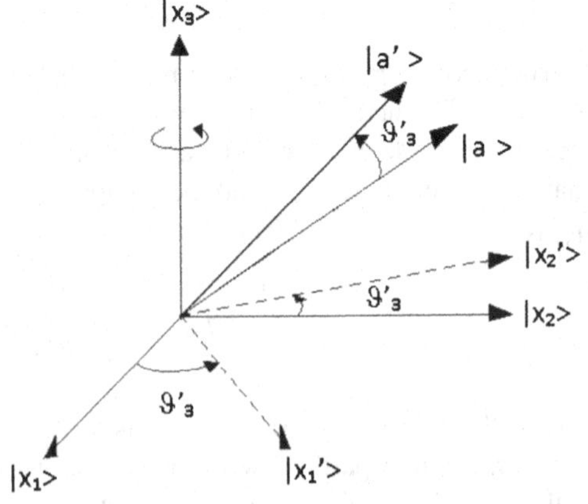

Fig. 4.2 Rotation of a basis. The basis vectors $\{|x_\mu(t)\rangle\}$ rotate around $|x_3\rangle$ which is fixed

$\{|x_\mu(t_2)\rangle\} = \{|x'_1\rangle, |x'_2\rangle, |x'_3\rangle\}$. If we consider the rotation angle ϑ'_3 to have a vector orientation along the axis $|x_3\rangle$ in accordance with the right hand rule we will have a positive direction of the angular velocity component $\omega'_3 = \mathrm{d}\vartheta'_3/\mathrm{d}t$.

We are seeking an expression for the rates of change of the components $\langle x_\mu(t)|\, a\rangle$ of the vector $|a\rangle$ resulting from the rotation of the basis alone. We want then to reference the vector $|a\rangle$ to the basis $\{|x_\mu(t)\rangle\}$ before and after the rotation of the basis without allowing the vector $|a\rangle$ to change in orientation or magnitude. Because our only reference during this process is to the basis $\{|x_\mu(t)\rangle\}$, we shall accomplish what we want by first holding $|a\rangle$ fixed in the basis $\{|x_\mu(t)\rangle\}$ while we rotate the basis and then projecting $|a\rangle$ back onto the original basis (as it was before rotation). We will then have components of the vector $|a\rangle$ that would be measured in the original basis as the basis itself changes. If we then make the rotation angle and the time taken infinitesimal we can obtain the rate of change of the vector resulting solely from the change in orientation of the basis.

Before the rotation we represent the vector $|a\rangle$ in the basis $\{|x_1\rangle, |x_2\rangle, |x_3\rangle\}$ as

$$|a\rangle = a_1|x_1\rangle + a_2|x_2\rangle + a_3|x_3\rangle . \tag{4.21}$$

and after the rotation we represent the vector $|a'\rangle$ in the basis $\{|x'_1\rangle, |x'_2\rangle, |x'_3\rangle\}$ as

$$|a'\rangle = a_1|x'_1\rangle + a_2|x'_2\rangle + a_3|x'_3\rangle . \tag{4.22}$$

The magnitudes of the components do not change.

Now we project the vector $|a'\rangle$ back onto the basis $\{|x_1\rangle, |x_2\rangle, |x_3\rangle\}$ by applying the projector

$$\mathbf{P}_\mathrm{x} = |x_1\rangle\langle x_1| + |x_2\rangle\langle x_2| + |x_3\rangle\langle x_3| \tag{4.23}$$

to the vector $|a'\rangle$ in (4.22). Carrying out the projection in detail, we have

$$\begin{aligned}\mathbf{P}_\mathrm{x}|a'\rangle &= (|x_1\rangle\langle x_1| + |x_2\rangle\langle x_2| + |x_3\rangle\langle x_3|)\cdots \\ &\quad\cdots\left(a_1|x'_1\rangle + a_2|x'_2\rangle + a_3|x'_3\rangle\right) \\ &= \left(a_1\langle x_1|x'_1\rangle + a_2\langle x_1|x'_2\rangle + a_3\langle x_1|x'_3\rangle\right)|x_1\rangle \\ &\quad + \left(a_1\langle x_2|x'_1\rangle + a_2\langle x_2|x'_2\rangle + a_3\langle x_2|x'_3\rangle\right)|x_2\rangle \\ &\quad + \left(a_1\langle x_3|x'_1\rangle + a_2\langle x_3|x'_2\rangle + a_3\langle x_3|x'_3\rangle\right)|x_3\rangle .\end{aligned} \tag{4.24}$$

The brackets $\langle x_\mu|x'_\nu\rangle$ are the scalar products between the basis vectors $|x_\mu\rangle$ and $|x'_\nu\rangle$ which are the cosines of the angles between these basis vectors. By inspection of Fig. 4.2 these are

$$\begin{aligned}\langle x_1|x'_1\rangle &= \cos\vartheta'_3 \\ \langle x_1|x'_2\rangle &= -\sin\vartheta'_3 \\ \langle x_1|x'_3\rangle &= 0 \\ \langle x_2|x'_1\rangle &= \sin\vartheta'_3\end{aligned}$$

$$
\begin{aligned}
\langle x_2 \left| x_2' \right\rangle &= \cos \vartheta_3' \\
\langle x_2 \left| x_3' \right\rangle &= 0 \\
\langle x_3 \left| x_1' \right\rangle &= 0 \\
\langle x_3 \left| x_2' \right\rangle &= 0 \\
\langle x_3 \left| x_3' \right\rangle &= 1.
\end{aligned}
\tag{4.25}
$$

Then (4.24) is

$$
\begin{aligned}
\mathbf{P}_{\mathrm{x}} \left| a' \right\rangle = & \left(a_1 \cos \vartheta_3' - a_2 \sin \vartheta_3' + 0 \right) \left| x_1 \right\rangle \\
& + \left(a_1 \sin \vartheta_3' + a_2 \cos \vartheta_3' + 0 \right) \left| x_2 \right\rangle \\
& + \left(0 + 0 + a_3 \right) \left| x_3 \right\rangle .
\end{aligned}
\tag{4.26}
$$

Written in matrix form (4.26) is

$$
\mathbf{P}_{\mathrm{x}} \left| a' \right\rangle = \begin{bmatrix} \cos \vartheta_3' & -\sin \vartheta_3' & 0 \\ \sin \vartheta_3' & \cos \vartheta_3' & 0 \\ 0 & 0 & 1 \end{bmatrix} \begin{bmatrix} a_1 \\ a_2 \\ a_3 \end{bmatrix} .
\tag{4.27}
$$

In (4.27) we have attained our mathematical goal. The square matrix in (4.27) is the *matrix form* of the operator $\mathrm{d}/\mathrm{d}t]_{\text{basis}}$ that carries out the rotation and projection. We shall now designate this rotation and projection operator about the axis $|x_3\rangle$ in *abstract form* as $\mathbf{R}_3\left(\vartheta_3'\right)$. To obtain a representation of $\mathbf{R}_3\left(\vartheta_3'\right)$ we must use projectors on both sides (see Sect. 4.2.3). The square matrix in (4.27) is, then, actually $\mathbf{P}_{\mathrm{x}}\mathbf{R}_3\left(\vartheta_3'\right)\mathbf{P}_{\mathrm{x}} = |\mu\rangle\langle\mu|\,\mathbf{R}_3\left(\vartheta_3'\right)|\nu\rangle\langle\nu|$. But we can avoid constantly writing the projectors if we agree that using an *arrow* instead of an equal sign indicates equality that would result if we were to include the projectors. That is

$$
\mathbf{R}_3\left(\vartheta_3'\right) \Longrightarrow \begin{bmatrix} \cos \vartheta_3' & -\sin \vartheta_3' & 0 \\ \sin \vartheta_3' & \cos \vartheta_3' & 0 \\ 0 & 0 & 1 \end{bmatrix} .
\tag{4.28}
$$

Our present interest is in the rate of change of the vector $|a\rangle$. We, therefore, need only an infinitesimal angle of rotation $\delta\vartheta_3'$ in (4.27). Holding only first order terms in the expansion of the sine and cosine the square matrix in (4.28) becomes

$$
\mathbf{R}_3\left(\delta\vartheta_3'\right) \Longrightarrow \begin{bmatrix} 1 & -\delta\vartheta_3' & 0 \\ \delta\vartheta_3' & 1 & 0 \\ 0 & 0 & 1 \end{bmatrix} .
\tag{4.29}
$$

Performing the same rotation and projection about the basis vectors $|x_1\rangle$ and $|x_2\rangle$ we find similar operators.

$$\mathbf{R}_1\left(\delta\vartheta_1'\right) \Longrightarrow \begin{bmatrix} 1 & 0 & 0 \\ 0 & 1 & -\delta\vartheta_1' \\ 0 & \delta\vartheta_1' & 1 \end{bmatrix}, \tag{4.30}$$

and

$$\mathbf{R}_2\left(\delta\vartheta_2'\right) \Longrightarrow \begin{bmatrix} 1 & 0 & \delta\vartheta_2' \\ 0 & 1 & 0 \\ -\delta\vartheta_2' & 0 & 1 \end{bmatrix}. \tag{4.31}$$

Generally matrices do not commute. However, the matrices (4.29)–(4.31) with unity on the diagonal, and off diagonal terms that are small enough that we can neglect all terms above first order in $\delta\vartheta_\mu'$, do commute. We may then obtain a general rotation operator $\mathbf{R}\left(\delta\vartheta'\right)$ for the arbitrary infinitesimal rotation $\delta\vartheta' = \left(\delta\vartheta_1', \delta\vartheta_2', \delta\vartheta_3'\right)$ in the basis $\{|x_1\rangle, |x_2\rangle, |x_3\rangle\}$ by multiplying matrix forms of the infinitesimal operators (4.29)–(4.31) in any order. The rotation operator for a general infinitesimal rotation is then, to first order in $\delta-$quantities,

$$\mathbf{R}\left(\delta\vartheta'\right) \Longrightarrow \begin{bmatrix} 1 & -\delta\vartheta_3' & \delta\vartheta_2' \\ \delta\vartheta_3' & 1 & -\delta\vartheta_1' \\ -\delta\vartheta_2' & \delta\vartheta_1' & 1 \end{bmatrix}. \tag{4.32}$$

The operator (4.32) is the sum of the identity and an operator that we shall call $\delta\mathbf{R}\left(\delta\vartheta'\right)$. That is

$$\begin{aligned} \mathbf{R}\left(\delta\vartheta'\right) &\Longrightarrow \begin{bmatrix} 1 & 0 & 0 \\ 0 & 1 & 0 \\ 0 & 0 & 1 \end{bmatrix} + \begin{bmatrix} 0 & -\delta\vartheta_3' & \delta\vartheta_2' \\ \delta\vartheta_3' & 0 & -\delta\vartheta_1' \\ -\delta\vartheta_2' & \delta\vartheta_1' & 0 \end{bmatrix} \\ &= \mathbf{1} + \delta\mathbf{R}\left(\delta\vartheta'\right) \end{aligned} \tag{4.33}$$

with

$$\delta\mathbf{R}\left(\delta\vartheta'\right) \Longrightarrow \begin{bmatrix} 0 & -\delta\vartheta_3' & \delta\vartheta_2' \\ \delta\vartheta_3' & 0 & -\delta\vartheta_1' \\ -\delta\vartheta_2' & \delta\vartheta_1' & 0 \end{bmatrix}. \tag{4.34}$$

Using (4.33) and (4.34) in (4.27) we find that a vector $|a\rangle$ of constant length becomes

$$\left|a'\right\rangle = \left(\mathbf{1} + \delta\mathbf{R}\left(\delta\vartheta'\right)\right)|a\rangle, \tag{4.35}$$

after a general infinitesimal rotation $\delta\vartheta' = \left(\delta\vartheta_1', \delta\vartheta_2', \delta\vartheta_3'\right)$ in the basis $\{|x_1\rangle, |x_2\rangle, |x_3\rangle\}$. From (4.35) we can then define the infinitesimal change in a vector of constant length resulting from an infinitesimal angular rotation $\delta\vartheta'$ as

$$\delta|a\rangle = \left|a'\right\rangle - |a\rangle = \delta\mathbf{R}\left(\delta\vartheta'\right)|a\rangle. \tag{4.36}$$

By dividing $\delta\,|a\rangle$ in (4.36) by the infinitesimal time δt during which the infinitesimal rotation takes place, and taking the limit $\delta t \to 0$, we obtain the third term in (4.18). That is

$$\lim_{\delta t\to 0}\frac{\delta\,|a\rangle}{\delta t} = \left(\lim_{\delta t\to 0}\frac{\delta\mathbf{R}\left(\delta\vartheta'\right)}{\delta t}\right)|a\rangle = \frac{\mathrm{d}\mathbf{R}\left(\vartheta'\right)}{\mathrm{d}t}\,|a\rangle$$
$$\Longrightarrow \left|x_\mu(t)\right\rangle\left(\left\langle x_\mu(t)\right|\left.\frac{\mathrm{d}}{\mathrm{d}t}\right]_{\text{basis}}\left|x_\nu(t)\right\rangle\right)\left\langle x_\nu(t)\right|a\rangle \tag{4.37}$$

where we have re-introduced the original form of the operator from (4.18). Equation (4.37) tells us that

$$\left.\frac{\mathrm{d}}{\mathrm{d}t}\right]_{\text{basis}} = \frac{\mathrm{d}\mathbf{R}\left(\vartheta'\right)}{\mathrm{d}t}. \tag{4.38}$$

From (4.34) we have $\mathrm{d}\mathbf{R}\left(\vartheta'\right)/\mathrm{d}t$ in matrix form as

$$\frac{\mathrm{d}\mathbf{R}\left(\vartheta'\right)}{\mathrm{d}t} \Longrightarrow \lim_{\delta t\to 0}\begin{bmatrix} 0 & -\delta\vartheta_3'/\delta t & \delta\vartheta_2'/\delta t \\ \delta\vartheta_3'/\delta t & 0 & -\delta\vartheta_1'/\delta t \\ -\delta\vartheta_2'/\delta t & \delta\vartheta_1'/\delta t & 0 \end{bmatrix}$$
$$= \begin{bmatrix} 0 & -\omega_3' & \omega_2' \\ \omega_3' & 0 & -\omega_1' \\ -\omega_2' & \omega_1' & 0 \end{bmatrix}, \tag{4.39}$$

where

$$\omega_\mu' = \lim_{\delta t\to 0}\frac{\delta\vartheta_\mu'}{\delta t}. \tag{4.40}$$

Combining (4.39) with (4.37) and (4.38) we have the final matrix form of the term we sought.

$$\left|x_\mu(t)\right\rangle\left(\left\langle x_\mu(t)\right|\left.\frac{\mathrm{d}}{\mathrm{d}t}\right]_{\text{basis}}\left|x_\nu(t)\right\rangle\right)\left\langle x_\nu(t)\right|a\rangle$$
$$= \left|x_\mu(t)\right\rangle\left(\left\langle x_\mu(t)\right|\frac{\mathrm{d}\mathbf{R}\left(\vartheta'\right)}{\mathrm{d}t}\left|x_\nu(t)\right\rangle\right)\left\langle x_\nu(t)\right|r_{\mathrm{i}}'\rangle$$
$$= \begin{bmatrix} 0 & -\omega_3' & \omega_2' \\ \omega_3' & 0 & -\omega_1' \\ -\omega_2' & \omega_1' & 0 \end{bmatrix}\begin{bmatrix} r_{\mathrm{i}1}' \\ r_{\mathrm{i}2}' \\ r_{\mathrm{i}3}' \end{bmatrix}. \tag{4.41}$$

And if we multiply the matrices appearing in (4.41) we have

$$\left|x_\nu(t)\right\rangle\left(\left\langle x_\mu(t)\right|\frac{\mathrm{d}\mathbf{R}\left(\vartheta'\right)}{\mathrm{d}t}\left|x_\nu(t)\right\rangle\right)\left\langle x_\mu(t)\right|r_{\mathrm{i}}'\rangle = \begin{bmatrix} -\omega_3'r_{\mathrm{i}2}' + \omega_2'r_{\mathrm{i}3}' \\ \omega_3'r_{\mathrm{i}1}' - \omega_1'r_{\mathrm{i}3}' \\ -\omega_2'r_{\mathrm{i}1}' + \omega_1'r_{\mathrm{i}2}' \end{bmatrix} \tag{4.42}$$

In standard vector notation the elements of the matrix in (4.42) are the components of the vector cross product $\boldsymbol{\omega}' \times \boldsymbol{r}_{\mathrm{i}}'$. That is

$$|x_\nu(t)\rangle \left(\langle x_\mu(t)| \frac{\mathrm{d}\mathbf{R}(\vartheta')}{\mathrm{d}t} |x_\nu(t)\rangle \right) \langle x_\mu(t)| r_{\mathrm{i}}' \rangle = \boldsymbol{\omega}' \times \boldsymbol{r}_{\mathrm{i}}'. \tag{4.43}$$

This standard vector form in (4.43) has advantages in some circumstances. We will use this when dealing with rolling constraints.

4.4.3 Combined Velocities

Combining (4.41) and (4.20) we have

$$\frac{\mathrm{d}}{\mathrm{d}t}|r_{\mathrm{i}}'(t)\rangle \Longrightarrow \begin{bmatrix} \dot{r}_{\mathrm{i}1}' \\ \dot{r}_{\mathrm{i}2}' \\ \dot{r}_{\mathrm{i}3}' \end{bmatrix} + \begin{bmatrix} 0 & -\omega_3' & \omega_2' \\ \omega_3' & 0 & -\omega_1' \\ -\omega_2' & \omega_1' & 0 \end{bmatrix} \begin{bmatrix} r_{\mathrm{i}1}' \\ r_{\mathrm{i}2}' \\ r_{\mathrm{i}3}' \end{bmatrix} \tag{4.44}$$

for the (total) time derivative of the vector $|r_{\mathrm{i}}'(t)\rangle$ represented in the rotating system with the basis $\{|x_\mu(t)\rangle\}$.

There was nothing specific about the identity of the vector $|r_{\mathrm{i}}'(t)\rangle$ that was used in the derivation of Eq. (4.44). The classical point particle to which the vector $|r_{\mathrm{i}}'(t)\rangle$ pointed was an arbitrary point represented in the basis $\{|x_\mu(t)\rangle\}$. Therefore Eq. (4.44) is the general formula for the time derivative of any vector quantity represented in the basis $\{|x_\mu(t)\rangle\}$, which is rotating with the arbitrary instantaneous angular velocity $|\omega'\rangle$ with components $(\omega_1', \omega_2', \omega_3')$.

With no motion of the classical point particles in the rigid body $\dot{r}_{\mathrm{i}\mu}' = \mathbf{0}$. For a rigid body (4.44) is then

$$\frac{\mathrm{d}}{\mathrm{d}t}|r_{\mathrm{i}}'(t)\rangle \Longrightarrow \begin{bmatrix} 0 & -\omega_3' & \omega_2' \\ \omega_3' & 0 & -\omega_1' \\ -\omega_2' & \omega_1' & 0 \end{bmatrix} \begin{bmatrix} r_{\mathrm{i}1}' \\ r_{\mathrm{i}2}' \\ r_{\mathrm{i}3}' \end{bmatrix}. \tag{4.45}$$

Combining (4.19) with (4.45) we have the total time derivative of $|r_{\mathrm{i}}(t)\rangle$ for the rigid body as

$$\begin{aligned} \frac{\mathrm{d}}{\mathrm{d}t}|r_{\mathrm{i}}(t)\rangle &= \frac{\mathrm{d}}{\mathrm{d}t}|R(t)\rangle + \frac{\mathrm{d}}{\mathrm{d}t}|r_{\mathrm{i}}'(t)\rangle \\ &\Longrightarrow \begin{bmatrix} V_1 \\ V_2 \\ V_3 \end{bmatrix} + \begin{bmatrix} 0 & -\omega_3' & \omega_2' \\ \omega_3' & 0 & -\omega_1' \\ -\omega_2' & \omega_1' & 0 \end{bmatrix} \begin{bmatrix} r_{\mathrm{i}1}' \\ r_{\mathrm{i}2}' \\ r_{\mathrm{i}3}' \end{bmatrix}, \end{aligned} \tag{4.46}$$

where the primes indicate quantities represented in the rotating basis $\{|x_\mu(t)\rangle\}$.

We may also write (4.46) in standard vector form as

$$\frac{\mathrm{d}}{\mathrm{d}t}\boldsymbol{r}_{\mathrm{i}} = \frac{\mathrm{d}}{\mathrm{d}t}\boldsymbol{R} + \boldsymbol{\omega}' \times \boldsymbol{r}_{\mathrm{i}}' \tag{4.47}$$

for the rigid body. If we consider the motion of a single classical particle referred to the moving basis, we have

$$\frac{\mathrm{d}}{\mathrm{d}t}\boldsymbol{r}_{\mathrm{i}} = \frac{\mathrm{d}}{\mathrm{d}t}\boldsymbol{R} + \boldsymbol{v}_{\mathrm{i}}' + \boldsymbol{\omega}' \times \boldsymbol{r}_{\mathrm{i}}', \tag{4.48}$$

which is the form frequently used in considering motion on the earth's surface or on turntables primarily because of the visual aid gained from our understanding of standard vectors [cf. e.g. [1], pp. 204–206].

4.4.4 General Rolling Constraint

In Sect. 2.5.1 we considered the rolling constraint as an example of a nonholonomic constraint. But there we only treated a rolling disk. Here we must consider the general motion of a solid (rigid) body moving on a general surface. Our previous treatment was insufficient to handle this.

When a rigid body is rolling on a surface the point on the body which contacts the surface does not move relatively to the surface. If we choose the point of contact to be located by the vector $\boldsymbol{r}_{\mathrm{i}}'$ appearing in (4.47) then the velocity of that point is the velocity of the point on the surface, i.e. $\mathrm{d}\boldsymbol{r}_{\mathrm{i}}/\mathrm{d}t = \boldsymbol{V}_{\text{surface}}$, which is represented in the basis $\{|X_\mu\rangle\}$. And the velocity of the center of mass (CM) of the rigid body $\mathrm{d}\boldsymbol{R}/\mathrm{d}t = \boldsymbol{V}_{\mathrm{CM}}$ is also represented in the fixed basis $\{|X_\mu\rangle\}$. Then, using (4.47), the rolling constraint for a rigid body becomes

$$\boldsymbol{V}_{\text{surface}} = \boldsymbol{V}_{\mathrm{CM}} + \boldsymbol{\omega}' \times \boldsymbol{r}_{\mathrm{i}}'. \tag{4.49}$$

The only difficulty here is the cross product term, which is represented in the basis $\{|x_\mu(t)\rangle\}$. For convenience we wish to represent all vectors in (4.49) in the fixed basis $\{|X_\mu\rangle\}$.

The transformation of the term $\boldsymbol{\omega}' \times \boldsymbol{r}'$ can be accomplished in a straightforward fashion if we return to the general form (4.43), which is

$$\boldsymbol{\omega}' \times \boldsymbol{r}_{\mathrm{i}}' = |x_\mu(t)\rangle\langle x_\mu(t)| \frac{\mathrm{d}\mathbf{R}(\vartheta')}{\mathrm{d}t} |x_\nu(t)\rangle\langle x_\nu(t)|\, r'\rangle. \tag{4.50}$$

We can transform this to the fixed basis $\{|X_\mu\rangle\}$ using the projector $\mathbf{P}_{\mathrm{X}} = |X_\mu\rangle\langle X_\mu| = \mathbf{1}$, which we may insert at any points we wish in the expression on the right hand side of (4.50). The result is

$$\boldsymbol{\omega}' \times \boldsymbol{r}_{\mathrm{i}}' = |X_\alpha\rangle \langle X_\alpha| \, x_\mu(t)\rangle \langle x_\mu(t)| \, X_\beta\rangle \langle X_\beta| \cdots$$
$$\cdots \frac{\mathrm{d}\mathbf{R}(\vartheta)}{\mathrm{d}t} |X_\gamma\rangle \langle X_\gamma| \, x_\nu(t)\rangle \langle x_\nu(t)| \, X_\lambda\rangle \langle X_\lambda| \, r'\rangle. \tag{4.51}$$

In (4.51) we have dropped the prime on ϑ in $\mathrm{d}\mathbf{R}(\vartheta)/\mathrm{d}t$ because the representation of the operator is now in the basis $\{|X_\mu\rangle\}$. We may also eliminate the projector $\mathbf{P}_{\mathrm{x}} = |x_\mu(t)\rangle\langle x_\mu(t)| = \mathbf{1}$ from the two places it appears in the expression on the right hand side in (4.51). And, since the basis vectors $\{|X_\mu\rangle\}$ are orthonormal, we have $\langle X_\alpha| \, X_\beta\rangle = \delta_{\alpha\beta}$ and $\langle X_\gamma| \, X_\lambda\rangle = \delta_{\gamma\lambda}$, which further compresses the expression on the right hand side of (4.51). Then (4.51) becomes

$$\boldsymbol{\omega}' \times \boldsymbol{r}_{\mathrm{i}}' = |X_\alpha\rangle \langle X_\alpha| \frac{\mathrm{d}\mathbf{R}(\vartheta)}{\mathrm{d}t} |X_\lambda\rangle \langle X_\lambda| \, r'\rangle. \tag{4.52}$$

That is two expressions (4.52) and (4.50) are of exactly the same form, although represented in different bases. We may then write the rolling constraint in the form (4.49) without the primes, which we had used to designate representation in the rotating basis $\{|x_\mu(t)\rangle\}$. In order to avoid any possible confusion with a particle position vector we will use $\boldsymbol{d}$ as the vector from the CM to the contact point with the surface on which the body rolls. That is

$$\boxed{\boldsymbol{V}_{\text{surface}} = \boldsymbol{V}_{\text{CM}} + \boldsymbol{\omega} \times \boldsymbol{d}} \tag{4.53}$$

where both $\boldsymbol{\omega}$ and $\boldsymbol{d}$ are now represented in the fixed basis $\{|X_\mu\rangle\}$, although they are still the angular velocity of the rigid body and a location vector from the CM of the rigid body. The expression (4.53) is now a completely general rolling constraint.

Example 4.4.1 As an example of the application of (4.53) we return to the disk rolling on a stationary plane. The situation is shown below.

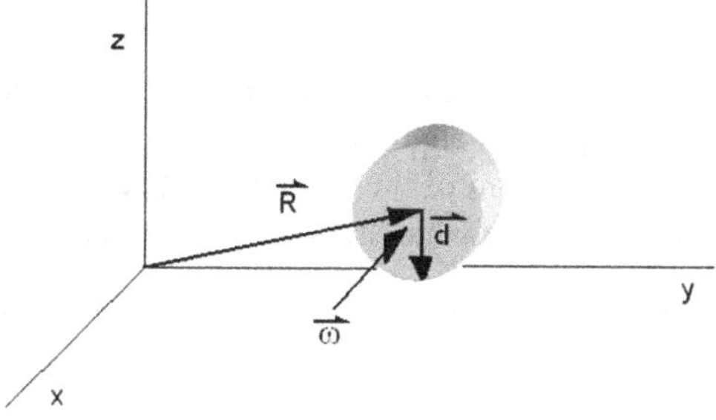

Disk rolling on a stationary plane

In this example we will use standard vector notation and write the fixed basis vectors as $(\hat{e}_x, \hat{e}_y, \hat{e}_z)$.

Since the floor is stationary in the fixed basis $\boldsymbol{V}_{\text{surface}} = \boldsymbol{0}$. The rolling constraint (4.53) is then

$$\boldsymbol{0} = \boldsymbol{V}_{\text{CM}} + \boldsymbol{\omega} \times \boldsymbol{d},$$

with

$$\boldsymbol{V}_{\text{CM}} = \frac{\mathrm{d}\boldsymbol{R}}{\mathrm{d}t} = \dot{X}\,\hat{e}_x + \dot{Y}\,\hat{e}_y.$$

The vector $\boldsymbol{d}$ from the CM to the surface and the angular momentum $\boldsymbol{\omega}$ are

$$\boldsymbol{d} = -a\,\hat{e}_z$$

and

$$\boldsymbol{\omega} = -\dot{\vartheta}\,\hat{e}_x.$$

With

$$\boldsymbol{\omega} \times \boldsymbol{d} = -a\dot{\vartheta}\,\hat{e}_y$$

the rolling constraint is

$$\boldsymbol{0} = \left(\dot{X}\right)\,\hat{e}_x + \left(\dot{Y} - a\dot{\vartheta}\right)\,\hat{e}_y,$$

or

$$\dot{X} = 0,\ \dot{Y} - a\dot{\vartheta} = 0.$$

This is the rolling constraint we would have intuitively written. Our general formulation makes the procedure automatic. This will be imperative in complex situations.

4.5 Rigid Body Dynamics

In this section we will use the results of our discussion on kinematics to obtain the kinetic and potential energies of a rigid body and the canonical equations for the motion of the body. The final form of the canonical equations for the components of the angular momentum will be what are known as the *Euler Equations* for the motion of the rigid body.

4.5.1 General Kinetic Energy

As in Sect. 4.4.1 we consider the solid (rigid) body to be made up of classical point particles of masses m_i located relative to the CM of the rigid body at $\left|r_i'(t)\right\rangle$ in the basis $\{|x_\mu(t)\rangle\}$, which we write using our arrow notation of Sect. 4.4.2 as

$$\left|r_{\mathrm{i}}'(t)\right\rangle \Longrightarrow \begin{bmatrix} r_{\mathrm{i}1}' \\ r_{\mathrm{i}2}' \\ r_{\mathrm{i}3}' \end{bmatrix} \tag{4.54}$$

in the standard basis. Because we have chosen the origin of this basis to be the CM,

$$\begin{aligned}
\sum_{\mathrm{i}} m_{\mathrm{i}} r_{\mathrm{i}1}' &= 0 \\
\sum_{\mathrm{i}} m_{\mathrm{i}} r_{\mathrm{i}2}' &= 0 \\
\sum_{\mathrm{i}} m_{\mathrm{i}} r_{\mathrm{i}3}' &= 0.
\end{aligned} \tag{4.55}$$

In Sect. 4.4.2 we obtained a general expression for the velocity of a particle making up a rigid body (see Sect. 4.4.3, Eq. (4.45)). If we carry out the matrix product indicated in (4.45) and write $d\left|r_{\mathrm{i}}'(t)\right\rangle/\mathrm{d}t = \left|\dot{r}_{\mathrm{i}}'(t)\right\rangle$ we have

$$\left|\dot{r}_{\mathrm{i}}'(t)\right\rangle \Longrightarrow \begin{bmatrix} V_1 - \omega_3' r_{\mathrm{i}2}' + \omega_2' r_{\mathrm{i}3}' \\ V_2 + \omega_3' r_{\mathrm{i}1}' - \omega_1' r_{\mathrm{i}3}' \\ V_3 - \omega_2' r_{\mathrm{i}1}' + \omega_1' r_{\mathrm{i}2}' \end{bmatrix} \tag{4.56}$$

From (4.56) the square of the velocity of the ith particle is

$$\begin{aligned}
\left\langle \dot{r}_{\mathrm{i}}'(t) \middle| \dot{r}_{\mathrm{i}}'(t)\right\rangle = &\left(V_1 - \omega_3' r_{\mathrm{i}2}' + \omega_2' r_{\mathrm{i}3}'\right)^2 \\
&+ \left(V_2 + \omega_3' r_{\mathrm{i}1}' - \omega_1' r_{\mathrm{i}3}'\right)^2 \\
&+ \left(V_3 - \omega_2' r_{\mathrm{i}1}' + \omega_1' r_{\mathrm{i}2}'\right)^2.
\end{aligned} \tag{4.57}$$

With (4.55) the kinetic energy of the rigid body is

$$\begin{aligned}
T = &\sum_\mu \frac{1}{2} m_\mu \left\langle \dot{r}_\mu'(t) \middle| \dot{r}_\mu'(t)\right\rangle \\
= &\frac{1}{2} M\left[V_1^2 + V_2^2 + V_3^2\right] \\
&+ \sum_{\mathrm{i}} \frac{1}{2} m_{\mathrm{i}} \left[\left(-\omega_3' r_{\mathrm{i}2}' + \omega_2' r_{\mathrm{i}3}'\right)^2 + \left(\omega_3' r_{\mathrm{i}1}' - \omega_1' r_{\mathrm{i}3}'\right)^2\right. \\
&\left. + \left(-\omega_2' r_{\mathrm{i}1}' + \omega_1' r_{\mathrm{i}2}'\right)^2\right].
\end{aligned} \tag{4.58}$$

The first term in (4.58) is the kinetic energy of translation of the CM.

$$T_{\mathrm{CM}} = \frac{1}{2} M \left[V_{\mathrm{X}}^2 + V_{\mathrm{Y}}^2 + V_{\mathrm{Z}}^2 \right]. \tag{4.59}$$

And the second term in (4.58) is the kinetic energy of rotation of the rigid body about its CM.

$$\begin{aligned} T'_{\mathrm{rot}} = \frac{1}{2} \sum_{\mathrm{i}} m_{\mathrm{i}} \{ & \left[\omega'_1 \left(r'^2_{\mathrm{i2}} + r'^2_{\mathrm{i3}} \right) - \omega'_2 r'_{\mathrm{i2}} r'_{\mathrm{i1}} - \omega'_3 r'_{\mathrm{i3}} r'_{\mathrm{i1}} \right] \omega'_1 \\ & + \left[\omega'_2 \left(r'^2_{\mathrm{i1}} + r'^2_{\mathrm{i3}} \right) - \omega'_1 r'_{\mathrm{i1}} r'_{\mathrm{i2}} - \omega'_3 r'_{\mathrm{i3}} r'_{\mathrm{i2}} \right] \omega'_2 \\ & + \left[\omega'_3 \left(r'^2_{\mathrm{i1}} + r'^2_{\mathrm{i2}} \right) - \omega'_1 r'_{\mathrm{i1}} r'_{\mathrm{i3}} - \omega'_2 r'_{\mathrm{i2}} r'_{\mathrm{i3}} \right] \omega'_3 \} . \end{aligned} \tag{4.60}$$

In (4.60) we have used the prime on T'_{rot} to indicate that this kinetic energy is represented in the body basis $\{|x_\mu(t)\rangle\}$.

The kinetic energy of the rigid body is then separated as

$$T = T_{\mathrm{CM}} + T'_{\mathrm{rot}}, \tag{4.61}$$

Provided we choose the origin of the body basis $\{|x_\mu(t)\rangle\}$ at the CM.

4.5.2 *Inertia Tensor*

The kinetic energy of rotation of the rigid body about the CM (4.60) has a particular symmetry. If we identify the elements of a matrix $I_{\alpha\beta}$ as

$$I_{\alpha\beta} = \sum_{\mathrm{i}} m_{\mathrm{i}} \left[\left(r'^2_{\mathrm{i1}} + r'^2_{\mathrm{i2}} + r'^2_{\mathrm{i3}} \right) \delta_{\alpha\beta} - r'_{\mathrm{i}\alpha} r'_{\mathrm{i}\beta} \right] \tag{4.62}$$

we can write T'_{rot} in (4.60) as

$$\begin{aligned} T'_{\mathrm{rot}} = \frac{1}{2} & \left[\left(\omega'_1 I_{11} + \omega'_2 I_{21} + \omega'_3 I_{31} \right) \omega'_1 \right. \\ & + \left(\omega'_2 I_{22} + \omega'_1 I_{12} + \omega'_3 I_{32} \right) \omega'_2 \\ & \left. + \left(I_{33} \omega'_3 + \omega'_1 I_{13} + \omega'_2 I_{23} \right) \omega'_3 \right] . \end{aligned} \tag{4.63}$$

The matrix $I_{\alpha\beta}$ is a property of the rigid body known as the *inertia tensor*.

With (4.62) we may write the kinetic energy of rotation about the CM as

$$T'_{\mathrm{rot}} = \frac{1}{2} \omega'_\alpha I_{\alpha\beta} \omega'_\beta, \tag{4.64}$$

using the summation convention. Written in matrix form in the basis $\{|x_\mu(t)\rangle\}$ the inertia tensor is

$$
\begin{aligned}
I &= \begin{bmatrix} I_{11} & I_{12} & I_{13} \\ I_{21} & I_{22} & I_{23} \\ I_{31} & I_{32} & I_{33} \end{bmatrix} \\
&= \begin{bmatrix} \sum_{\mathrm{i}} m_{\mathrm{i}} \left(r_{\mathrm{i}2}'^2 + r_{\mathrm{i}3}'^2\right) & -\sum_{\mathrm{i}} m_{\mathrm{i}} r_{\mathrm{i}1}' r_{\mathrm{i}2}' & -\sum_{\mathrm{i}} m_{\mathrm{i}} r_{\mathrm{i}1}' r_{\mathrm{i}3}' \\ -\sum_{\mathrm{i}} m_{\mathrm{i}} r_{\mathrm{i}2}' r_{\mathrm{i}1}' & \sum_{\mathrm{i}} m_{\mathrm{i}} \left(r_{\mathrm{i}1}'^2 + r_{\mathrm{i}3}'^2\right) & -\sum_{\mathrm{i}} m_{\mathrm{i}} m_\mu r_{\mathrm{i}2}' r_{\mathrm{i}3}' \\ -\sum_{\mathrm{i}} m_{\mathrm{i}} r_{\mathrm{i}3}' r_{\mathrm{i}1}' & -\sum_{\mathrm{i}} m_{\mathrm{i}} m_\mu r_{\mathrm{i}3}' r_{\mathrm{i}2}' & \sum_{\mathrm{i}} m_{\mathrm{i}} m_\mu \left(r_{\mathrm{i}1}'^2 + r_{\mathrm{i}2}'^2\right) \end{bmatrix},
\end{aligned}
\tag{4.65}
$$

which has the symmetry $I_{\alpha\beta} = I_{\beta\alpha}$. We can always diagonalize a symmetric matrix [see e.g. [21], p. 54]. For most bodies of interest there is no need to carry out the diagonalization separately because the axes in which the inertia tensor is diagonal are the axes of symmetry, which are usually obvious. We shall, therefore, simply write the general inertia tensor for a rigid body as

$$
I = \begin{bmatrix} I_1 & 0 & 0 \\ 0 & I_2 & 0 \\ 0 & 0 & I_3 \end{bmatrix}, \tag{4.66}
$$

designating the elements with single indices. If any two of these elements are equal we have a *body of revolution*. If all three are equal we have a *sphere*.

The kinetic energy of rotation of a rigid body is then

$$
T_{\mathrm{rot}}' = \frac{1}{2} \sum_{\alpha}^{3} \omega_\alpha' I_\alpha \omega_\alpha'. \tag{4.67}
$$

4.5.3 Rotational Potential Energy

We shall assume that the potential energy is a scalar function of spatial coordinates alone. Because we are considering our rigid body to be made up of classical point particles i, with arbitrary masses, it is convenient to consider the potential energy per unit mass as the basis for our discussion. In the case of an electric field we would consider, equivalently, the potential energy per unit charge.

The potential energy of a classical point particle located at the point designated by the vector $|r_{\mathrm{i}}'(t)\rangle$ in Fig. 4.1 may always be separated into a value at the location of the CM and an added value resulting from the location of the classical point particle relatively to the CM. If we designate the value of the potential energy (per unit mass) at the CM, i.e. at the point $|R(t)\rangle = (R_1, R_2, R_3)$, as φ_{CM} and the potential energy (per unit mass) of the ith classical point particle at the point $|r_{\mathrm{i}}'(t)\rangle = \left(r_{\mathrm{i}1}', r_{\mathrm{i}2}', r_{\mathrm{i}3}'\right)$, referenced to the value at the CM, as φ_{i}' then the potential energy of the rigid body is $M\varphi_{\mathrm{CM}}(R_1, R_2, R_3) + \sum_{\mathrm{i}} m_{\mathrm{i}} \varphi_{\mathrm{i}}' \left(r_{\mathrm{i}1}', r_{\mathrm{i}2}', r_{\mathrm{i}3}'\right)$. The potential energy is then separated into a portion dependent on the CM coordinates (R_1, R_2, R_3) and a portion dependent on the body coordinates $\left(r'_{i1}, r'_{i2}, r'_{i3}\right)$, expressed in the basis $\{|x_\mu(t)\rangle\}$, which is stationary in the body.

We now define

$$\Phi_{CM} \equiv M\varphi_{CM}(R_1, R_2, R_3) \tag{4.68}$$

and

$$\Phi'_{rot} \equiv \sum_i m_i \varphi'_i\left(r'_{i1}, r'_{i2}, r'_{i3}\right), \tag{4.69}$$

where we use the prime to indicate that Φ'_{rot} is represented in the body basis.

With (4.61), (4.68), and (4.69) the Lagrangian for a rigid body is

$$L = (T_{CM} - \Phi_{CM}) + \left(T'_{rot} - \Phi'_{rot}\right). \tag{4.70}$$

The Lagrangian for a rigid body then separates into a part involving the motion of the CM,

$$L_{CM} = T_{CM} - \Phi_{CM}, \tag{4.71}$$

which depends only on $(R_1, R_2, R_3, V_1, V_2, V_3)$, and a part involving rotation about the CM

$$L'_{rot} = T'_{rot} - \Phi'_{rot}, \tag{4.72}$$

which depends only on the body coordinates and angular velocities $\left(r'_{i1}, r'_{i2}, r'_{i3}, \omega'_1, \omega'_2, \omega'_3\right)$. The CM of a rigid body moves as a point of mass M under the influence of the potential $\Phi_{CM}(R_1, R_2, R_3)$. And the motion of the rigid body about the CM is determined by the potential $\Phi'_{rot}\left(r'_{i1}, r'_{i2}, r'_{i3}\right)$.

There remains, however, one problem. The kinetic energy of the rotating body is a function of angular velocities $\omega'_\mu = \mathrm{d}\vartheta'_\mu/\mathrm{d}t$, while the potential energy is a function of positions $r'_{i\mu}$. But the Lagrangian must be a function of coordinates q and their time derivatives $\dot{q}$. We cannot have a Lagrangian with a mixed dependence on positions $r'_{i\mu}$ and angular velocities $\omega'_\mu = \mathrm{d}\vartheta'_\mu/\mathrm{d}t$. We must, therefore, convert the functional dependence of the potential energy Φ'_{rot} from positions to angles of rotation. We can resolve this difficulty in a natural manner if we carefully consider the motion of the ith classical point particle as the basis $\{|x_\mu(t)\rangle\}$ rotates. Then the rotational equivalence of a force, which is known as a *torque*,[3] will also emerge from the derivatives with respect to the angles.

As in our derivation of T'_{rot}, we again consider first only rotations about a single axis $|x_3(t)\rangle$ of the basis $\{|x_\mu(t)\rangle\}$. Here, however, it will be more convenient to actually consider the components of the position vector $|r'_i(t)\rangle$ locating the ith point particle directly, rather than introducing a separate vector $|a\rangle$.

[3]The torque is a product of force and distance, which results in a change in angular momentum. The torque is also sometimes called a moment.

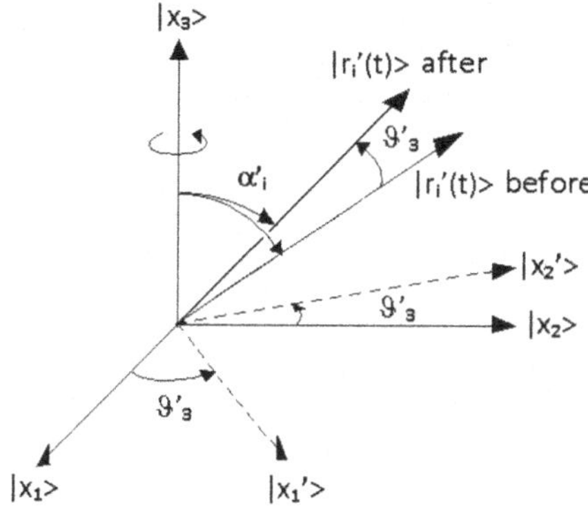

Fig. 4.3 $\{|x_\mu(t)\rangle\}$ rotate around $|x_3\rangle$ which is fixed

In Fig. 4.3 we have shown the rotation of the position vector $|r'_{\mathrm{i}}(t)\rangle$ as the basis rotates through an angle ϑ'_3 about $|x_3(t)\rangle$.

In Fig. 4.3 the $|x_1\rangle$ and $|x_2\rangle$components of the vector $|r'_{\mathrm{i}}(t)\rangle$ are

$$\begin{aligned} r'_{\mathrm{i}1} &= \left(r'_{\mathrm{i}} \sin\alpha'_{\mathrm{i}}\right)\cos\vartheta'_3 \\ r'_{\mathrm{i}2} &= \left(r'_{\mathrm{i}} \sin\alpha'_{\mathrm{i}}\right)\sin\vartheta'_3, \end{aligned} \tag{4.73}$$

where r'_{i} is the magnitude of the position vector to the ith particle. During an infinitesimal change $\mathrm{d}\vartheta'_3$ in the angle ϑ'_3 the components $r'_{\mathrm{i}1}$ and $r'_{\mathrm{i}2}$ change by the infinitesimal amounts

$$\begin{aligned} \mathrm{d}r'_{\mathrm{i}1} &= -\left(r'_{\mathrm{i}} \sin\alpha'_{\mathrm{i}}\right)\sin\vartheta'_3 \mathrm{d}\vartheta'_3 = -r'_{\mathrm{i}2}\mathrm{d}\vartheta'_3 \\ \mathrm{d}r'_{\mathrm{i}2} &= \left(r'_{\mathrm{i}} \sin\alpha'_{\mathrm{i}}\right)\cos\vartheta'_3 \mathrm{d}\vartheta'_3 = r'_{\mathrm{i}1}\mathrm{d}\vartheta'_3. \end{aligned} \tag{4.74}$$

We then identify the partial derivatives of $r_{\mathrm{i}2}$ and $r_{\mathrm{i}3}$ with respect to ϑ'_3 as

$$\begin{aligned} \frac{\partial r'_{\mathrm{i}1}}{\partial\vartheta'_3} &= -r'_{\mathrm{i}2} \\ \frac{\partial r'_{\mathrm{i}2}}{\partial\vartheta'_3} &= r'_{\mathrm{i}1}. \end{aligned} \tag{4.75}$$

The partial derivative of the potential per unit mass φ'_{i} at the location of the ith particle, with respect to the angle ϑ'_3, is then

$$\begin{aligned} \frac{\partial\varphi'_{\mathrm{i}}}{\partial\vartheta'_3} &= \frac{\partial\varphi'_{\mathrm{i}}}{\partial r'_{\mathrm{i}1}}\frac{\partial r'_{\mathrm{i}1}}{\partial\vartheta'_3} + \frac{\partial\varphi'_{\mathrm{i}}}{\partial r'_{\mathrm{i}2}}\frac{\partial r'_{\mathrm{i}2}}{\partial\vartheta'_3} \\ &= r_{\mathrm{i}2}\frac{\partial\varphi'_{\mathrm{i}}}{\partial r'_{\mathrm{i}3}} - r_{\mathrm{i}3}\frac{\partial\varphi'_{\mathrm{i}}}{\partial r'_{\mathrm{i}2}}. \end{aligned} \tag{4.76}$$

We shall now designate the gradient of φ'_{i} evaluated at the location of the ith particle as $\mathrm{grad}'_{\mathrm{i}}\,\varphi'_{\mathrm{i}}$. In column matrix form

$$\mathrm{grad}'_{\mathrm{i}}\,\varphi'_{\mathrm{i}} \Longrightarrow \begin{bmatrix} \partial\varphi'_{\mathrm{i}}/\partial r'_{\mathrm{i}1} \\ \partial\varphi'_{\mathrm{i}}/\partial r'_{\mathrm{i}2} \\ \partial\varphi'_{\mathrm{i}}/\partial r'_{\mathrm{i}3} \end{bmatrix}. \tag{4.77}$$

With (4.77) we recognize that (4.76) is the $|x_3\rangle$–component of the cross product of the vector $\left|r'_{\mathrm{i}}(t)\right\rangle$ with $\mathrm{grad}'_{\mathrm{i}}\,\varphi'_{\mathrm{i}}$ evaluated at the location of the ith particle. That is

$$\frac{\partial\varphi'_{\mathrm{i}}}{\partial\vartheta'_3} = \left[\boldsymbol{r}'_{\mathrm{i}} \times \mathrm{grad}'_{\mathrm{i}}\,\varphi'_{\mathrm{i}}\right]_3. \tag{4.78}$$

The other partial derivatives with respect to the angles produce similar terms. Specifically

$$\frac{\partial\varphi'_{\mathrm{i}}}{\partial\vartheta'_1} = \left[\boldsymbol{r}'_{\mathrm{i}} \times \mathrm{grad}'_{\mathrm{i}}\,\varphi'_{\mathrm{i}}\right]_1. \tag{4.79}$$

and

$$\frac{\partial\varphi'_{\mathrm{i}}}{\partial\vartheta'_2} = \left[\boldsymbol{r}'_{\mathrm{i}} \times \mathrm{grad}'_{\mathrm{i}}\,\varphi'_{\mathrm{i}}\right]_2 \tag{4.80}$$

The product $m_{\mathrm{i}}\left(-\mathrm{grad}'_{\mathrm{i}}\varphi'_{\mathrm{i}}\right)$ is the (external) force acting on the ith particle. And the cross product of the position vector $\boldsymbol{r}'_{\mathrm{i}}$ of the ith particle with this force is the *torque*, about the CM, due to this force. We shall designate this torque as $\vec{\gamma}_{\mathrm{i}}$. Then

$$\vec{\gamma}_{\mathrm{i}} = -\boldsymbol{r}'_{\mathrm{i}} \times \mathrm{grad}'_{\mathrm{i}}\,\left(m_{\mathrm{i}}\varphi'_{\mathrm{i}}\right). \tag{4.81}$$

If we sum over all parts making up the rigid body we have the total torque, from *external conservative forces*, acting on the rigid body as

$$\vec{\gamma} = \sum_{\mathrm{i}} \vec{\gamma}_{\mathrm{i}} = -\sum_{\mathrm{i}} \boldsymbol{r}'_{\mathrm{i}} \times \mathrm{grad}'_{\mathrm{i}}\,\left(m_{\mathrm{i}}\varphi'_{\mathrm{i}}\right). \tag{4.82}$$

With (4.78)–(4.80) and (4.69), Eq. (4.82) becomes

$$\vec{\gamma} = -\left[|x_1\rangle\frac{\partial}{\partial\vartheta'_3} + |x_2\rangle\frac{\partial}{\partial\vartheta'_2} + |x_3\rangle\frac{\partial}{\partial\vartheta'_3}\right]\Phi'_{\mathrm{rot}}. \tag{4.83}$$

Then the components of $\vec{\gamma}$ are

$$\gamma_\lambda = -\frac{\partial\Phi'_{\mathrm{rot}}}{\partial\vartheta_\lambda} \text{ for } \lambda = 1, 2, 3. \tag{4.84}$$

We have, therefore succeeded in our task. In the Eq. (4.73) and the $|x_3\rangle$ – component $r'_{i3} = r'_i \cos\alpha'_i$ we have the transformation of the potential φ'_i from position to angular coordinates. And from (4.84) we have the components of the torque, which is the equivalent of the force for rotational motion. We now have the Lagrangian $L'_{rot}(\vartheta, \dot{\vartheta})$ and can proceed to obtain the Hamiltonian and the canonical equations.

4.5.4 *Rotational Canonical Equations*

From (4.67) and (4.72) the Lagrangian for rotation of a rigid body about the CM is

$$L'_{rot} = \frac{1}{2} I_\mu \dot{\vartheta}'^2_\mu - \Phi'_{rot}(\vartheta), \tag{4.85}$$

and the components of the canonical (angular) momenta P'_{ϑ_μ} are

$$P'_{\vartheta_\mu} = \frac{\partial L'_{rot}}{\partial \dot{\vartheta}'_\mu} = I_\mu \dot{\vartheta}'_\mu = I_\mu \omega'_\mu \tag{4.86}$$

with no sum on μ. In (4.86) we have used the definition of ω'_μ from (4.40). The Hamiltonian $\mathcal{H}'_{rot}$, in the body system, is then

$$\mathcal{H}'_{rot} = \frac{1}{2} \sum_\mu \frac{P'^2_{\vartheta_\mu}}{I_\mu} + \Phi'_{rot}(\vartheta), \tag{4.87}$$

where we have included the summation sign because the index μ appears three times in the sum.

From (4.87) the canonical equations are

$$\frac{\partial}{\partial P'_{\vartheta_\mu}} \mathcal{H}'_{rot} = \frac{P'_{\vartheta_\mu}}{I_\mu} = \frac{d}{dt} \vartheta'_\mu, \tag{4.88}$$

with no sum on μ, and

$$\frac{\partial}{\partial \vartheta'_\mu} \mathcal{H}'_{rot} = -\frac{d}{dt} P'_{\vartheta_\mu}. \tag{4.89}$$

Then (4.88) is

$$\frac{d}{dt} \vartheta'_\mu = \frac{P'_{\vartheta_\mu}}{I_\mu}, \tag{4.90}$$

or

$$P'_{\vartheta_\mu} = I_\mu \omega'_\mu, \tag{4.91}$$

which is simply the definition of components of the canonical angular momenta in terms of the angular velocities. Because the I_μ are generally distinct for each μ, the angular momentum is generally not parallel to the angular velocity $\mathrm{d}\vartheta'_\mu/\mathrm{d}t$.

Using (4.44) the time derivative in Eq. (4.89) is

$$\begin{aligned}\frac{\mathrm{d}}{\mathrm{d}t}P'_{\vartheta_\mu} &= \begin{bmatrix} \dot{P}'_{\vartheta 1} \\ \dot{P}'_{\vartheta 2} \\ \dot{P}'_{\vartheta 3} \end{bmatrix} + \begin{bmatrix} -\omega'_3 I_2 \omega'_2 + \omega'_2 I_3 \omega'_3 \\ \omega'_3 I_1 \omega'_1 - \omega'_1 I_3 \omega'_3 \\ -\omega'_2 I_1 \omega'_1 + \omega'_1 I_2 \omega'_2 \end{bmatrix} \\ &= \begin{bmatrix} \dot{P}'_{\vartheta 1} - \omega'_3 I_2 \omega'_2 + \omega'_2 I_3 \omega'_3 \\ \dot{P}'_{\vartheta 2} + \omega'_3 I_1 \omega'_1 - \omega'_1 I_3 \omega'_3 \\ \dot{P}'_{\vartheta 3} - \omega'_2 I_1 \omega'_1 + \omega'_1 I_2 \omega'_2 \end{bmatrix}. \end{aligned} \tag{4.92}$$

And with (4.84) we have the left hand side of (4.89) as

$$\frac{\partial}{\partial \vartheta'_\mu}\mathcal{H}'_{\mathrm{rot}} = \frac{\partial}{\partial \vartheta'_\mu}\Phi'_{\mathrm{rot}} = -\gamma'_\mu .$$

The canonical equation (4.92) is then

$$\frac{\mathrm{d}}{\mathrm{d}t}P'_{\vartheta_\mu} = \begin{bmatrix} \dot{P}'_{\vartheta 1} - \omega'_3 I_2 \omega'_2 + \omega'_2 I_3 \omega'_3 \\ \dot{P}'_{\vartheta 2} + \omega'_3 I_1 \omega'_1 - \omega'_1 I_3 \omega'_3 \\ \dot{P}'_{\vartheta 3} - \omega'_2 I_1 \omega'_1 + \omega'_1 I_2 \omega'_2 \end{bmatrix} = \begin{bmatrix} \gamma'_1 \\ \gamma'_2 \\ \gamma'_3 \end{bmatrix}. \tag{4.93}$$

If we now introduce (4.91) into (4.93) we have

$$\boxed{\begin{bmatrix} \gamma'_1 - I_1 \dot{\omega}'_1 + \omega'_3 I_2 \omega'_2 - \omega'_2 I_3 \omega'_3 \\ \gamma'_2 - I_2 \dot{\omega}'_2 - \omega'_3 I_1 \omega'_1 + \omega'_1 I_3 \omega'_3 \\ \gamma'_3 - I_3 \dot{\omega}'_3 + \omega'_2 I_1 \omega'_1 - \omega'_1 I_2 \omega'_2 \end{bmatrix} = \begin{bmatrix} 0 \\ 0 \\ 0 \end{bmatrix}}. \tag{4.94}$$

The set of Eq. (4.94) is traditionally referred to simply as the *Euler Equations* for rotational motion.

Our use of the indices 1, 2, 3 for the basis vectors has been for our use of the Dirac transformation methods. We may always revert to the more common Cartesian coordinates using X, Y, Z and x, y, z if we choose.

The sets of Eqs. (4.90) and (4.93) are the canonical equations for the angles and the angular momenta. The Euler Equations (4.94) are a combination of these canonical equations. These are all written without constraints. We realize that the canonical equations come from Hamilton's variational principle and that the constraints may, therefore, be easily introduced into these equations through use of Lagrange multipliers, which may be functions of the time.

We must be careful in treating any problem involving rotational motion. Because of the nonlinear coupling among the angular velocities revealed in the Euler Equations, we cannot expect our intuition to be reliable regarding the motion of rotating bodies. And in the Euler Equations the angular velocities are referenced to the basis fixed in the rotating body, rather than to the external system fixed in space. We must be aware of this in any direct application of the Euler Equations. Our measurements are seldom made directly on these angular velocities (see exercises).

In many applications to simple geometrical shapes, such as disks with fixed axes or spheres, it may be simplest to begin with the Hamiltonian referred to a fixed external basis and obtain the canonical equations as we have in the preceding chapter. We must only keep our wits about us. Thinking in terms of equations is a form of intuition.

Example 4.5.1 (*Ball rolling on a rotating Turntable*) A ball of radius a and mass m rolls on a turntable, which has a constant angular velocity $\vec{\Omega}$ about the vertical axis of the fixed system. We have illustrated the situation in Fig. 4.4. What is the trajectory of the ball as seen in the stationary system? That is, if we look vertically downward on the table and ignore the fact that the turntable is rotating, what is the geometric figure that we see? It is easiest to consider that the turntable is made of some dark material making it possible to ignore the motion. The ball may then be a white (super) ball that rolls without slipping on this turntable.

This problem would be notoriously difficult if it were not for some major simplifications. The torque on the ball comes from the reaction to the rolling constraint. We, therefore, do not require a separate evaluation of this torque. The rolling constraint may also, of course, be written in the fixed basis. The external gravitational force is only in the vertical direction, so the external potential is a constant and may be taken to be zero. And, finally, for the spherical ball, we may write the kinetic energy in the fixed basis (see below).

We choose the origin of our fixed basis (coordinate system) to be the center of the turntable. If we designate the point on the turntable which is in contact with the ball as $\boldsymbol{r}_{\mathrm{p}}$ the rolling constraint is

$$\frac{\mathrm{d}}{\mathrm{d}t}\vec{r}_{\mathrm{p}} = \frac{\mathrm{d}}{\mathrm{d}t}\vec{R} + \vec{\omega} \times \vec{d},$$

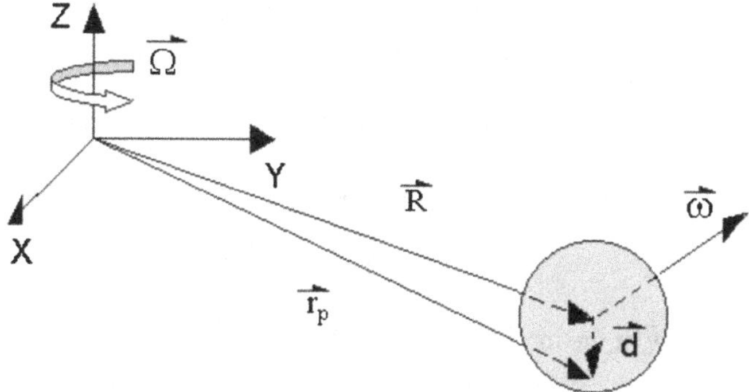

Fig. 4.4 Basic vectors describing a ball rolling without slipping on a rotating turntable

in which the term $\vec{\omega} \times \vec{d}$ is represented in the fixed system (see (4.53)). The point $\vec{r}_{\mathrm{p}}$ is not fixed in space, but is a point on the turntable and has the velocity of the point on the turntable. Noting that the point

$$\vec{r}_{\mathrm{p}} = \vec{R} + \vec{d}$$

is, simultaneously, a point on the rigid body and a point on the turntable, the velocity of $\vec{r}_{\mathrm{p}}$ is $\vec{\Omega} \times \vec{r}_{\mathrm{p}}$. That is

$$\begin{aligned}\frac{\mathrm{d}}{\mathrm{d}t}\vec{r}_{p} &= \vec{\Omega} \times \left(\vec{R} + \vec{d}\right)\\ &= \vec{\Omega} \times \vec{R},\end{aligned}$$

since $\vec{d}$ is parallel to the turntable angular velocity $\vec{\Omega}$. The rolling constraint is then

$$\vec{\Omega} \times \vec{R} = \frac{\mathrm{d}}{\mathrm{d}t}\vec{R} + \vec{\omega} \times \vec{d}$$

represented in the fixed system.

The abstract form of the kinetic energy of rotation is

$$T_{\mathrm{rot}} = \frac{1}{2}\langle\omega| I |\omega\rangle .$$

For a sphere the elements of the inertia tensor are $I_0\delta_{\alpha\beta}$. For the sphere then

$$T_{\mathrm{rot}} = \frac{1}{2} I_0 \langle\omega |x_\mu\rangle \langle x_\mu| \omega\rangle ,$$

which we can show is

$$T_{\mathrm{rot}} = \frac{1}{2} I_0 \langle\omega |X_\lambda\rangle \langle X_\lambda| \omega\rangle$$

in the fixed basis (see exercises). We then have the complete representation of the dynamics in the fixed system.

Written in the stationary system the terms of interest are the inertia tensor

$$I = \begin{bmatrix} I_0 & 0 & 0 \\ 0 & I_0 & 0 \\ 0 & 0 & I_0 \end{bmatrix},$$

the angular velocity of the turntable

$$\vec{\Omega} = \begin{bmatrix} 0 \\ 0 \\ \Omega \end{bmatrix},$$

the angular velocity of the ball

$$\vec{\omega} = \begin{bmatrix} \omega_1 \\ \omega_2 \\ \omega_3 \end{bmatrix},$$

the point of contact between the ball and the turntable

$$\vec{d} = \begin{bmatrix} 0 \\ 0 \\ -a \end{bmatrix},$$

and the vector to the CM

$$\vec{R} = \begin{bmatrix} X \\ Y \\ Z \end{bmatrix}.$$

The rolling constraint is then

$$\begin{bmatrix} -\Omega Y \\ \Omega X \\ 0 \end{bmatrix} = \begin{bmatrix} \dot{X} - a\omega_2 \\ \dot{Y} + a\omega_1 \\ 0 \end{bmatrix}, \tag{4.95}$$

which is two constraint equations

$$\begin{aligned} -\Omega Y &= \dot{X} - a\omega_2 \\ \Omega X &= \dot{Y} + a\omega_1. \end{aligned} \tag{4.96}$$

These may be written as time derivatives of functions $g_1 = g_2 = 0$. Then

$$\begin{aligned} \frac{\mathrm{d}g_1}{\mathrm{d}t} &= 0 = \Omega Y + \dot{X} - a\omega_2 \\ \frac{\mathrm{d}g_2}{\mathrm{d}t} &= 0 = -\Omega X + \dot{Y} + a\omega_1. \end{aligned} \tag{4.97}$$

Calling the rotation angle of the turntable Θ_3,

$$\Omega = \dot{\Theta}_3 = \frac{\mathrm{d}\Theta_3}{\mathrm{d}t},$$

and writing

$$\omega_\mu = \frac{\mathrm{d}\vartheta_\mu}{\mathrm{d}t},$$

the Eq. (4.97) become

$$\begin{aligned} \mathrm{d}g_1 &= 0 = \mathrm{d}X + Y\mathrm{d}\Theta_3 - a\mathrm{d}\vartheta_2 \\ \mathrm{d}g_2 &= 0 = \mathrm{d}Y - X\mathrm{d}\Theta_3 + a\mathrm{d}\vartheta_1. \end{aligned} \tag{4.98}$$

We now associate the Lagrange multipliers λ_1 and λ_2 with the constraints $g_1 = 0$ and $g_2 = 0$ respectively. We can pick off the partial derivatives of g_1 and g_2 with respect to the coordinates from (4.98). When introduced into the canonical equations the Lagrange multipliers λ_1 and λ_2 will appear as variable (time dependent) forces resulting in torques acting on the ball.

Since there is no potential energy, we can begin by immediately writing the Hamiltonian as

$$\begin{aligned} \mathcal{H} &= T \\ &= T_{\mathrm{CM}} + T_{\mathrm{Rot}} + T_{\mathrm{Turntable}} \\ &= \frac{1}{2m}\left(P_1^2 + P_2^2\right) + \frac{1}{2I_0}\left(P_{\vartheta_1}^2 + P_{\vartheta_2}^2 + P_{\vartheta_3}^2\right) + \frac{1}{2I_{\mathrm{T}}} P_{\Theta}^2, \end{aligned} \tag{4.99}$$

where $T_{\mathrm{Turntable}}$ is the kinetic energy of the turntable. The components of the linear momenta of the CM of the ball are P_1 and P_2. P_{ϑ_μ} is the angular momentum of the ball for rotation about the axis $|X_\mu\rangle$, since we have expressed the kinetic energy of the ball in the fixed basis. And P_Θ is the angular momentum of the rotating turntable. I_{T} is the moment of inertia of the turntable.

Because the angular velocity of the turntable is constant the kinetic energy of the turntable is constant.

The canonical equations with constraints are then

$$\begin{aligned} \dot{P}_1 &= \lambda_1, \\ \dot{P}_2 &= \lambda_2, \\ \dot{P}_{\vartheta_1} &= a\lambda_2, \\ \dot{P}_{\vartheta_2} &= -a\lambda_1, \\ \dot{P}_{\vartheta_3} &= 0, \\ \dot{P}_{\Theta} &= \lambda_1 Y - \lambda_2 X, \end{aligned} \tag{4.100}$$

and

$$\begin{aligned} \dot{X} &= \frac{1}{m} P_1, \\ \dot{Y} &= \frac{1}{m} P_2, \\ \dot{\vartheta}_1 &= \frac{1}{I_0} P_{\vartheta_1}, \end{aligned}$$

$$\dot{\vartheta}_2 = \frac{1}{I_0} P_{\vartheta_2},$$
$$\dot{\vartheta}_3 = \frac{1}{I_0} P_{\vartheta_3},$$
$$\dot{\Theta}_3 = \frac{1}{I_T} P_{\Theta}, \tag{4.101}$$

and the constraint Ω = constant is

$$\dot{\Theta}_3 = \Omega. \tag{4.102}$$

We note that the equations $\dot{P}_{\vartheta_1} = a\lambda_2$ and $\dot{P}_{\vartheta_2} = -a\lambda_1$ are, in fact, the first and second of the Euler Equations (4.94). All components of the inertia tensor are identical for the sphere. so the coupling terms in (4.94) vanish and the torques are $a\lambda_2$ and $-a\lambda_1$.

We can eliminate the λs among the first four equations of (4.100) resulting in

$$\dot{P}_1 + \frac{1}{a}\dot{P}_{\vartheta_2} = 0$$
$$\dot{P}_2 - \frac{1}{a}\dot{P}_{\vartheta_1} = 0. \tag{4.103}$$

Multiplying the original rolling constraint equations by I_0 we get

$$P_{\vartheta_2} = I_0\omega_2 = \frac{I_0}{a}\Omega Y + \frac{I_0}{a}\dot{X}$$
$$P_{\vartheta_1} = I_0\omega_1 = \frac{I_0}{a}\Omega X - \frac{I_0}{a}\dot{Y} \tag{4.104}$$

The Eqs. (4.103) and (4.104) are of a form which makes them soluble in the complex plane. The procedure we follow here is somewhat standard.

We begin by defining the complex valued quantities

$$Z = X + iY$$
$$P_Z = P_1 + iP_2$$
$$L_Z = P_{\vartheta_1} + iP_{\vartheta_2} \tag{4.105}$$

We then introduce the imaginary quantity $i = \sqrt{-1}$ into the Eq. (4.103) to obtain.

$$\dot{P}_1 - i\frac{1}{a}i\dot{P}_{\vartheta_2} = 0$$
$$i\dot{P}_2 - i\frac{1}{a}\dot{P}_{\vartheta_1} = 0. \tag{4.106}$$

And we then add the Eq. (4.106) to get

$$\dot{P}_Z - i\frac{1}{a}\dot{L}_Z = 0. \tag{4.107}$$

In a similar fashion we introduce the imaginary quantity i into Eq. (4.104) to obtain

$$\begin{aligned} iP_{\vartheta_2} &= \frac{I_0}{a}\Omega iY + i\frac{I_0}{a}\dot{X} \\ P_{\vartheta_1} &= \frac{I_0}{a}\Omega X + i\frac{I_0}{a}i\dot{Y}, \end{aligned} \tag{4.108}$$

which we add to get

$$L_Z = \frac{I_0}{a}\Omega Z + i\frac{I_0}{a}\dot{Z}. \tag{4.109}$$

We then have the differential equations (4.103), (4.107) and (4.109) along with the definitions (4.105) as the equations we must solve.

We may integrate (4.107) immediately to give

$$P_Z - i\frac{1}{a}L_Z = \text{constant}. \tag{4.110}$$

With the first two equations of (4.101) and of (4.105), the Eq. (4.110) becomes

$$\dot{Z} = i\frac{1}{ma}L_Z + K, \tag{4.111}$$

where K is a complex valued constant. With (4.109) Eq. (4.111) becomes

$$\begin{aligned} \dot{Z} &= i\left(\frac{I_0\Omega}{ma^2 + I_0}\right)Z + \left(\frac{ma^2}{ma^2 + I_0}\right)K \\ &= i\Omega_0 Z + \left(\frac{ma^2\Omega_0}{\Omega I_0}\right)K, \end{aligned} \tag{4.112}$$

where

$$\Omega_0 = \frac{I_0\Omega}{ma^2 + I_0} \tag{4.113}$$

The solution to (4.112) is

$$Z = \widetilde{Z}\exp(i\Omega_0 t) + C, \tag{4.114}$$

where $\widetilde{Z}$ and C are complex constants. Inserting (4.114) into (4.112) we find

$$C = i\frac{ma^2}{I_0\Omega}K.$$

Then

$$Z = \widetilde{Z} \exp(i\Omega_0 t) + i\frac{ma^2}{I_0\Omega}K. \tag{4.115}$$

The coordinates X and Y are the real and imaginary parts of Z (see (4.105)). Identifying the real and imaginary parts of (4.115) we have

$$X = \widetilde{Z}_\mathrm{r} \cos\Omega_0 t - \widetilde{Z}_\mathrm{i} \sin\Omega_0 t - K_\mathrm{i}\frac{ma^2}{I_0\Omega} \tag{4.116}$$

$$Y = \widetilde{Z}_\mathrm{r} \sin\Omega_0 t + \widetilde{Z}_\mathrm{i} \cos\Omega_0 t + K_\mathrm{r}\frac{ma^2}{I_0\Omega}. \tag{4.117}$$

We now have four arbitrary constants to evaluate. We find these from the initial values of position and velocity of the ball. The velocities are

$$\dot{X} = -\widetilde{Z}_\mathrm{r}\Omega_0 \sin\Omega_0 t - \widetilde{Z}_\mathrm{i}\Omega_0 \cos\Omega_0 t \tag{4.118}$$

$$\dot{Y} = \widetilde{Z}_\mathrm{r}\Omega_0 \cos\Omega_0 t - \widetilde{Z}_\mathrm{i}\Omega_0 \sin\Omega_0 t \tag{4.119}$$

Let us choose to release the ball from the point $Y = Y_0$ and $X = 0$ with the velocities $\dot{X} = V_0$ and $\dot{Y} = 0$ in the fixed basis. Then we have the equations

$$0 = \widetilde{Z}_\mathrm{r} - K_\mathrm{i}\frac{ma^2}{I_0\Omega} \text{ and } Y_0 = \widetilde{Z}_\mathrm{i} + K_\mathrm{r}\frac{ma^2}{I_0\Omega}$$

and

$$V_0 = -\widetilde{Z}_\mathrm{i}\Omega_0 \text{ and } 0 = \widetilde{Z}_\mathrm{r}\Omega_0$$

to solve. The solution to these four equations is

$$Z_\mathrm{r} = 0,\ Z_\mathrm{i} = -\frac{V_0}{\Omega_0} = \frac{\Omega}{\Omega_0}Y_0,\ K_\mathrm{r} = \frac{I_0\Omega}{ma^2}\left(Y_0 + \frac{\Omega}{\Omega_0}Y_0\right) \text{ and } K_\mathrm{i} = 0.$$

Then

$$X = \frac{V_0}{\Omega_0}\sin\Omega_0 t \tag{4.120}$$

and

$$Y = -\frac{V_0}{\Omega_0}\cos\Omega_0 t + \left(Y_0 + \frac{\Omega}{\Omega_0}Y_0\right). \tag{4.121}$$

Since the ball never slips, we must release the ball at rest with respect to the turntable. Then $V_0 = \Omega Y_0$, which is the (tangential) velocity of the point on the turntable at which we release the ball. Then (4.120) and (4.121) become

$$X = \frac{\Omega}{\Omega_0} Y_0 \sin \Omega_0 t \tag{4.122}$$

and

$$Y = -\frac{\Omega}{\Omega_0} Y_0 \cos \Omega_0 t + \left(Y_0 + \frac{\Omega}{\Omega_0} Y_0 \right). \tag{4.123}$$

From (4.122) and (4.123) here we find that

$$X^2 + \left(Y - Y_0 - \frac{\Omega}{\Omega_0} Y_0 \right)^2 = \left(\frac{\Omega}{\Omega_0} Y_0 \right)^2. \tag{4.124}$$

Using the moment of inertia of a sphere $I_0 = 2ma^2/\ 5$ we have $\Omega/\Omega_0 = 7/2$ and the trajectory (4.124) becomes

$$X^2 + \left(Y - Y_0 - \frac{7Y_0}{2} \right)^2 = \left(\frac{7Y_0}{2} \right)^2, \tag{4.125}$$

which is a circle with radius $7Y_0/2$ centered at the point with coordinates $X = 0$ and $Y = Y_0 + 7Y_0/2$. The velocities are

$$\dot{X} = \Omega Y_0 \cos \Omega_0 t \tag{4.126}$$

$$\dot{Y} = \Omega Y_0 \sin \Omega_0 t. \tag{4.127}$$

The speed of the ball on the circle is then a constant equal to ΩY_0. And at time $t = 0$ the velocity in the $Y-$direction is zero while the velocity in the $X-$direction is ΩY_0.

To observe this motion we must choose values of Ω and Y_0 such that Y_0 plus twice the radius of the trajectory or $8Y_0$ is less than the radius of the turntable. We require then a reasonable turntable angular velocity Ω and a low enough value of Y_0 that a full trajectory of the ball will remain on the turntable.

In Fig. 4.5 we have plotted the trajectory of the ball in the fixed basis. The initial conditions are those we chose above. We have also shown the directions of motion of the turntable and of the ball.

As we pointed out in this example, although it seems as though we did not require use of the full Euler Equations in (4.94), they are present in terms based on our understanding of the torques in terms of undetermined multipliers. In this example these torques are functions of the time, varying as the ball moves on the turntable. As we may also have noted, the Hamiltonian in (4.99) separates into a CM contribution $(1/2m)\left(P_1^2 + P_2^2\right)$ and a contribution from the motion about the CM $(1/2I_0)\left(P_{\vartheta_1}^2 + P_{\vartheta_2}^2 + P_{\vartheta_3}^2\right)$.[4]

[4]This example particularly reveals the power of the Hamiltonian formulation. Application of Newton's Laws directly in vector form place the unknown force between ball and table as central.

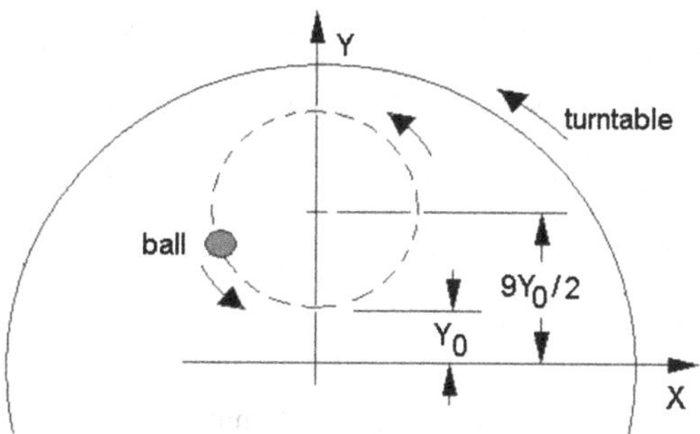

Fig. 4.5 Trajectory of a ball rolling without slipping on a rotating turntable

4.6 Summary

In this chapter we developed the principles necessary for us to treat real physical problems that are more complex than those involving classical point particles, rods or pendula. Any treatment of Analytical Mechanics would be incomplete without this study. And for the engineer this is a particularly important step.

To simplify the treatment of the rotating coordinate system in the solid body and the development of the final form of the canonical equations for the solid body, as well as to transform some critical terms to the fixed basis, we introduced the vector notation and the transformation theory of Dirac. Many readers may have already encountered this vector formulation in studies of the quantum theory and some will encounter it later. This is, at a minimum, a very important part of the mathematics of physics. And the use of projectors in the transformation of coordinates, as we have done here, introduces a simplicity that is invaluable in the treatment of solid bodies.

We presented as brief a development of the Dirac vectors and transformation theory as possible, while still providing what was needed for our work. This appears to differ from the original introduction by Dirac [see [18], Sects. 6 and 16], in which the development was directed toward state vectors in quantum theory. But the difference is only in the application. The wave function of quantum mechanics is a space and time representation of the (abstract) state vector of the quantum system.

Our derivation of the equations of motion for the rigid solid body was entirely in the context of the canonical equations of Hamilton and, therefore, based on a variational principle. In the final analysis we presented the canonical equations for the coordinates and the momenta, which we then combined to give the Euler Equations for rotational motion. There is nothing new in the physics here. But there is a naturally occurring nonlinear coupling among the angular velocities, which is the source of counterintuitive results. The practitioner should always seek understanding based on the mathematical laws, rather than intuition based on linear motion.

We have avoided many of the standard related topics in rigid body motion. For example we have neglected the Euler angles, which are helpful, although unnecessary in applications. And we have chosen not to discuss the motion of the top, which is interesting as an application. We have also avoided any explicit discussion of the observational consequences of the term $\boldsymbol{\omega}' \times \boldsymbol{r}'_{\mathrm{i}}$ which appears in (4.48) and in matrix form in (4.44). This term results in a kinetic reaction contribution to the acceleration, which is termed the Coriolis force. We have simply decided to concentrate on a presentation of the kinematics and Analytical Mechanics of solid and rigid bodies.

4.7 Exercises

4.1. In the text we obtained the matrices (operators) for infinitesimal rotations about three basis vectors as operators.

$$\mathbf{R}_1\left(\delta\vartheta'_1\right) \Longrightarrow \begin{bmatrix} 1 & 0 & 0 \\ 0 & 1 & -\delta\vartheta'_1 \\ 0 & \delta\vartheta'_1 & 1 \end{bmatrix},$$

$$\mathbf{R}_2\left(\delta\vartheta'_2\right) \Longrightarrow \begin{bmatrix} 1 & 0 & \delta\vartheta'_2 \\ 0 & 1 & 0 \\ -\delta\vartheta'_2 & 0 & 1 \end{bmatrix},$$

and

$$\mathbf{R}_3\left(\delta\vartheta'_3\right) \Longrightarrow \begin{bmatrix} 1 & -\delta\vartheta'_3 & 0 \\ \delta\vartheta'_3 & 1 & 0 \\ 0 & 0 & 1 \end{bmatrix}.$$

Sow that these commute by carrying out the calculation. Pick any two matrices you wish for this demonstration.

4.2. The kinetic energy of rotation of a rigid body has the abstract form

$$T_{\mathrm{rot}} = \frac{1}{2}\left\langle\omega\right| I \left|\omega\right\rangle$$

If we project this onto the basis fixed in the body using

$$P_{\mathrm{x}} = \left|x_\mu\right\rangle\left\langle x_\mu\right| = \mathbf{1}$$

we have

$$T_{\mathrm{rot}} = \frac{1}{2}\left\langle\omega\right.\left|x_\mu\right\rangle\left\langle x_\mu\right| I \left|x_\nu\right\rangle\left\langle x_\nu\right|\left.\omega\right\rangle.$$

Show that if the rigid body is a sphere this kinetic energy has the same form represented in the fixed system with basis $\left\{\left|X_\mu\right\rangle\right\}$.

4.3. A spinning disk of radius a is lowered onto a table. The disk is rotating about the axis parallel to the table at an initial angular velocity ω_0. The disk begins slipping and eventually starts to roll. Study the motion and determine the point at which rolling begins.

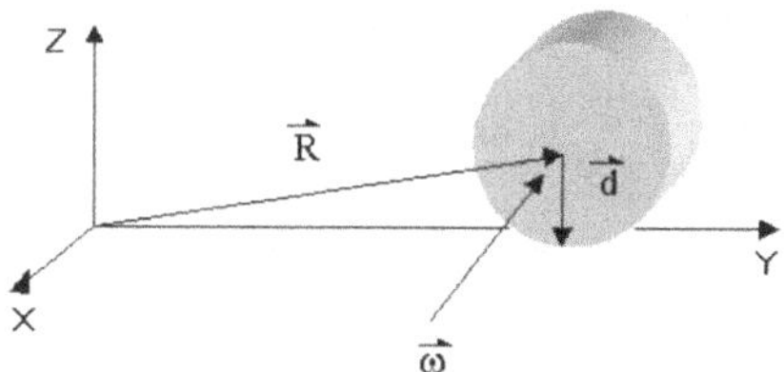

Disk released onto table with initial angular velocity $\omega = \omega_0$

Recall that the horizontal kinetic frictional force has a magnitude

$$f_{\text{friction}} = \mu_{\text{k}} mg.$$

This is the horizontal force while the disk is slipping. In the Hamiltonian formulation this appears as a Lagrange multiplier. That is

$$\dot{p}_\ell = -\frac{\partial \mathcal{H}}{\partial q_\ell} + \sum_{\text{k}=1}^{\text{n}} \lambda_{\text{k}} (t) \frac{\partial g_{\text{k}}}{\partial q_\ell}.$$

4.4. Consider the disk of mass M and radius a rolling down a hill, as shown below.

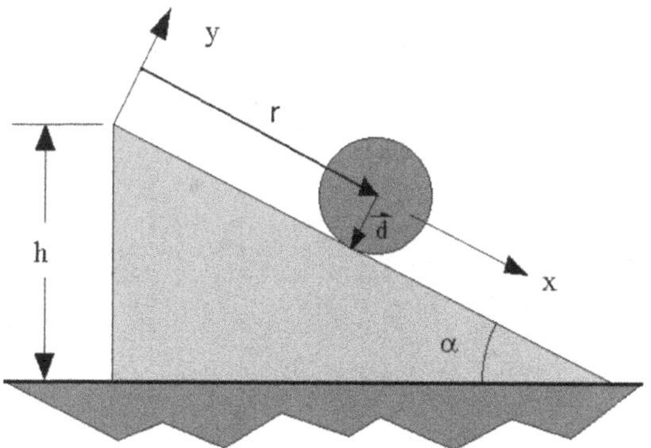

Disk rolling down hill

Obtain the description of the motion by solving the canonical equations for $r = x(t)$.

4.5. Consider a small ball of mass m is rolling inside of a cone with axis along the $z-$axis of cylindrical coordinates and defined by the angle α from the central axis. The situation is shown here

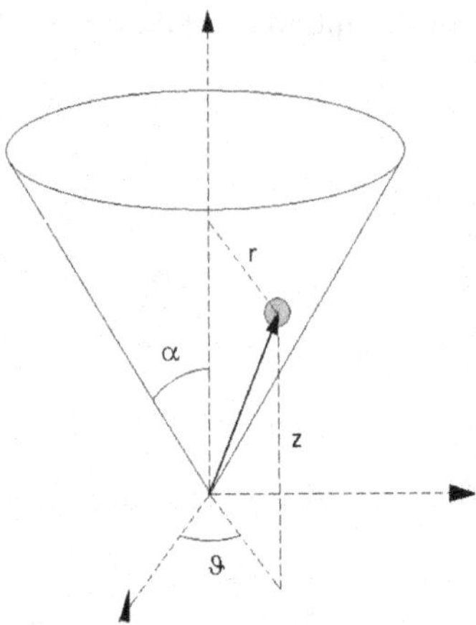

A small ball rolling without slipping inside a cone

Obtain the canonical equations for the motion of the ball. The fact that the ball moves on the inner surface of the cone introduces a constraint. And there will be three rolling constraints corresponding to the three cylindrical coordinates.

4.6. In the figure below we have drawn a small right circular cylinder of radius a and length ℓ rolling without slipping on a larger right circular cylinder with radius $A > a$ and length $\geq \ell$. The larger cylinder is fastened to the laboratory bench and does not move.

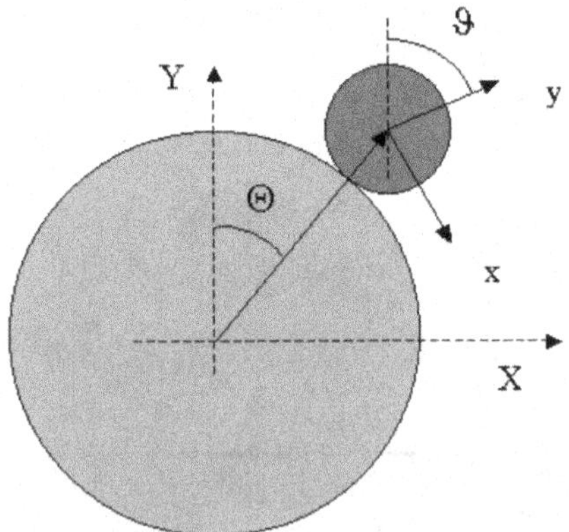

Small solid right circular cylinder of radius a and length ℓ rolling on large right circular cylinder of radius $A > a$ and length $> \ell$

The fixed coordinate system is (X, Y, Z). The unit vector $\hat{e}_Z$ is, according to the right hand system, along the axis of the larger (fixed) cylinder and oriented out of the figure. The system (x, y, z) is fixed in the smaller cylinder with the unit vector

$\hat{e}_Z$ along the axis of the small cylinder and out of the figure. The unit vectors $\hat{e}_\Theta$ and $\hat{e}_\vartheta$ are then parallel and positive in the direction of increasing Θ and ϑ.

We carefully balance the smaller cylinder along the top of the larger cylinder at $X = 0$ and then set it in motion with a very small nudge. At what point (value of Θ) does the smaller cylinder lose contact with the larger?

4.7. In the figure below we have drawn a small solid sphere of radius a rolling without slipping on a larger sphere of radius $A > a$. The larger sphere is fastened to the laboratory bench and is stationary.

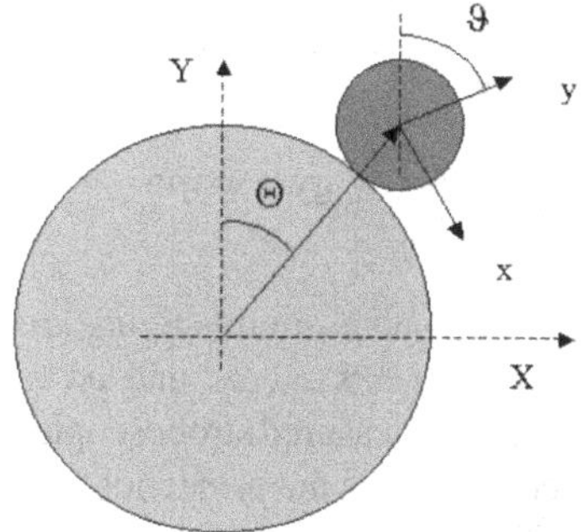

Solid sphere of radius a rolling on a fixed sphere of radius $A > a$

We carefully balance the smaller sphere at the top of the larger sphere and then set it in motion with a very small nudge. The smaller sphere then rolls down a great circle of the larger sphere, which is in the plane of the figure. We choose spherical coordinate systems for both spheres. The angles Θ and ϑ are the azimuthal angles. The polar angle $\phi = \pi/2$, which is the plane of motion. The unit vectors $\hat{e}_\Theta$ and $\hat{e}_\vartheta$ are then parallel and positive in the direction of increasing Θ and ϑ.

At what point (value of Θ) does the smaller sphere lose contact with the larger?

4.8. Here we shall seek an understanding of the rather mysterious motion of the toy gyroscope. This toy is not really a gyroscope. A real gyroscope pivots about a fixed point at the CM of the gyroscope. The toy gyroscope is actually a top, because it pivots about a point which is not the CM. In seeking an understanding we shall approach the problem by inserting the motion we have observed and asking whether or not this is consistent with Euler's Equations for rotational motion.

We have drawn a toy gyroscope in the figure below. The mass of the gyroscope flywheel is m. The torque about the pivot point is γ and is equal to the product of mg and the moment arm from the center of mass of the flywheel to the the pivot point. The flywheel of the toy gyroscope rotates around the z axis with an angular velocity ω_0 and the precession velocity is ω_p. The fixed coordinates are (X, Y, Z). The body coordinates are (x, y, z). Initially the axes Z and z are aligned.

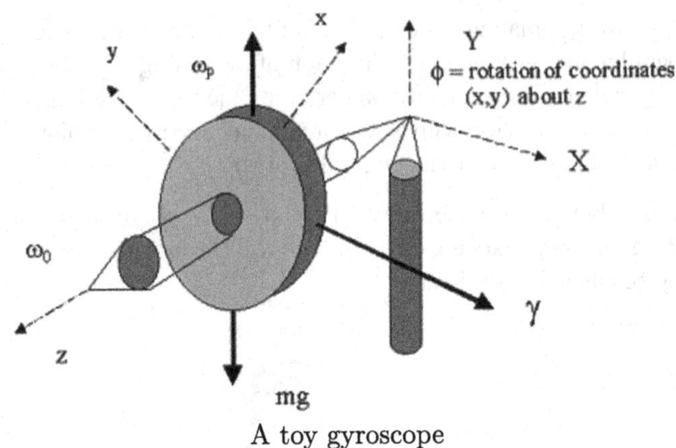

A toy gyroscope

The two angular velocities: ω_0 and ω_p in this figure are measurable. The angular velocity of the flywheel has components ω_x, ω_y, and ω_z, which we have identified as ω_0. The angular velocity ω_0 may be measured stroboscopically before the experiment. The angular velocity of precession ω_p is the projection of ω_x and ω_y on the fixed axis Y as we have shown in the drawing here.

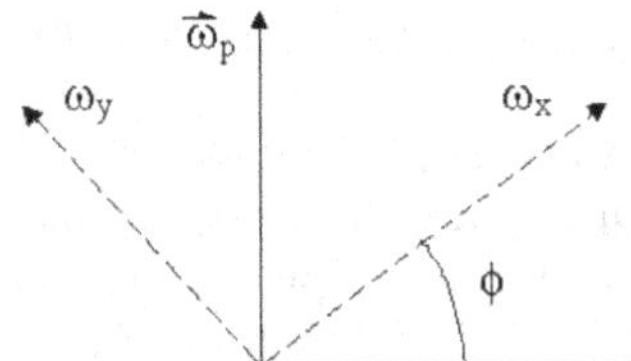

Combination of angular velocities of the flywheel into ω_p

In the experiment we observe that the toy gyroscope precesses around the vertical axis Y. This is counter-intuitive. It seems to float as it precesses rather than falling over as we may expect.

Obtain the relationship between the angular velocity of precession ω_p and the angular velocity of the flywheel about the $Z-$axis ω_0.

4.9. The rolling ball pendulum is a pendulum in which a bowling ball (of mass M and radius a) may be either suspended by a wire of length ℓ and allowed to swing through an arc, or removed from the suspending wire and allowed to roll on a circular track of radius $R = \ell$ constructed to follow the same path as that of the swinging pendulum. We have drawn the rolling ball pendulum in the figure here.

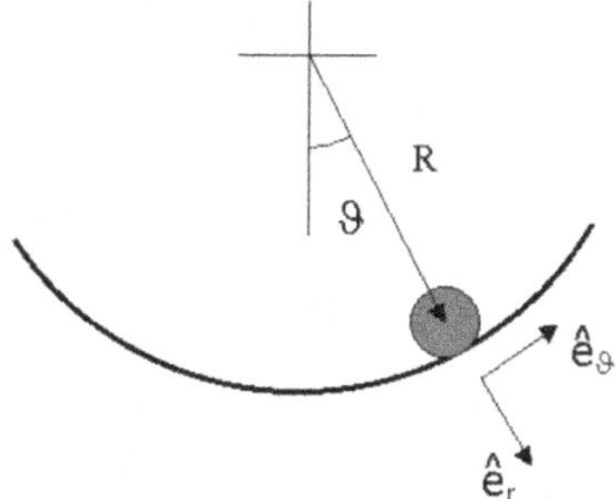

The rolling ball pendulum. We have shown here only the track with the ball rolling on it. The ball (a bowling ball) may also be suspended to result in a simple pendulum following the same path

Experimentally we can measure the period in each case and compare them. The fact of the matter is that we find two periods. The rolling period is slightly longer than that of the simple pendulum. Almost all students initially claim that the difference is the result of friction. But this is not the case if the ball rolls on the track, because rolling on a smooth surface is frictionless.

Analyze the two situations and find the difference in the two periods. show that there is no friction from rolling.

4.10. In the figure below we have a disk, which is free to rotate without friction about the central axis, and has a rigid-rod pendulum affixed to the rim of the disk.

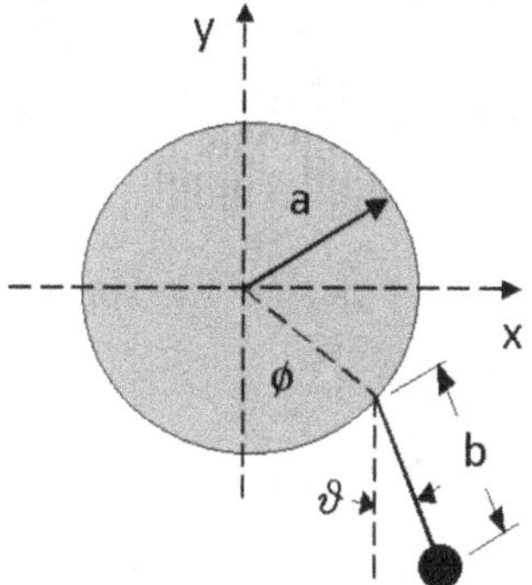

Disk with a pendulum

Obtain the angular velocities in terms of the canonical momenta. Then turn to the Euler–Lagrange Equations for the study of small vibrations. In the final analysis simplify to the case in which $a = b$.

4.11. We have devised a sort of toy we designate as a spring pendulum, which we have drawn in the figure here.

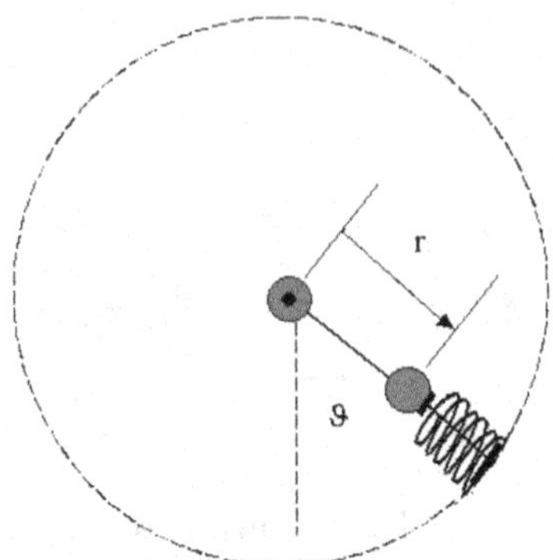

The spring pendulum

In the spring pendulum we have a rigid rod of negligible mass and of length R mounted on a bearing about which it rotates without friction. A metal ball of mass m with a hole drilled through it slides also without friction on the rod. To the ball we have affixed a spring, which is solidly mounted to the end of the rod.

Study the motion of this toy. Obtain the canonical and the Euler–Lagrange equations. Seek a numerical solution if software is available.

In the figures below we have plotted results from the numerical solution of the nonlinear canonical equations describing the spring pendulum.

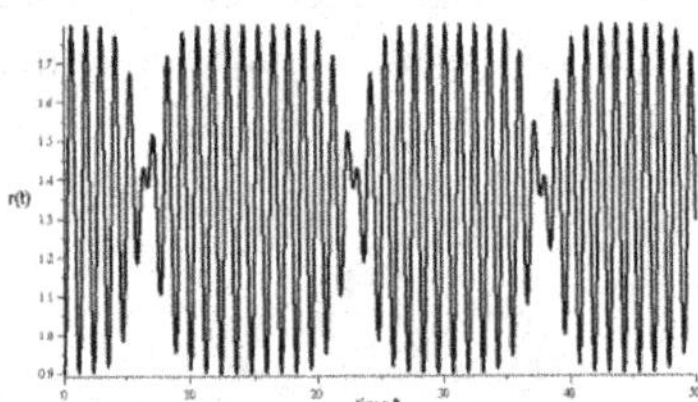

Radial motion $r(t)$ vs t for the spring pendulum

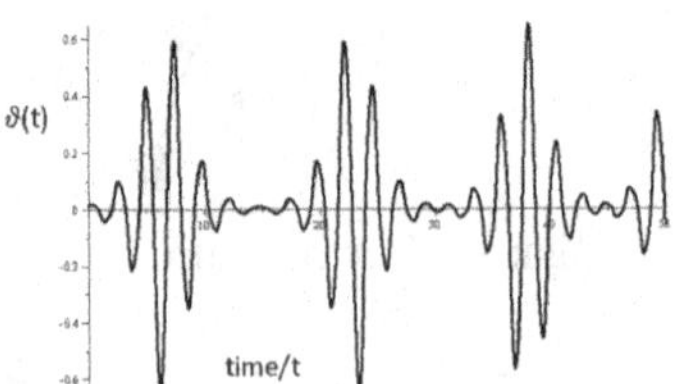

Angular motion $\vartheta(t)$ vs t for the spring pendulum

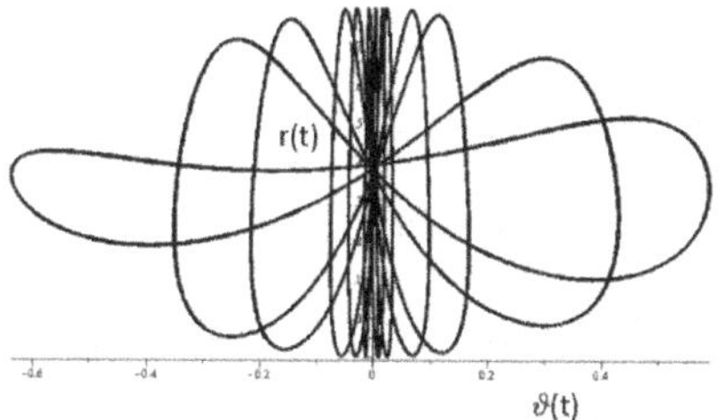

Plot of $r(t)$ vs $\vartheta(t)$ for the spring pendulum

4.12. In American football the forward pass is a critical part of the game. And many children, playing touch football in the backyard or on school playgrounds, dream of learning how to throw that ideal spiral pass, in which the football does not tumble awkwardly but seems to move like a bullet in slow motion to the receiver. Here we shall analyze the motion of the passed football. This is an interesting problem not only for those who have tried to throw that splendid spiral pass. It is also interesting for the spectator in the stands, who believes a certain motion was observed, which is not physically possible.

Below is a picture of an American football and the body coordinates.

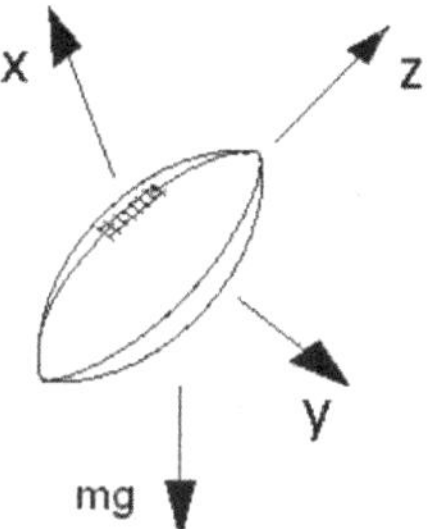

American football after passing

The moments of inertia are chosen as

$$I_x = I_y = I'$$
$$I_z = I$$

Study the motion of the spiral pass, and the possible wobble that destroys the beautiful spiral.

Chapter 5
Hamilton–Jacobi Approach

> One of the principal objects of theoretical research in any department of knowledge is to find the point of view from which the subject appears in its greatest simplicity.[1]
>
> Josiah Willard Gibbs

5.1 Introduction

In our first chapter on the history of Analytical Mechanics we discussed the two essays by Hamilton in which he introduced the Principal Function and showed that Analytical Mechanics can be expressed in terms of a pair of partial differential equations and a subsequent set of algebraic equations. This was unquestionably a mathematical *tour de force*. This is an elegant and simply beautiful mathematical formulation of Analytical Mechanics.

Jacobi was critical of Hamilton's treatment, but not of the basic idea, which he held in high esteem. He generalized Hamilton's approach to include dissipative systems. He reduced Hamilton's two partial differential equations to a single equation. And he pointed to the simplicity of ordinary differential equations, which could usually be formulated from a partial differential equation. He found simplicity without losing the generality.

In this chapter we will follow what was essentially Jacobi's path to a simplification of Hamilton's idea. Our approach will be more modern than Jacobi's original and, we hope, more easily understood.

---

[1] We introduce this chapter with the quote from Gibbs regarding simplicity, because that is the subject of this chapter. Lanczos begins his chapter with a quote from Exodus 3:5, which is appropriate, but intimidating. Exodus 3:5 begins with "Do not come near".

C.S. Helrich, *Analytical Mechanics*, Undergraduate Lecture Notes in Physics, DOI 10.1007/978-3-319-44491-8_5

We will not reduce the Hamilton–Jacobi approach to formulae that can be memorized. Rather we will keep Jacobi's generator before us in each application and work carefully with the separation of the Hamiltonian in applications.

We begin with a review of Hamilton's idea in modern notation.

5.2 Hamilton's Idea

In the first of his two essays, that we outlined in our historical discussion, Hamilton introduced the Principal Function S

$$\begin{aligned} S &= \int_0^{\mathrm{t}} L\,(q,p)\,\mathrm{d}t \\ &= \int_0^{\mathrm{t}} \left[\sum_{\mathrm{i}}^{\mathrm{n}} p_{\mathrm{i}}\dot{q}_{\mathrm{i}} - \mathcal{H}\,(q,p) \right] \mathrm{d}t, \end{aligned} \tag{5.1}$$

using modern notation q_{j} for the generalized coordinates at an arbitrary time. This is also known as the *canonical integral* [see [65], pp. 168–169]. In the second essay he obtained the two partial differential equations

$$\frac{\partial S}{\partial t} + \mathcal{H}\left(\frac{\partial S}{\partial q_1}, \ldots, q_1, \ldots \right) = 0 \tag{5.2}$$

and

$$\frac{\partial S}{\partial t} + \mathcal{H}\left(\frac{\partial S}{\partial e_1}, \ldots, e_1, \ldots \right) = 0, \tag{5.3}$$

from a variation of (5.1). Here e_{i} are the values of the generalized coordinates q_{i} at the initial time.[2] The Function $\mathcal{H}$, which we now call the Hamiltonian, appeared in the second essay as the result of a Legendre transformation of the Lagrangian.

The solution for $S\,(q_1, \ldots, e_1, \ldots, t)$ could then be used to obtain algebraic equations for the canonical momenta as

$$p_{\mathrm{i}} = \frac{\partial S}{\partial q_{\mathrm{i}}} \tag{5.4}$$

and the Hamiltonian as

$$\mathcal{H} = -\frac{\partial S}{\partial t}. \tag{5.5}$$

[2] These equations actually appeared in the first essay in a slightly different form.

With this approach Hamilton had shown that the Analytical Mechanics of Lagrange could be reduced to obtaining the solution of a pair of partial differential equations for the Principal Function S and a subsequent set of algebraic equations.

In his derivation, however, Hamilton required that $\mathcal{H} =$ constant (see (1.14) and (1.15)). So his formulation applied only to conservative systems.

5.3 Jacobi's Contribution

Jacobi had no intention of detracting from Hamilton's beautiful theory. But he was rather critical of Hamilton's failure to include nonconservative systems, for which $\mathcal{H}$ is not constant, from the outset. And he found it curious that Hamilton had claimed that the Principal Function S must satisfy two, essentially identical, partial differential equations, since it was easy to show that only one was sufficient. He also pointed out that little was gained through reduction to a single partial differential equation, since the theory of partial differential equations showed that a partial differential equation for a function such as S can often be reduced to a system of ordinary differential equations, which do not contain the original function S itself. This set of ordinary differential equations is easier to solve than the original partial differential equation.

We must, however, be careful in our evaluation of any shortcomings in Hamilton's essays. Michiyo Nakane and Craig Fraser point out that Jacobi may not have read the entirety of Hamilton's essays, concentrating on the theory and not the applications [85]. Nevertheless, buried in Jacobi's analysis and critique of Hamilton's theory lies a way around the difficulties in application of the theory and a deeper appreciation of the philosophical basis of the science of mechanics.

In Jacobi's first paper of 1837 dealing with Hamilton's theory he showed that the first of Hamilton's equations (5.2) follows from a variation of the Principal Function even for nonconservative systems [52]. In the second paper of 1837 he showed that only a single partial differential equation for the Principal Function is required [53]. The argument, which we outline in Sect. 1.10.2, Jacobi presents as a theorem, known now as *Jacobi's Theorem*. It is this theorem that frequently appears alone in textbooks as the *Hamilton–Jacobi method*.

Jacobi traditionally introduced the new methods into his lectures as soon as practicable. Jacobi's treatment of the second partial differential equation for S (5.3) we find in his lectures on dynamics given in the winter semester of 1842–43 at the University of Königsberg [[54], Lecture 19, pp. 153–157]. This is carried out with clarity and Jacobi's traditional rigor. The proof that we can drop the second Eq. (5.3) appears here and is actually not difficult. The final solution for S then depends on the initial conditions through the required arbitrary constants.

After showing that Hamilton's theory was not limited to forms of the potential (then known as the force function) that were independent of the time, and that only one partial differential equation (5.25) sufficed, Jacobi wrote (retaining the 19th century spelling) [53]

Hamilton scheint mir dadurch seine schöne Entdeckung in ein falsches Licht gesetzt zu haben, ausserdem dass sie dadurch zu gleicher Zeit unnöthig complicirt und beschränkt wird. (It seems to me that, through these limitations, Hamilton has presented his beautiful theory in a false light, which has also simultaneously made the theory unnecessarily complicated and limited.)

5.4 Time Dependence

Here we will show that the form of the Principal Function (canonical integral), when the Lagrangian depends explicitly on the time, results also in Hamilton's theory. The theory is then not limited to a constant Hamiltonian.

If we have a system with n generalized coordinates and a Lagrangian that depends explicitly on the time t we may, for mathematical convenience, choose to treat the time simply as another generalized coordinate [see [65], pp. 185–189].[3] That is we consider that the time t is the $(n+1)^{\text{st}}$ generalized coordinate $q_{\text{n+1}}$. And in the place of t we introduce a variable of integration t'. That is $\mathrm{d}t = \left(\mathrm{d}t/\mathrm{d}t'\right)\mathrm{d}t'$ and $\dot{q}_{\text{m}} = \left(\mathrm{d}q_{\text{m}}/\mathrm{d}t'\right) / \left(\mathrm{d}t/\mathrm{d}t'\right) = q'_{\text{m}}/q'_{\text{n+1}}$, where the prime indicates differentiation with respect to t'. The Principal Function (5.1) is then

$$S = \int_0^{t'} L\left(q_1,\ldots,q_{\text{n+1}},\frac{q'_1}{q'_{\text{n+1}}},\ldots,\frac{q'_{\text{n}}}{q'_{\text{n+1}}}\right) q'_{\text{n+1}}\mathrm{d}t'. \tag{5.6}$$

If we define a new Lagrangian as

$$L_{\text{n+1}} = L\left(q_1,\ldots,q_{\text{n+1}},\frac{q'_1}{q'_{\text{n+1}}},\ldots,\frac{q'_{\text{n}}}{q'_{\text{n+1}}}\right) q'_{\text{n+1}}, \tag{5.7}$$

then (5.6) becomes

$$S_{\text{n+1}} = \int_0^{t'} L_{\text{n+1}}\left(q_1,\ldots,q_{\text{n+1}},q'_1,\ldots,q'_{\text{n+1}}\right)\mathrm{d}t', \tag{5.8}$$

where the designation $S_{\text{n+1}}$ indicates dependence of the Lagrangian on $q_{\text{n+1}}$ and integration with respect to t'. The corresponding Hamiltonian, which we call $\mathcal{H}_{\text{n+1}}$, has then $n+1$ generalized coordinates. Since $L_{\text{n+1}}$ does not depend explicitly on t', we have $\mathrm{d}\mathcal{H}/\mathrm{d}t' = 0$.

Because $S_{\text{n+1}}$ in (5.8) has the same mathematical form as (5.1), a variation of (5.8) will produce the same results as a variation of (5.1) except that now the time variable is t' and there are additional canonical variables $q_{\text{n+1}}$ and $p_{\text{n+1}}$.

[3]Time is another coordinate in a more fundamental sense relativistically. Here we are considering only a mathematical identification.

The canonical momentum p_{n+1} is defined as

$$p_{n+1} = \frac{\partial L_{n+1}}{\partial q'_{n+1}}$$

From (5.7) this canonical momentum is

$$\begin{aligned} p_{n+1} &= L\left(q_1, \ldots, q_{n+1}, \frac{q'_1}{q'_{n+1}}, \ldots, \frac{q'_n}{q'_{n+1}}\right) + q'_{n+1} \sum^{n+1} \frac{\partial L}{\partial \dot{q}_i} \frac{\partial}{\partial q'_{n+1}} \left(\frac{q'_i}{q'_{n+1}}\right) \\ &= L\left(q_1, \ldots, q_n, \dot{q}_1, \ldots, \dot{q}_n, t\right) - \sum^{n} p_i \dot{q}_i, \end{aligned} \tag{5.9}$$

where we have used $q_{n+1} = t$ in the Lagrangian. Then

$$p_{n+1} = -\mathcal{H}\left(q, p, t\right), \tag{5.10}$$

the original Hamiltonian of n variables and the time.

We may obtain the Hamiltonian written for $n + 1$ coordinates and momenta using a Legendre transformation. The result is

$$\begin{aligned} \mathcal{H}_{n+1} &= \sum_{i}^{n+1} \left(\frac{\partial L_{n+1}}{\partial q'_i}\right) q'_i - L_{n+1} \\ &= \sum_{i}^{n+1} p_i q'_i - L_{n+1} \end{aligned} \tag{5.11}$$

Since the basic form of the Lagrangian is the difference of kinetic and potential energies, we are able to also obtain a general expression for L_{n+1}. The kinetic energy is quadratic in the velocities $\dot{q}_m$. Therefore the Lagrangian L_{n+1} involves products of the form $\left(q'_m/q'_{n+1}\right)^2 q'_{n+1} = q'_m \left(q'_m/q'_{n+1}\right) = q'_m \dot{q}_m$ for $m \leq n$ and products of the potential function $V(q)$, which is independent of the velocities, and q'_{n+1}. That is L_{n+1} is of the form

$$L_{n+1} = \sum_{i}^{n} a_i \dot{q}_i q'_i - V(q)\, q'_{n+1}, \tag{5.12}$$

where the coefficients a_i are possibly functions of the coordinates q_i. From (5.12) we see that

$$\begin{aligned} \sum_{i}^{n+1} \left(\frac{\partial L_{n+1}}{\partial q'_i}\right) q'_i &= \sum_{i}^{n} a_i \dot{q}_i q'_i - V(q)\, q'_{n+1} \\ &= L_{n+1}, \end{aligned} \tag{5.13}$$

which is also a result of Euler's theorem for homogeneous functions. Since $p_i = \partial L_{n+1}/\partial q_i'$ for all i, (5.13) is

$$\sum_{i}^{n+1} p_i q_i' = L_{n+1}. \tag{5.14}$$

With (5.14) Eq. (5.11) becomes

$$\mathcal{H}_{n+1} = 0. \tag{5.15}$$

Our transformation of the time t to the coordinate q_{n+1} then results in a Hamiltonian equal to zero and a canonical integral

$$S_{n+1} = \int_0^{t'} \sum_{i}^{n+1} p_i q_i' dt'. \tag{5.16}$$

If (5.15) stood alone all coordinates and momenta q_i and p_i for $1 \leq i \leq n+1$ would be constants, which is not generally possible. But (5.15) is constrained by (5.10). If we write this constraint as

$$\mathcal{K} = p_{n+1} + \mathcal{H}(q, p, t) = 0 \tag{5.17}$$

with

$$\int_0^{t'} \mathcal{K} dt' = 0, \tag{5.18}$$

our canonical integral (5.16) becomes

$$S_{n+1} = \int_0^{t'} \left(\sum_{i}^{n+1} p_i q_i' - \mathcal{K} \right) dt'. \tag{5.19}$$

This is the same reasoning as used in the introduction of Lagrange undetermined multipliers, except that we have no need for a multiplier here. The form we have obtained for the canonical integral (5.19) is the same as the original canonical integral (5.1) with $\mathcal{K} = p_{n+1} + \mathcal{H}(q, p, t)$ taking the role of the Hamiltonian.

The variable t' is a dummy of integration. If we change t' to τ, retain the prime to indicate now differentiation with respect to τ, and drop the subscript on S, (5.19) becomes

$$S = \int_0^{\tau} \left(\sum_{i}^{n+1} p_i q_i' - \mathcal{K} \right) d\tau. \tag{5.20}$$

This form of the canonical integral is termed an *extended form* of the canonical integral and is considered to be the most advanced form of the canonical integral. The variation of (5.20) results in the standard canonical equations

$$\left.\begin{aligned} \mathrm{d}q_{\mathrm{i}}/\mathrm{d}\tau &= \partial\mathcal{K}/\partial p_{\mathrm{i}} \\ \mathrm{d}p_{\mathrm{i}}/\mathrm{d}\tau &= -\partial\mathcal{K}/\partial q_{\mathrm{i}} \end{aligned}\right\} \text{ for } 1 \leq i \leq n+1 \tag{5.21}$$

But there are now $2n+2$ canonical variables. With the constraint (5.17) the canonical equations (5.21) are

$$\left.\begin{aligned} \mathrm{d}q_{\mathrm{i}}/\mathrm{d}\tau &= \partial\mathcal{H}/\partial p_{\mathrm{i}} \\ \mathrm{d}p_{\mathrm{i}}/\mathrm{d}\tau &= -\partial\mathcal{H}/\partial q_{\mathrm{i}} \end{aligned}\right\} \text{ for } 1 \leq i \leq n \tag{5.22}$$

and if we normalize the time τ such that $\mathrm{d}t/\mathrm{d}\tau = 1$,

$$\left.\begin{aligned} \mathrm{d}q_{\mathrm{n+1}}/\mathrm{d}\tau &= 1 \\ \mathrm{d}p_{\mathrm{n+1}}/\mathrm{d}\tau &= -\partial\mathcal{H}/\partial q_{\mathrm{n+1}} \end{aligned}\right\}. \tag{5.23}$$

Our treatment of the time as an independent canonical coordinate has then produced a canonical integral in which the Hamiltonian $\mathcal{K}$ does not contain the time τ. The treatment in terms of the $2n+2$ coordinates of this extended formulation is the same as that of a conservative system. This results in Hamilton's beautiful theory as well, which was Jacobi's point.

5.5 Hamilton–Jacobi Equation

In the Appendix we provide a detailed derivation of (5.2) including the possibility of time dependence in the Hamiltonian and, therefore, in the Principal Function S. Our approach there is to first find the total differential of the Principal Function $\mathrm{d}S$. In this we consider variations in the values of the coordinates and the velocities δq and $\delta\dot{q}$ at the time end points t_1 and t_2 as well as in the Lagrangian. We also consider variations in the end point times, which we designate as Δt. The result is

$$\mathrm{d}S = \left[\sum_{\mathrm{i}} (p_{\mathrm{i}}\Delta q_{\mathrm{i}}) - \mathcal{H}\Delta t\right]_{t_1}^{t_2}, \tag{5.24}$$

with

$$\Delta q_{\mathrm{i}} = \delta q_{\mathrm{i}} + \dot{q}_{\mathrm{i}}\Delta t$$

at each of the times t_1 and t_2. Because $\mathrm{d}S$ is a Pfaffian the terms Δq_{i} and Δt are differentials,

$$\mathrm{d}q_{\mathrm{i}1} = \Delta q_{\mathrm{i}}\,(t_1)$$
$$\mathrm{d}q_{\mathrm{i}2} = \Delta q_{\mathrm{i}}\,(t_2)$$

and

$$\mathrm{d}t_1 = \Delta t\,(t_1)$$
$$\mathrm{d}t_2 = \Delta t\,(t_2)\,.$$

From here the equation for the Principal Function S follows as

$$\boxed{\partial S/\partial t + \mathcal{H}\,(\partial S/\partial q, q, t) = 0,} \tag{5.25}$$

with the initial values of the coordinates identified as constants α

$$q_{\mathrm{i}1} = \alpha_{\mathrm{i}}$$

in S and the initial momenta as

$$p_{\mathrm{i}1} = -\partial S/\partial \alpha_{\mathrm{i}}.$$

The nonlinear partial differential equation (5.25) is known as the *Hamilton–Jacobi Equation*.

As we indicated above, we accept Jacobi's proof that only a single partial differential equation is necessary. This emerges logically in our derivation, although we make no pretense at Jacobi's rigor.

We then have a complete formulation of Analytical Mechanics based on a single partial differential equation, which is valid for time dependent Lagrangians and, therefore, for time dependent Hamiltonians. These were Jacobi's first points regarding Hamilton's development.

5.6 Canonical Transformation

To arrive at Jacobi's point, that the partial differential equation (5.25) can be reduced to a simpler set of ordinary differential equations, will take us down a longer path.

The first step on this path is to consider that the dynamical evolution of the system may be considered as a transformation of the coordinates and momenta. We call this a *canonical transformation* because it is defined by the requirement that the canonical equations are valid for each step. This will lead us to the concept of a *generating function*, or *generator*, for the transformation, which satisfies a partial differential equation. And our attention will become focused on obtaining the generator. For many problems of interest in physics the equation for the generator is separable and we obtain the set of ordinary differential equations we are seeking.

In our development we use Edmund T. Whittaker's notation [[125], p. 305] identifying the final coordinates and momenta with capital letters Q and P.

5.6.1 The Generating Function

The sets of coordinates and momenta before and after a canonical transformation, (q, p) and (Q, P) respectively, will satisfy the canonical equations provided variations of the canonical integrals written for the initial and final coordinates and momenta vanish. That is provided

$$\delta \int_0^t \left[\sum_i p_i \dot{q}_i - \mathcal{H}(q, p, t) \right] dt = 0 \tag{5.26}$$

and

$$\delta \int_0^t \left[\sum_i P_i \dot{Q}_i - \mathcal{H}(Q, P, t) \right] dt = 0. \tag{5.27}$$

In the second canonical integral $\mathcal{H}(Q, P, t)$ is the final Hamiltonian. The integrands in (5.26) and (5.27) are not equal. The variations will still vanish, however, if the integrands differ by the time derivative of an arbitrary function, which we shall call F_1 [cf. [80], pp. 233–236]. That is

$$\sum_i p_i \dot{q}_i - \mathcal{H}(q, p, t) = \sum_i P_i \dot{Q}_i - \mathcal{H}(Q, P, t) + \frac{dF_1}{dt}, \tag{5.28}$$

or

$$dF_1 = \sum_i p_i dq_i - \sum_i P_i dQ_i + [\mathcal{H}(Q, P, t) - \mathcal{H}(q, p, t)]\, dt \tag{5.29}$$

The function F_1 is the *generating function* or *generator* of the canonical transformation. From the form of the differential of F_1 in (5.29) we see that F_1 depends on (q, Q, t). The Pfaffian of F_1 is

$$dF_1 = \sum_i \frac{\partial F_1}{\partial q_i} dq_i + \sum_i \frac{\partial F_1}{\partial Q_i} dQ_i + \frac{\partial F_1}{\partial t} dt. \tag{5.30}$$

The generator $F_1(q, Q, t)$ then links the initial configuration of the system, identified by the coordinates q, to the configuration at a later time identified by the coordinates Q, with the requirement that the coordinates q and Q and their corresponding canonical momenta p and P satisfy the canonical equations. And we obtain equations for canonical momenta p_i, P_i, and for $\partial F_1/\partial t$ by identifying the coefficients of the differentials in (5.29) and (5.30). These equations are

$$\frac{\partial F_1}{\partial q_i} = p_i \tag{5.31}$$

$$\frac{\partial F_1}{\partial Q_i} = -P_i \tag{5.32}$$

$$\frac{\partial F_1}{\partial t} = \mathcal{H}(Q, P, t) - \mathcal{H}(q, p, t). \tag{5.33}$$

The fact that the canonical equations

$$\dot{Q}_i = \frac{\partial \mathcal{H}(Q, P, t)}{\partial P_i} \text{ and } \dot{P}_i = -\frac{\partial \mathcal{H}(Q, P, t)}{\partial Q_i} \tag{5.34}$$

are valid is a requirement of the canonical transformation and follows from (5.27).

We may use the *Legendre transformation* [see [44], pp. 47–51] to obtain generating functions of the form $F_2(q, P, t)$, $F_3(p, Q, t)$, and $F_4(p, P, t)$.[4] For example, we obtain $F_2(q, P, t)$ by transforming out Q from $F_1(q, Q, t)$ in favor of $P_i = -\partial F_1/\partial Q_i$ (see (5.32)). That is

$$\begin{aligned} F_2(q, P, t) &= F_1(q, Q, t) - \sum_i Q_i \frac{\partial F_1}{\partial Q_i} \\ &= F_1 + \sum_i Q_i(q, P) P_i, \end{aligned} \tag{5.35}$$

where the dependence on P in Q is obtained from (5.32). Then

$$\mathrm{d}F_2 = \sum_i p_i \mathrm{d}q_i + \sum_i Q_i \mathrm{d}P_i + \frac{\partial F_1}{\partial t}\mathrm{d}t \tag{5.36}$$

with

$$\frac{\partial F_2}{\partial q_i} = p_i, \ \frac{\partial F_2}{\partial P_i} = Q_i, \text{ and } \frac{\partial F_2}{\partial t} = \frac{\partial F_1}{\partial t}. \tag{5.37}$$

We leave the remaining two transformations as exercises.

Because the time t does not enter the Legendre transformations among F_1, F_2, F_3, and F_4, the forms of the partial derivatives with respect to time remain unchanged and are all equal to the difference between final and initial Hamiltonians (see (5.33)). From (5.33) and (5.31) we have a partial differential equation for $F_1(q, Q, t)$, which is

$$\frac{\partial F_1(q, Q, t)}{\partial t} = \mathcal{H}\left[Q, \frac{\partial F_1(q, Q, t)}{\partial Q}, t\right] - \mathcal{H}\left[q, \frac{\partial F_1(q, Q, t)}{\partial q}, t\right]. \tag{5.38}$$

[4]We have chosen this subscript notation to coincide with that of Goldstein.

Similarly with (5.37), we have for $F_2(q, P, t)$

$$\frac{\partial F_2(q, P, t)}{\partial t} = \mathcal{H}\left[\frac{\partial F_2(q, P, t)}{\partial P}, P, t\right] - \mathcal{H}\left[q, \frac{\partial F_2(q, P, t)}{\partial q}, t\right]. \tag{5.39}$$

Once we have the generator $F_1(q, Q, t)$ we may obtain the momenta p_j from (5.31) and the momenta P_j from (5.32), which are algebraic equations. Similarly, once we have $F_2(q, P, t)$ we may obtain the momenta p_j and the momenta P_j from (5.37), which are also algebraic equations. We are then faced with solving a single partial differential equation and then algebraic equations. And, as Jacobi pointed out, we may often be able to reduce the partial differential equation to a set of ordinary differential equations. In any case we realize that if we have found the generator we have a solution for the mechanical problem. The generator has taken center stage from the canonical equations and even from the Hamilton–Jacobi Equation itself.

From (5.31), (5.32) and (5.37) we see that we only require partial derivatives of the generator to obtain equations for the canonical momenta from F_1 or the coordinates from F_2. Therefore we may drop any additive integration constants that may appear in a solution for the generator as unimportant.

5.6.2 Conservative Systems

For *conservative systems* we drop the time dependence in the Hamiltonian $\mathcal{H}$ and the value of the final Hamiltonian $\mathcal{H}(Q, P)$ will be the same as that of $\mathcal{H}(q, p)$. Both will be equal to the total mechanical energy $\mathcal{E}$. The partial differential equation for the generator is then obtained from

$$\mathcal{H}(q, p) = \mathcal{E}. \tag{5.40}$$

If we choose the generator to be $F_{1,2}$ the canonical momenta in (5.40) are replaced by $\partial F_{1,2}/\partial q_i$ and if we choose $F_{3,4}$ the coordinates are replaced by $-\partial F_{3,4}/\partial p_i$. Specifically for the choice $F_{1,2}$, the partial differential equation for the generator is

$$\mathcal{H}\left(q, \frac{\partial F_{1,2}}{\partial q}\right) = \mathcal{E}. \tag{5.41}$$

and for the choice $F_{3,4}$ the equation for the generator is

$$\mathcal{H}\left(\frac{\partial F_{3,4}}{\partial p}, p\right) = \mathcal{E}. \tag{5.42}$$

We are completely free to choose any of these forms of the generator, although the most common choice for the initial dependence is the coordinate q. So we will choose the form of the generator here to be $F_{1,2}$.

The partial differential equations (5.41) or (5.42) are first order in the n coordinates q. So the solution will require n constants, one of which is the energy $\mathcal{E}$. We choose, in keeping with tradition, to designate these constants as the set $\alpha = (\alpha_1, \ldots, \alpha_n)$. And we shall designate $\alpha_1 = \mathcal{E}$. Our generator F_1 or F_2 may then be written generally as $F_{1,2}(q, a, t)$. Whether we identify the set of constants α as $(\mathcal{E}, Q)$, for F_1, or as $(\mathcal{E}, P)$, for F_2, is actually of no consequence. As we shall see in our examples the units on Q and P are not normally the units we may expect for coordinates and momenta. The only mathematical requirement is that Q and P satisfy the canonical equations, which is guaranteed by (5.27) and results in (5.32) and (5.37).

Let us consider that we have chosen $F_1(q, Q, t)$ to be the form of our generator. Then we identify the set α to be the final coordinates Q with one of these Qs as the energy $\mathcal{E}$. Which of the final coordinates Q (or momenta in the case of $\alpha = P$) we choose to be equal to the energy $\mathcal{E}$ is arbitrary, since the numbering of the final coordinates is arbitrary. To simplify bookkeeping we shall (arbitrarily) choose $Q_1 = \mathcal{E} = \mathcal{H}(Q, P)$ [cf. [65], p. 231]. Then, from the canonical equations for (Q, P),

$$\dot{P}_1 = -\frac{\partial \mathcal{H}(Q, P)}{\partial Q_1} = -1 \text{ for } i = 1 \tag{5.43}$$

and

$$P_1 = \tau - t, \tag{5.44}$$

where τ is a reference time. That is the final momentum P_1 is the time.[5] The remaining final momenta $P_2, \ldots, P_n$ satisfy the canonical equations

$$\dot{P}_i = -\frac{\partial \mathcal{H}(Q, P)}{\partial Q_i} = \partial Q_1 / \partial Q_i = 0 \text{ for } i > 1, \tag{5.45}$$

which means that

$$P_i = \text{constant for } i > 1. \tag{5.46}$$

The final dynamical state of our system, represented in the phase space $(Q_1, \ldots, Q_n, P_1, \ldots, P_n)$, is a point moving along the P_1 axis at a rate $\dot{P}_1 = -\partial \mathcal{H}(Q, P)/\partial Q_1 = -1$. Had we chosen P_1 to be the energy $\mathcal{E}$ we would only have reversed the roles of Q_1 and P_1 but the physics would not have been affected. The only simpler representation of our system would be obtained if we could reduce the momentum P_1 to a constant.

We now choose the one dimensional harmonic oscillator to demonstrate the application of this approach.

Example 5.6.1 The Hamiltonian for a one dimensional harmonic oscillator written in terms of the initial coordinates and momenta is

[5] This is not a trivial observation. We recall that identification of q_{n+1} as the time above resulted in p_{n+1} as (negative of) the Hamiltonian (see (5.10)).

$$\mathcal{H}(q,p) = \frac{1}{2m}p^2 + \frac{1}{2}kq^2 = \mathcal{E}. \tag{5.47}$$

We choose q as our initial variable. Then from (5.31) the partial differential equation for the generator $F_{1,2}(q,\mathcal{E})$ is

$$\frac{1}{2m}\left(\frac{\partial F_{1,2}}{\partial q}\right)^2 + \frac{1}{2}kq^2 = \mathcal{E}. \tag{5.48}$$

We can integrate this first order equation immediately to give

$$F_{1,2}(q,\mathcal{E}) = \pm\int\sqrt{2m\mathcal{E} - mkq^2}dq, \tag{5.49}$$

where we have dropped the additive integration constant as unimportant.

It is not necessary, and often not desirable, to actually perform the integration in (5.49) to obtain the generator. Occasionally the expression obtained by differentiating the generator in integral form is easier to integrate than the original. However, in the case of (5.49) the integral is not difficult. The result is

$$F_{1,2}(q,\mathcal{E}) = \pm\frac{1}{2}\sqrt{mk}\left[q\sqrt{(2\mathcal{E}/k) - q^2} + \frac{2\mathcal{E}}{k}\sin^{-1}\frac{q}{\sqrt{2\mathcal{E}/k}}\right]. \tag{5.50}$$

According to our preceding discussion, we may now choose $\mathcal{E}$ to be either the final coordinate Q or the final momentum P. We shall choose $Q = \mathcal{E}$. From the canonical equation the final canonical momentum is then

$$\dot{P} = -\frac{\partial}{\partial Q}\mathcal{H}(Q,P) = -\frac{\partial}{\partial \mathcal{E}}\mathcal{E} = -1. \tag{5.51}$$

And, as above, we introduce the constant τ to specify an initial time so that

$$P = \tau - t. \tag{5.52}$$

The generator then is F_1. From (5.32) we can obtain the relation between the final canonical momentum P and the initial and final coordinates q and $Q = \mathcal{E}$ as

$$P = -\frac{\partial}{\partial \mathcal{E}}F_1(q,\mathcal{E}) = \mp\sqrt{\frac{m}{k}}\sin^{-1}\frac{q}{\sqrt{2\mathcal{E}/k}}. \tag{5.53}$$

And from either the Hamiltonian (5.47) or (5.31) we can obtain the relationship between the initial canonical momentum and the initial and final coordinates as

$$p = \frac{\partial}{\partial q}F_1(q,\mathcal{E}) = \pm\sqrt{mk}\sqrt{(2\mathcal{E}/k) - q^2}. \tag{5.54}$$

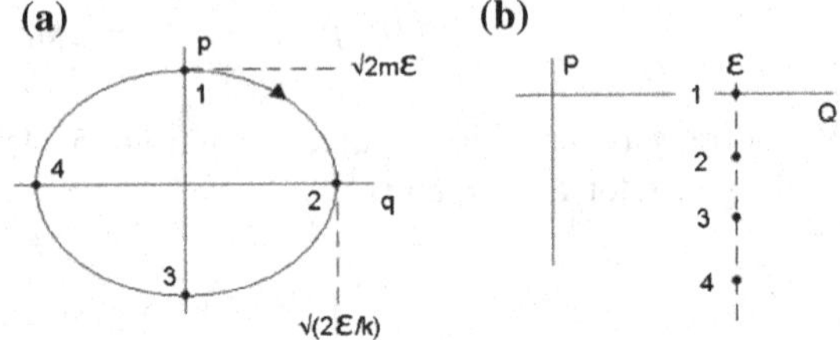

Fig. 5.1 Phase space representations of the harmonic oscillator. The initial (q, p) representation is in panel (**a**) and the final (Q, P) representation is in panel (**b**)

This completes our solution in terms of the Hamilton–Jacobi approach. Indeed the difficult aspect of the solution was completed once we had the generator (5.50), since only differentiation and algebra remained. To make our solution appear familiar we need only use (5.52) for the final momentum. If we do so the initial coordinate and momentum become

$$q = \pm\sqrt{2\mathcal{E}/k}\,\sin\sqrt{\frac{k}{m}}\,(t-\tau) \tag{5.55}$$

and

$$p = \pm\sqrt{2m\mathcal{E}}\,\cos\sqrt{\frac{k}{m}}\,(t-\tau)\,. \tag{5.56}$$

In (q, p) and (Q, P) we have two phase space representations of the harmonic oscillator. From the constant value of the Hamiltonian (5.47) we know that the trajectory of the representative point for the oscillator in the initial phase space (q, p) is an ellipse. This we plot in Fig. 5.1 panel (a). The final generalized coordinate Q is equal to the constant system energy $\mathcal{E}$. Therefore the motion of the representative point in the final phase space (Q, P) is along a straight line parallel to the P axis. And the final momentum P is $\tau - t$, which decreases from a value of zero when $t = \tau$. The final phase plot we have shown in Fig. 5.1 panel (b). The points 1, 2, 3, and 4 correspond in each panel.

5.6.3 *Time Dependent Hamiltonian*

A system for which the Hamiltonian is time dependent is not conservative. In Sect. 5.4 we considered these systems. There we chose to treat the time t as the $(n+1)^{\text{st}}$ canonical variable and arrived at an extended canonical integral (5.20) with the constraint (5.17). We shall now consider this extended theory in the context of the canonical transformation. For continuity we repeat the extended canonical integral (5.20) and the constraint (5.17) from Sect. 5.4 here.

$$S = \int_0^T \left(\sum_{\mathrm{i}}^{\mathrm{n+1}} p_{\mathrm{i}} q_{\mathrm{i}}' - \mathcal{K} \right) \mathrm{d}\tau \tag{5.57}$$

and

$$\mathcal{K} = p_{n+1} + \mathcal{H}(q, p, t) = 0. \tag{5.58}$$

For the sake of easy comparison with what we have done in the preceding section we shall again chose the generator to be $F_1(q, Q, t)$. From (5.31) with $q_{n+1} = t$ we have $p_{n+1} = \partial F_1/\partial t$ and the constraint equation (5.58) becomes

$$\frac{\partial F_1}{\partial t} + \mathcal{H}\left(q, \frac{\partial F_1}{\partial q}, t\right) = 0, \tag{5.59}$$

which is the partial differential equation for the generator F_1 in the time dependent case.

In the final form, (5.57), with the constraint $\mathcal{K} = 0$ included explicitly, is

$$S = \int_0^{\mathcal{T}} \sum_{i}^{n+1} P_i Q_i' d\tau, \tag{5.60}$$

which is the canonical integral for a system with a Hamiltonian equal to zero. For such a system the canonical equations require that the coordinates Q_i and momenta P_i are constants. Therefore our use of the extended canonical integral as the basis of our analysis permits us to define the final phase of the system to be a single point. The time now appears in the partial differential equation for the generator (5.59).

The partial differential equation for F_1 (5.59) contains $n+1$ first order partial derivatives. There are then $n+1$ arbitrary constants required for the solution of (5.59). There are also $n+1$ arbitrary constant coordinates Q_i resulting from the variation of (5.60). And the $n+1$ momenta P_i are obtained from the algebraic equations (5.32). As we pointed out in Sect. 5.6.1 we may ignore any additive constants of integration that may appear in F_1.

Designating the constant final coordinates as $Q_i = \alpha_i$, the differential equation for the generating function (5.59) may be written as

$$\frac{\partial F_1(q, \alpha, t)}{\partial t} + \mathcal{H}\left[q, \frac{\partial F_1(q, \alpha, t)}{\partial q}, t\right] = 0, \tag{5.61}$$

which has the form of the Hamilton–Jacobi Equation (5.25). The difference between Eq. (5.61) and the Hamilton–Jacobi Equation is that the generator $F_1(q, \alpha, t)$ links the initial system configuration (q, p) to a final configuration (Q, P) in which Q and P are constants, while the Principal Function $S(q, p, t)$ does not require this limitation on the final generalized coordinates or canonical momenta. The generator $F_1(q, \alpha, t)$ is then a subset of the solutions to the Hamilton–Jacobi Equation. Because they are linked by the Legendre transformation, the same may be said of the other forms of the generator as well.

This extended method includes the time independent Hamiltonian as a special case. This is also the method treated by numerous authors as simply the Hamilton–Jacobi method [cf. [34], pp. 445–449].

Example 5.6.2 We will now consider the one dimensional harmonic oscillator using the extended approach we have just introduced. We will consider only the generator $F_1(q, Q, t)$. The partial differential equation for the generator is (5.61), which becomes

$$\frac{\partial F_1}{\partial t} + \frac{1}{2m}\left(\frac{\partial F_1}{\partial q}\right)^2 + \frac{1}{2}kq^2 = 0 \tag{5.62}$$

for the one dimensional harmonic oscillator.

Equation (5.62) is *separable*. That is if we try the solution

$$F_1(q, t) = F_t(t) + F_q(q), \tag{5.63}$$

Equation (5.62) becomes

$$\frac{1}{2m}\left(\frac{dF_q}{dq}\right)^2 + \frac{1}{2}kq^2 = -\frac{dF_t}{dt}. \tag{5.64}$$

Because the left hand side of (5.64) is a function only of the independent variable q and is equal to a function only of the independent variable t on the right hand side, both sides of (5.64) must be equal to a constant, which we choose to call α_1. This is what we term a separation constant. We then have two separate ordinary differential equations

$$\frac{1}{2m}\left(\frac{dF_q}{dq}\right)^2 + \frac{1}{2}kq^2 = \alpha_1, \tag{5.65}$$

$$\frac{dF_t}{dt} = -\alpha_1. \tag{5.66}$$

Integrating (5.65) we have (see (5.50))

$$F_q(q, \alpha_1) = \pm\frac{1}{2}\sqrt{mk}\left[q\sqrt{(2\alpha_1/k) - q^2} + \frac{2\alpha_1}{k}\sin^{-1}\frac{q}{\sqrt{2\alpha_1/k}}\right], \tag{5.67}$$

and integrating (5.66),

$$F_t(t, \alpha_1) = -\alpha_1 t. \tag{5.68}$$

Then the generator is

$$F_1(q, \alpha_1, t) = \pm\frac{1}{2}\sqrt{mk}\left[q\sqrt{(2\alpha_1/k) - q^2} + \frac{2\alpha_1}{k}\sin^{-1}\frac{q}{\sqrt{2\alpha_1/k}}\right] - \alpha_1 t, \tag{5.69}$$

where we have dropped the additive integration constants.

In the extended method the final Q is a constant, which we choose to be the separation constant α_1. From (5.32) The final canonical momentum is then

$$\begin{aligned} P &= -\partial F_1/\partial\alpha_1 \\ &= \mp\sqrt{\frac{m}{k}}\sin^{-1}\frac{q}{\sqrt{2\alpha_1/k}}+t, \end{aligned} \tag{5.70}$$

which we choose to be the constant β_1. Then

$$\beta_1 = \mp\sqrt{\frac{m}{k}}\sin^{-1}\frac{q}{\sqrt{2\alpha_1/k}}+t \tag{5.71}$$

And from (5.31) the initial canonical momentum is

$$\begin{aligned} p &= \frac{\partial}{\partial q}\frac{1}{2}\sqrt{mk}\left(q\sqrt{(2\alpha_1/k)-q^2}+\frac{2\alpha_1}{k}\sin^{-1}\frac{q}{\sqrt{2\alpha_1/k}}\right) \\ &= \sqrt{m\left(2\alpha_1-kq^2\right)}. \end{aligned} \tag{5.72}$$

From (5.71) the initial coordinate is

$$q = \pm\sqrt{2\alpha_1/k}\sin\sqrt{\frac{k}{m}}\left(t-\beta_1\right). \tag{5.73}$$

Then, from (5.72), we have

$$p = \pm\sqrt{2m\alpha_1}\cos\sqrt{\frac{k}{m}}\left(t-\beta_1\right) \tag{5.74}$$

With the constants $\mathcal{E}$ and τ identified as α_1 and β_1 Eqs. (5.73) and (5.74) are identical to (5.55) and (5.56). Our results from the extended method are then identical to those obtained for a standard treatment of the linear harmonic oscillator.

In our Example 5.6.2 here the initial phase plot is identical to that in Fig. 5.1 panel (a). But the final phase plot is a single stationary point. The time has been carried by the generator and appears in the solutions for the initial coordinate and momentum in (5.73) and (5.74).

5.6.4 Separation of Variables

In this section we consider only the extended method, which includes both time dependent and time independent Hamiltonians. The equation for the generator $F_1(q,\alpha,t)$ or $F_2(q,\alpha,t)$ then takes the form

$$\frac{\partial F_{1,2}(q,\alpha,t)}{\partial t} + \mathcal{H}\left[q, \frac{\partial F_{1,2}(q,\alpha,t)}{\partial q}, t\right] = 0. \tag{5.75}$$

If the Hamiltonian is either independent of time or if the time dependence of the Hamiltonian is entirely contained in an additive function $f(t)$ so that

$$\mathcal{H}(q,p,t) = \mathcal{H}(q,p) + f(t), \tag{5.76}$$

then we can always separate the generator into the sum of a part dependent only on the time and a part dependent only on spatial variable(s) as

$$F_{1,2}(q,\alpha,t) = F_{\mathrm{t}}(\alpha,t) + F_{\mathrm{q}}(q,\alpha), \tag{5.77}$$

as we did in Example 5.6.2. That is if $H(q,p,t)$ has the form (5.76) then, using (5.77), Eq. (5.75) becomes

$$\frac{\mathrm{d}F_{\mathrm{t}}}{\mathrm{d}t} + f(t) = -\mathcal{H}\left(q, \frac{\partial F_{\mathrm{q}}(q,\alpha)}{\partial q}\right). \tag{5.78}$$

Because the left and right hand sides of (5.78) depend on distinct independent variables, and are always equal to one another, they must be equal to a constant. Then

$$\mathcal{H}\left(q, \frac{\partial F_{1,2}}{\partial q}\right) = \alpha_1 \tag{5.79}$$

and

$$\frac{\mathrm{d}F_{\mathrm{t}}(t,\alpha)}{\mathrm{d}t} + f(t) = -\alpha_1. \tag{5.80}$$

The separation of the Hamiltonian $\mathcal{H}(q,p)$ itself is, however, not required for this step. If $f(t) = 0$ we have the situation we treated in Example 5.6.2.

If the spatial portion of the generator can also be separated into a sum of functions dependent separately on each of the n coordinates q_{i}, that is if

$$F_{\mathrm{q}}(q,\alpha) = F_{\mathrm{q}_1}(q_1,\alpha) + \cdots + F_{\mathrm{q}_n}(q_{\mathrm{n}},\alpha), \tag{5.81}$$

then the Hamiltonian $\mathcal{H}(q,p)$ is separable and the generator becomes

$$F_{1,2}(q,\alpha,t) = F_{\mathrm{t}}(t,\alpha) + F_{\mathrm{q}_1}(q_1,\alpha) + \cdots + F_{\mathrm{q}_n}(q_{\mathrm{n}},\alpha). \tag{5.82}$$

If the Hamiltonian is separable and the generator has the form (5.82) then the initial canonical momenta are

$$p_{\mathrm{j}} = \frac{\partial F_{1,2}(q,\alpha,t)}{\partial q_{\mathrm{j}}} = \frac{\mathrm{d}F_{\mathrm{q}_j}(q_{\mathrm{j}},\alpha)}{\mathrm{d}q_{\mathrm{j}}} \tag{5.83}$$

(see (5.31) and (5.37)). Therefore we are able to obtain a separated solution of the form (5.81) for the spatial contribution to the generator provided (5.79) results in separate equations of the form (5.83).

The great Italian mathematician Tullio Levi-Civita developed a general method to test the separability of a Hamiltonian [referenced in [65], p. 240]. In practice, however, it is easier to assume that the spatial dependence of the generator has the form (5.81) and see if a separation results. In Example 5.6.3 we show this for a spherically symmetric potential centered on the origin.

Example 5.6.3 We consider the motion of a mass m in a spherically symmetric potential $-K/r^{\mathrm{n}}$ and choose the spherical coordinate system as the basis of our description.

We assume a separation of variables by writing the generator as

$$F_{1,2}(q,\alpha,t) = F_{\mathrm{t}}(\alpha,t) + F_{\mathrm{r}}(r,\alpha) + F_{\vartheta}(\vartheta,\alpha) + F_{\phi}(\phi,\alpha). \tag{5.84}$$

Then Eq. (5.78) for this problem is

$$\frac{\mathrm{d}F_{\mathrm{t}}}{\mathrm{d}t} = -\frac{1}{2m}\left[\left(\frac{\mathrm{d}F_{\mathrm{r}}}{\mathrm{d}r}\right)^2 + \frac{1}{r^2\sin^2\phi}\left(\frac{\mathrm{d}F_{\vartheta}}{\mathrm{d}\vartheta}\right)^2 + \frac{1}{r^2}\left(\frac{\mathrm{d}F_{\phi}}{\mathrm{d}\phi}\right)^2\right] + \frac{K}{r^{\mathrm{n}}}. \tag{5.85}$$

We must, of course, avoid the points $\phi = 0, \pi$ at which $\sin\phi = 0$. The time separation produces the constant α_1 as

$$\frac{\mathrm{d}F_{\mathrm{t}}}{\mathrm{d}t} = -\alpha_1 \tag{5.86}$$

$$\frac{1}{2m}\left[\left(\frac{\mathrm{d}F_{\mathrm{r}}}{\mathrm{d}r}\right)^2 + \frac{1}{r^2\sin^2\phi}\left(\frac{\mathrm{d}F_{\vartheta}}{\mathrm{d}\vartheta}\right)^2 + \frac{1}{r^2}\left(\frac{\mathrm{d}F_{\phi}}{\mathrm{d}\phi}\right)^2\right] - \frac{K}{r^{\mathrm{n}}} = \alpha_1. \tag{5.87}$$

From the form of the Hamiltonian in (5.87) we have

$$\begin{aligned}\left(\frac{\mathrm{d}F_{\vartheta}}{\mathrm{d}\vartheta}\right)^2 &= 2m\alpha_1 r^2\sin^2\phi + \frac{2mK}{r^{\,\mathrm{n-2}}}\sin^2\phi \\ &\quad -r^2\sin^2\phi\left(\frac{\mathrm{d}F_{\mathrm{r}}}{\mathrm{d}r}\right)^2 - \sin^2\phi\left(\frac{\mathrm{d}F_{\phi}}{\mathrm{d}\phi}\right)^2.\end{aligned} \tag{5.88}$$

Because the independent coordinate ϑ appears only on the left hand side of (5.88) the left and right hand sides of (5.88) are equal to a constant. That is

$$\frac{\mathrm{d}F_{\vartheta}}{\mathrm{d}\vartheta} = \alpha_2$$

and the Hamiltonian (5.87) becomes

$$\frac{1}{2m}\left[\left(\frac{\mathrm{d}F_{\mathrm{r}}}{\mathrm{d}r}\right)^2+\frac{\alpha_2^2}{r^2\sin^2\phi}+\frac{1}{r^2}\left(\frac{\mathrm{d}F_{\phi}}{\mathrm{d}\phi}\right)^2\right]-\frac{K}{r^{\mathrm{n}}}=\alpha_1. \tag{5.89}$$

From the form of the Hamiltonian in (5.89) we have

$$\frac{\alpha_2^2}{\sin^2\phi}+\left(\frac{\mathrm{d}F_{\phi}}{\mathrm{d}\phi}\right)^2=2m\alpha_1 r^2+\frac{2mK}{r^{\mathrm{n}\text{-}2}}-r^2\left(\frac{\mathrm{d}F_{\mathrm{r}}}{\mathrm{d}r}\right)^2. \tag{5.90}$$

Because the independent coordinate ϕ appears only on the left hand side of (5.90) the left and right hand sides of (5.90) are equal to a constant. That is

$$\frac{\alpha_2^2}{\sin^2\phi}+\left(\frac{\mathrm{d}F_{\phi}}{\mathrm{d}\phi}\right)^2=\alpha_3^2 \tag{5.91}$$

or

$$\frac{\mathrm{d}F_{\phi}}{\mathrm{d}\phi}=\pm\sqrt{\alpha_3^2-\alpha_2^2/\sin^2\phi}, \tag{5.92}$$

provided $\alpha_3^2>\alpha_2^2/\sin^2\phi$. Then the Hamiltonian in (5.90) becomes

$$\frac{\mathrm{d}F_{\mathrm{r}}}{\mathrm{d}r}=\pm\sqrt{2m\alpha_1+2mK/r^{\mathrm{n}}-\alpha_3^2/r^2}. \tag{5.93}$$

And we have separated the partial differential equation for the generator into the set of four ordinary differential equations

$$\frac{\mathrm{d}F_{\mathrm{t}}}{\mathrm{d}t}=-\alpha_1 \tag{5.94}$$

$$\frac{\mathrm{d}F_{\vartheta}}{\mathrm{d}\vartheta}=\alpha_2, \tag{5.95}$$

$$\frac{\mathrm{d}F_{\phi}}{\mathrm{d}\phi}=\pm\sqrt{\alpha_3^2-\alpha_2^2/\sin^2\phi}, \tag{5.96}$$

$$\frac{\mathrm{d}F_{\mathrm{r}}}{\mathrm{d}r}=\pm\sqrt{2m\alpha_1+2mK/r^{\mathrm{n}}-\alpha_3^2/r^2}. \tag{5.97}$$

The generator (5.84) is then

$$\begin{aligned}F_{1,2}(r,\vartheta,\phi,t,\alpha)=&-\alpha_1 t+\alpha_2\vartheta\pm\int \mathrm{d}\phi\sqrt{\alpha_3^2-\alpha_2^2/\sin^2\phi}\\&\pm\int \mathrm{d}r\sqrt{2m\alpha_1+2mK/r^{\mathrm{n}}-\alpha_3^2/r^2},\end{aligned} \tag{5.98}$$

dropping all additive constants of integration as unimportant.

The constants $\alpha_1, \ldots, \alpha_3$ are *separation constants* that we obtained in the separation process. We can evaluate these from the initial conditions. For example, we see in (5.79) that α_1 is the value of the Hamiltonian, which is the system energy $\mathcal{E}$, and α_2 is the value of the constant azimuthal momentum. And, if we choose to work with the generator as F_1, the α_{i} are the final coordinates Q. From (5.32) the final momenta are $P_{\mathrm{i}} = -\partial F_1/\partial Q_{\mathrm{i}}$, which are also constants. The final phase is then a point in the space (Q, P).

If we identify the constants $\alpha_{1,2,3}$ as $\mathcal{E}$, Θ, and Φ the generator (5.98) is

$$F_1(r, \vartheta, \phi, t, \mathcal{E}, \Theta, \Phi) = -\mathcal{E}t + \Theta\vartheta \pm \int \mathrm{d}\phi \sqrt{\Phi^2 - \Theta^2/\sin^2\phi} \\ \pm \int \mathrm{d}r \sqrt{2m\mathcal{E} + 2mK/r^{\mathrm{n}} - \Phi^2/r^2}. \tag{5.99}$$

Using (5.32) the final (constant) canonical momenta obtained from (5.99) are

$$P_{\mathcal{E}}(r, t, R, \Phi) = -\frac{\partial F_1}{\partial \mathcal{E}} \\ = t \mp m \int \mathrm{d}r \frac{1}{\sqrt{2m\mathcal{E} + 2mK/r^{\mathrm{n}} - \Phi^2/r^2}} \tag{5.100}$$

$$P_{\Theta}(\vartheta, \phi, t, \Theta, \Phi) = -\frac{\partial F_1}{\partial \Theta} \\ = -\vartheta \pm \Theta \int \mathrm{d}\phi \frac{1}{\sqrt{\Phi^2 - \Theta^2/\sin^2\phi}} \tag{5.101}$$

$$P_{\Phi}(r, \phi, t, R, \Theta, \Phi) = -\frac{\partial F_1}{\partial \Phi} = \mp\frac{1}{2}\int \mathrm{d}\phi \frac{1}{\sqrt{\Phi^2 - \Theta^2/\sin^2\phi}} \\ \pm \int \frac{\mathrm{d}r}{r^2} \frac{1}{\sqrt{2mR + 2mK/r^{\mathrm{n}} - \Phi^2/r^2}}, \tag{5.102}$$

which are algebraic equations. In (5.100) we have used the subscript $\mathcal{E}$ to indicate that this momentum is conjugate to the energy $\mathcal{E}$ and note that $P_{\mathcal{E}} \propto t$. Using (5.31) the initial canonical momenta obtained from (5.99) are

$$p_{\mathrm{r}}(r, t, R, \Phi) = \frac{\partial F_1}{\partial r} = \pm\sqrt{2mR + 2mK/r^{\mathrm{n}} - \Phi^2/r^2} \tag{5.103}$$

$$p_{\vartheta}(\Theta) = \frac{\partial F_1}{\partial \vartheta} = \Theta \tag{5.104}$$

$$p_{\phi}(\phi, t, \Theta, \Phi) = \frac{\partial F_1}{\partial \phi} = \pm\sqrt{\Phi^2 - \Theta^2/\sin^2\phi}, \tag{5.105}$$

which are also algebraic equations.

Our present solution requires, however, that we avoid the polar angles $\phi = 0, \pi$ as we pointed out when we first wrote the Hamiltonian above. We must now require that $\Phi^2 > \Theta^2/\sin^2\phi$ as well. Otherwise the final momenta P_Θ and P_Φ are no longer real. The mathematical solution would be to accept a limited range of ϕ. But such a limitation would require a confining force and would destroy the symmetry of the potential that permitted the separation of the Hamiltonian. We are left then with the only alternative that $\phi =$ constant, which symmetry requires to be $\phi = \pi/2$. The angular momentum $p_\phi = \mathrm{d}F_\phi/\mathrm{d}\phi$ is then zero and motion takes place in a plane. From (5.91) $\alpha_3^2 = \alpha_2^2$ or $\Phi = \Theta^2$. This is a well-known result for motion in spherical potentials and is normally introduced initially in more elementary texts.

We have then only the time and ϑ separations. Our set of ordinary differential equations is then

$$\frac{\mathrm{d}F_\mathrm{t}}{\mathrm{d}t} = -\alpha_1 \tag{5.106}$$

$$\frac{\mathrm{d}F_\vartheta}{\mathrm{d}\vartheta} = \alpha_2, \tag{5.107}$$

$$\frac{\mathrm{d}F_\mathrm{r}}{\mathrm{d}r} = \pm\sqrt{2m\alpha_1 + 2mK/r^\mathrm{n} - \alpha_2^2/r^2}. \tag{5.108}$$

and the generator (5.84) is

$$F_1(r, \vartheta, t, \mathcal{E}, \Theta) = -\mathcal{E}t + \Theta\vartheta \pm \int \mathrm{d}r\sqrt{2m\mathcal{E} + 2mK/r^\mathrm{n} - \Theta^2/r^2}, \tag{5.109}$$

dropping additive constants.

Using (5.32) the final (constant) canonical momenta obtained from (5.109) are

$$P_\mathcal{E}(r, t, \mathcal{E}, \Theta) = -\frac{\partial F_1}{\partial \mathcal{E}} = t \mp m\int \mathrm{d}r \frac{1}{\sqrt{2m\mathcal{E} + 2mK/r^\mathrm{n} - \Theta^2/r^2}} \tag{5.110}$$

$$P_\Theta(\vartheta) = -\frac{\partial F_1}{\partial \Theta} = -\vartheta \pm \Theta\int \frac{\mathrm{d}r}{r^2}\frac{1}{\sqrt{2m\mathcal{E} + 2mK/r^\mathrm{n} - \Theta^2/r^2}} \tag{5.111}$$

which are algebraic equations. Using (5.31) the initial canonical momenta obtained from (5.99) are

$$p_\mathrm{r}(r, t, R, \Phi) = \frac{\partial F_1}{\partial r} = \pm\sqrt{2m\mathcal{E} + 2mK/r^\mathrm{n} - \Theta^2/r^2} \tag{5.112}$$

$$p_\vartheta(\Theta) = \frac{\partial F_1}{\partial \vartheta} = \Theta \tag{5.113}$$

Our example here of motion in the symmetrical central potential has been general enough to include most of the aspects of the Hamilton–Jacobi approach to the solution of mechanical systems. We have based our solution on the extended method, which is applicable to either conservative systems or those with time dependent Hamiltonians. We have chosen an example for which the Hamiltonian is separable in order to obtain the ordinary differential equations to which Jacobi pointed. And our solution has resulted in final coordinates and momenta that are constants so that the final phase representation (Q, P) is a point.

We have left three integrals unevaluated. These are in the generator (5.109) and the final canonical momenta $P_{\mathcal{E}}$ and P_Θ in (5.110) and (5.111). These are tabulated [35]. If we choose to evaluate the integral in (5.109) we will have a closed algebraic expression for the generator from which we can obtain the final canonical momenta $P_{\mathcal{E}}$ and P_Θ. We may also choose to leave the generator in the form (5.109) and evaluate the two integrals in (5.110) and (5.111). This option we may prefer in the event that the partial derivatives of the integrated form of the generator become unduly complicated. The choice is ours to make.

5.7 Action and Angle Variables

The analysis of a complex dynamical system is considerably simplified if some of the coordinates are periodic. In the general situation, however, a coordinate q_i and the corresponding conjugate momentum p_i are coupled nonlinearly to other coordinates and momenta. Even when the orbit in the phase space (q_i, p_i) is closed, the coordinate q_i is not generally a periodic function of time. That is the velocity $\dot{q}_i$ at which the representative point begins a subsequent passage on the orbit may not be the same as that for the preceding passage. But it may be multiply periodic and representable in a Fourier series with multiple frequencies [cf. [65], p. 247; [34], p. 466]. Therefore a primary goal in the study of complex systems is to discover the basic periodicities of the motion even when these may be obscured.

The French astronomer Charles–Eugène Delaunay (1816–1872) recognized that the periodicities of a separable system could be discovered in a combination of the Hamiltonian and the generator for the canonical transformation. His work on lunar motion in 1846 brought the power of Hamiltonian methods to the attention of physicists for application to astrophysical problems [[94]; [65], pp. 245–254; [34], pp. 528–530]. There Delaunay wrote the Hamiltonian as the sum of an unperturbed term and a perturbation, which he developed in a Fourier series [[76], p. 503]. Delauney's work inspired what became the approach based on the action and angle variables.

The action variables are constants that take the place of the final coordinates or momenta (Q or P) in the Hamilton–Jacobi approach. They are a more natural choice for the set of constants α than are values determined from combinations of the initial values of coordinates, since they are based on the phase space orbits of the separable coordinates. The angle variables carry the time.

To introduce the action and angle variables we begin with the extended approach and assume that the Hamiltonian is separable. Then we can write the generator in the form (5.82). The separation produces canonical momenta $p_{\mathrm{i}} = p_{\mathrm{i}}(q_{\mathrm{i}}, \alpha)\,(= \mathrm{d}F/\mathrm{d}q_{\mathrm{i}})$ as functions of the corresponding conjugate coordinate q_{i} and separation constants α. There is generally a range of values that can be taken on by the coordinates q_{i}, which produce ranges of the canonical momenta p_{i} (see, e.g., Fig. 5.1 panel (a)). If we integrate each of the canonical momenta p_{i} over a cycle of the conjugate coordinate q_{i} we obtain quantities that are functions only of the constants α. These quantities are the *action variables* J_{i} defined as

$$\begin{aligned} J_{\mathrm{i}}(\alpha) &\equiv \oint p_{\mathrm{i}} \mathrm{d}q_{\mathrm{i}} \\ &= \oint \frac{\mathrm{d}F_{\mathrm{i}}(q_{\mathrm{i}}, \alpha)}{\mathrm{d}q_{\mathrm{i}}} \mathrm{d}q_{\mathrm{i}}, \end{aligned} \tag{5.114}$$

where we have used (5.83).

We consider, for example, the spherically symmetric problem (Example 5.6.3) with motion in the equatorial plane, $\phi = \text{constant} = \pi/2$. The separable Hamiltonian in that situation is

$$\mathcal{H}(q, p) = \frac{1}{2m}\left(p_{\mathrm{r}}^2 + \frac{p_{\vartheta}^2}{r^2}\right) - \frac{K}{r^{\mathrm{n}}} = \alpha_1 .$$

Noting that $\alpha_1 = \mathcal{E}$, we have the action variables

$$\begin{aligned} J_{\mathrm{r}}(\mathcal{E}, \Phi) &= \oint \frac{\mathrm{d}F_{\mathrm{r}}(r, \mathcal{E}, \Phi)}{\mathrm{d}r} \mathrm{d}r \\ &= \pm \oint \mathrm{d}r \sqrt{2m\mathcal{E} + 2mK/r^{\mathrm{n}} - \Theta^2/r^2}, \end{aligned} \tag{5.115}$$

and

$$J_{\vartheta}(\Theta) = \oint \mathrm{d}\vartheta \frac{\mathrm{d}F_{\vartheta}(\vartheta, \Theta)}{\mathrm{d}\vartheta} = 2\pi\Theta. \tag{5.116}$$

We have plotted the phase space orbit

$$p_{\mathrm{r}}(r, \mathcal{E}, \Theta) = \pm\sqrt{2m\mathcal{E} + 2m\frac{K}{r} - \frac{\Theta^2}{r^2}} \tag{5.117}$$

for the for the mass m moving under a central force with $n = 1$ in Fig. 5.2 using data for an electron moving under the Coulomb force from a nucleus without the

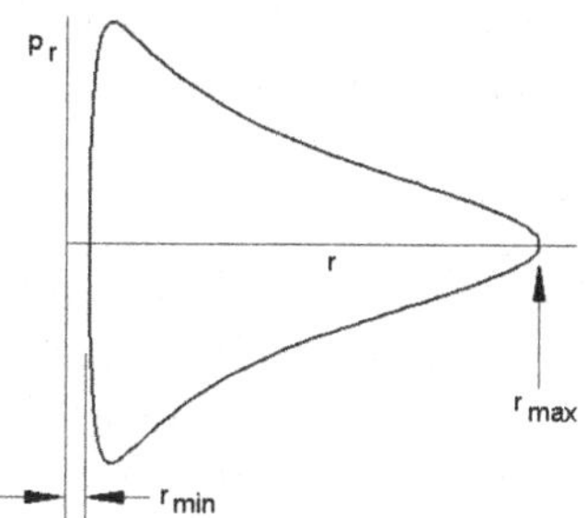

Fig. 5.2 Phase space orbit p_r as a function of r for the motion of a charged particle under a central Coulomb force

Bohr restriction to a circular orbit and relaxing the quantum requirement on angular momentum. The two parameters $(\mathcal{E}, \Theta)$ specify the orbit. The integration in (5.115) is from r_{min} back to r_{min} around the phase space orbit. We integrate first from $r_{min} \to r_{max}$ with $dr > 0$ and then from $r_{max} \to r_{min}$ with $dr < 0$. In (5.116) the integral is simply from $\vartheta = 0 \to 2\pi$.

In principle these $J_i(\alpha)$ can be inverted algebraically to obtain the constants α_i as functions of J. That is

$$\alpha_i = \alpha_i(J), \text{ which includes } \mathcal{E} = \mathcal{E}(J). \tag{5.118}$$

For the separable Hamiltonian the generator of the canonical transformation (5.82) then becomes

$$F_{1,2}(q, J) = \sum_j^n F_j(q_j, J) \tag{5.119}$$

which is a function of the initial coordinates q and the (constant) action variables J.

Our choice of q as the initial coordinate specifies the generator as either F_1 or F_2. The second (constant) coordinate may then be either Q or P. That is, if we wish to do so, we may identify the action variables J as either coordinates or canonical momenta. For example, Lanczos identifies the action variables as Q [[65], p. 248] while Goldstein prefers to identify them as (angular) momenta based on their dimensions $[pq]$ [[34], p. 461]. But neither identification of J as P or Q is necessarily compelling or intuitive. At this point the action variables are simply constant (canonical) variables associated with the final state of the system. And their dimension $[pq]$ is now called simply *action*.[6]

From (5.32) and (5.37) we have the relations between the canonical variables at the initial and final points for our system as

$$P_i = -\frac{\partial F_1(q, Q)}{\partial Q_i} \text{ and } Q_i = \frac{\partial F_2(q, P)}{\partial P_i} \tag{5.120}$$

[6]For example the units of Planck's *quantum of action* $h = 6.6260755 \times 10^{-34}(\text{kgms}^{-1})\text{m}$ are normally expressed as the product of energy and time $\text{Js} = (\text{kg m}^2\text{s}^{-2})\text{s}$.

In this sense we may then define new final point canonical variables *conjugate to the action variables* as

$$\omega_{\mathrm{i}}(q, J) \equiv \frac{\partial F(q, J)}{\partial J_{\mathrm{i}}}. \tag{5.121}$$

These new canonical variables are not constants of the motion, since they carry the relation between the initial coordinates q and the action variables. In defining $\omega_{\mathrm{i}}(q, J)$ as a positive partial derivative of the generator $F(q, J)$ we have chosen the action variables J to be the *analogs* of the final canonical momenta P and the new variables ω to be the *analogs* of the final coordinates Q.

The ω_{i} were first called *angle variables* in 1916 by the German astronomer Karl Schwarzschild [106]. This was Schwarzschild's last publication. He had been director of the Astrophysical Observatory at Potsdam, but volunteered for service in the German Army in 1914 and died in May of 1916 of a disease he contracted in Russia. He was 42 and a lieutenant of artillery [[6], vol 20, p. 3; [112]].

Since the variables (ω, J) are conjugate canonical variables, they must satisfy canonical equations obtained from the Hamiltonian $\mathcal{H}(J)$. From (5.118) we realize that the final Hamiltonian $\mathcal{H}(J) = \mathcal{E}(J)$ is functionally dependent only on the action variables J and independent of the angle variables ω. Therefore the canonical equations for the action variables are

$$\dot{J}_{\mathrm{i}} = -\frac{\partial \mathcal{H}(J)}{\partial \omega_{\mathrm{i}}} = 0, \tag{5.122}$$

and the canonical equations for the angle variables are

$$\dot{\omega}_{\mathrm{i}} = \frac{\partial \mathcal{H}(J)}{\partial J_{\mathrm{i}}} = \nu_{\mathrm{i}}(J), \tag{5.123}$$

where ν_{i} is a constant dependent on only the action variables. The signs in the Eqs. (5.122) and (5.123) are consistent with the sign choice in the definition of $\omega_{\mathrm{i}}(q, J)$ in (5.121). Since we already know that the action variables are constants of the motion, (5.122) contains no additional information. Integrating (5.123) we have

$$\omega_{\mathrm{i}} = \nu_{\mathrm{i}}(J)\, t + \delta_{\mathrm{i}}, \tag{5.124}$$

where the δ_{i} is an integration constant.

In the action and angle variables we have a complete solution to the dynamical problem for a system with a separable Hamiltonian. But our objective was not to simply obtain a solution. Our objective was to obtain an analysis of the periodicities in the motion of a complex system. Since the angle variables are the analogs of the coordinates, we may attain our goal by considering the changes in the angle variables during the natural motion of the system. From (5.121) the differential change of the angle variable ω_{i} with differential changes in the generalized coordinates q_{j} is

$$\mathrm{d}\omega_{\mathrm{i}} = \sum_{\mathrm{j}} \frac{\partial^2 F(q, J)}{\partial q_{\mathrm{j}} \partial J_{\mathrm{i}}} \mathrm{d}q_{\mathrm{j}} = \frac{\partial}{\partial J_{\mathrm{i}}} \sum_{\mathrm{j}} \frac{\partial F(q, J)}{\partial q_{\mathrm{j}}} \mathrm{d}q_{\mathrm{j}}, \tag{5.125}$$

since the order of partial differentiation is makes no difference. Using (5.83) we have

$$\mathrm{d}\omega_{\mathrm{i}} = \frac{\partial}{\partial J_{\mathrm{i}}} \sum_{\mathrm{j}} p_{\mathrm{j}} \mathrm{d}q_{\mathrm{j}}. \tag{5.126}$$

If we integrate (5.126) over all phase space orbits $(q_{\mathrm{j}}, p_{\mathrm{j}})$, we have the change in the angle variable ω_{i} resulting from a passage of the system through a time encompassing all of the characteristic motions defined in terms of the original generalized coordinates q_{j}.

$$\Delta\omega_{\mathrm{i}} = \frac{\partial}{\partial J_{\mathrm{i}}} \sum_{\mathrm{j}} \oint p_{\mathrm{j}} \mathrm{d}q_{\mathrm{j}} = \frac{\partial}{\partial J_{\mathrm{i}}} \sum_{\mathrm{j}} J_{\mathrm{j}} = \sum_{\mathrm{j}} \delta_{\mathrm{ij}} = 1. \tag{5.127}$$

The angle variable ω_{i} then changes by a value of unity as the generalized coordinate q_{i} undergoes a cycle in its motion. And the change $\Delta\omega_{\mathrm{i}}$ is unaffected by cycles in the other generalized coordinates $q_{\mathrm{j}\neq\mathrm{i}}$.

From (5.124) we see that if τ_{i} is the characteristic time taken for the phase space orbit $(q_{\mathrm{i}}, p_{\mathrm{i}})$ then the change $\Delta\omega_{\mathrm{i}} = 1$ in a system cycle is

$$\Delta\omega_{\mathrm{i}} = 1 = \nu_{\mathrm{i}}\tau_{\mathrm{i}} = \frac{\partial \mathcal{H}(J)}{\partial J_{\mathrm{i}}} \tau_{\mathrm{i}}, \tag{5.128}$$

using (5.123). We may then obtain the characteristic periods of motion for a complex system once we have the Hamiltonian as a function of the action variables.

This result, expressed in (5.128), is the basis for the analysis of the system periodicities that we sought. We have here a window that allows us to pick out the periodicities of the individual generalized coordinates independently of the complexities resulting from the relations among the generalized coordinates and momenta without solving for the dependence of each coordinate on the time. Mathematically all we require is a solution in terms of the action variables $J_{\mathrm{i}}(\alpha)$, which are integrals involving only the separation constants $p_{\mathrm{i}}(q_{\mathrm{i}}, \alpha) = \mathrm{d}F_{1,2}/\mathrm{d}q_{\mathrm{i}}$. None of this is, however, necessarily easy for the general problem, since we require an algebraic solution for the separation constants α in terms of the action variables J (see (5.118)). So our windows on the motion may be difficult to open.

Although the action variables may twist our intuition, the simplicity of the contour integrals, as opposed to the indefinite integrals required to obtain the components of the generator, make this approach attractive for complicated problems.

As our first example for which we shall obtain an action-angle variable solution, we turn again to the one dimensional harmonic oscillator.

Example 5.7.1 The Hamiltonian for the one-dimensional oscillator of mass m is (5.47), which we rewrite here with the natural frequency $\omega_0 = \sqrt{k/m}$

$$\mathcal{H}(q, p) = \frac{p^2}{2m} + \frac{1}{2} m\omega_0^2 q^2 = \mathcal{E}. \tag{5.129}$$

This is the algebraic equation for the phase space orbit, which we have shown in Fig. 5.1 panel (a). The limits of the coordinate $q_{\text{max/min}} = \pm\sqrt{2\mathcal{E}/m\omega_0^2}$ occur when the momentum vanishes and the limits of the momentum $p_{\text{max/min}} = \pm\sqrt{2m\mathcal{E}}$ occur when the coordinate vanishes. The action variable is

$$\begin{aligned} J_{\text{q}}(\mathcal{E}) &= \oint p\text{d}q \\ &= \oint \text{d}q\sqrt{2m\mathcal{E} - m^2\omega_0^2 q^2}. \end{aligned} \tag{5.130}$$

The integral over a cycle of the coordinate q is from q_{min} to q_{max} with $\text{d}q > 0$ and returning from q_{max} to q_{min} with $\text{d}q < 0$. Then

$$J_{\text{q}}(\mathcal{E}) = 2m\omega_0 \int_{-\sqrt{2\text{E}/\text{m}\omega_0^2}}^{\sqrt{2\text{E}/\text{m}\omega_0^2}} \text{d}q\sqrt{2\mathcal{E}/m\omega_0^2 - q^2}. \tag{5.131}$$

Integrating (5.131) we have

$$\begin{aligned} J_{\text{q}}(\mathcal{E}) &= 2m\omega_0 \left[\frac{q}{2}\sqrt{2\mathcal{E}/m\omega_0^2 - q^2}\right. \\ &\quad \left. + \frac{\mathcal{E}}{m\omega_0^2}\sin^{-1}\left(\frac{q}{\sqrt{2\mathcal{E}/m\omega_0^2}}\right)\right]_{-\sqrt{2\mathcal{E}/\text{m}\omega_0^2}}^{\sqrt{2\mathcal{E}/\text{m}\omega_0^2}} \\ &= \frac{2\pi\mathcal{E}}{\omega_0}. \end{aligned} \tag{5.132}$$

In terms of the action J_{q} the Hamiltonian is

$$\mathcal{H}\left(J_{\text{q}}\right) = \mathcal{E} = \frac{\omega_0}{2\pi} J_{\text{q}}. \tag{5.133}$$

We find the rate of change of the angle variable from (5.133) as

$$\dot{\omega}_{\text{q}} = \frac{\partial \mathcal{H}\left(J_{\text{q}}\right)}{\partial J_{\text{q}}} = \nu_{\text{q}} = \frac{\omega_0}{2\pi}, \tag{5.134}$$

and the angle variable is

$$\omega_{\mathrm{q}} = \frac{\omega_0}{2\pi} t + \delta_{\mathrm{q}}. \tag{5.135}$$

Using (5.128) the period of motion is

$$\tau_{\mathrm{q}} = 1/\left[\partial \mathcal{H}\left(J_{\mathrm{q}}\right)/\partial J_{\mathrm{q}}\right] = \frac{2\pi}{\omega_0}. \tag{5.136}$$

We have then the action and angle variables for the one dimensional harmonic oscillator. There is a single constant of integration δ_{q} in the result for the angle variable. If we identify δ_{q} as $-\omega_0 \tau_{\mathrm{q}}/2\pi$, then

$$2\pi\omega_{\mathrm{q}} = \omega_0\left(t - \tau_{\mathrm{q}}\right), \tag{5.137}$$

and the action and angle solution has revealed the periodicity of the motion.

From our solution here in terms of the action variable $J_{\mathrm{q}} = 2\pi\mathcal{E}/\omega_0$, the angle variable $\omega_{\mathrm{q}} = \omega_0\left(t - \tau\right)/2\pi$, and the separation constant $\alpha_1 = \mathcal{E} = p^2/2m + m\omega_0^2 q^2/2$ we have a complete description of the motion of the harmonic oscillator in terms of the periodicity of the coordinate and the phase space orbit. If we wish we may convert this to the standard elementary solution for q in terms of t. But the information contained in that solution is less than that contained in the action and angle variables and the phase space plot.

As a slightly more complicated example we return to the spherical potential, written again in three dimensions. We shall now choose $n = 1$, which is appropriate for the gravitational force or the electrical (Coulomb) force. The orbit resulting from the gravitational force is the planetary orbit studied by Johannes Kepler and this problem is normally referred to as the Kepler problem.

Example 5.7.2 The Hamiltonian for the Kepler potential $-K/r$ is

$$\mathcal{H}(q,p) = \frac{1}{2m}\left(p_{\mathrm{r}}^2 + \frac{p_\vartheta^2}{r^2\sin^2\phi} + \frac{p_\phi^2}{r^2}\right) - \frac{K}{r} \tag{5.138}$$

Using the extended method with the time as a coordinate in the generator, and assuming a separation of the generator as

$$F_{1,2}(q,\alpha,t) = F_{\mathrm{t}}(\alpha,t) + F_{\mathrm{r}}(r,\alpha) + F_\vartheta(\vartheta,\alpha) + F_\phi(\phi,\alpha),$$

the equation for the generator (5.75) for the Kepler problem is

$$0 = \frac{\mathrm{d}F_{\mathrm{t}}}{\mathrm{d}t} + \frac{1}{2m}\left[\left(\frac{\mathrm{d}F_{\mathrm{r}}}{\mathrm{d}r}\right)^2\right.$$

$$+\frac{1}{r^2\sin^2\phi}\left(\frac{\mathrm{d}F_\vartheta}{\mathrm{d}\vartheta}\right)^2+\frac{1}{r^2}\left(\frac{\mathrm{d}F_\phi}{\mathrm{d}\phi}\right)^2\Bigg]-\frac{K}{r}.$$

The first separation results in $\mathrm{d}F_\mathrm{t}/\mathrm{d}t=-\alpha_1$ and

$$\alpha_1=\frac{1}{2m}\Bigg[\left(\frac{\mathrm{d}F_\mathrm{r}}{\mathrm{d}r}\right)^2 +\frac{1}{r^2\sin^2\phi}\left(\frac{\mathrm{d}F_\vartheta}{\mathrm{d}\vartheta}\right)^2+\frac{1}{r^2}\left(\frac{\mathrm{d}F_\phi}{\mathrm{d}\phi}\right)^2\Bigg]-\frac{K}{r}. \tag{5.139}$$

The first separation constant α_1, in this conservative case, is equal to the mechanical energy $\mathcal{E}<0$.

Our second separation results in α_2, which is the angular momentum

$$\frac{\mathrm{d}F_\vartheta}{\mathrm{d}\vartheta}=\text{constant}=\alpha_2. \tag{5.140}$$

We separate the ϕ dependence in (5.139) as

$$\frac{\alpha_2^2}{\sin^2\phi}+\left(\frac{\mathrm{d}F_\phi}{\mathrm{d}\phi}\right)^2=2m\mathcal{E}r^2+2mKr-\left(\frac{\mathrm{d}F_\mathrm{r}}{\mathrm{d}r}\right)^2r^2. \tag{5.141}$$

From (5.141) we identify the third separation constant as α_3^2 with

$$\left(\frac{\mathrm{d}F_\phi}{\mathrm{d}\phi}\right)^2+\frac{\alpha_2^2}{\sin^2\phi}=\text{constant}=\alpha_3^2. \tag{5.142}$$

We are then left with

$$\left(\frac{\mathrm{d}F_\mathrm{r}}{\mathrm{d}r}\right)^2=2m\mathcal{E}+\frac{2mK}{r}-\frac{\alpha_3^2}{r^2}. \tag{5.143}$$

From our previous study of the spherically symmetrical potential we recall that there is actually no ϕ dependence and that $F_\phi=0$. We shall here, however, hold F_ϕ as unequal to zero for as long as we can. We will gain some insight by doing so.

The total angular momentum $\boldsymbol{L}_\mathrm{p}$ of the particle moving in the spherical potential is

$$\begin{aligned}\boldsymbol{L}_\mathrm{p}&=m\boldsymbol{r}\times\frac{\mathrm{d}}{\mathrm{d}t}\boldsymbol{r}=mr\hat{e}_\mathrm{r}\times\left[\dot{r}\hat{e}_\mathrm{r}+(r\sin\phi)\,\dot{\vartheta}\hat{e}_\vartheta+r\dot{\phi}\hat{e}_\phi\right]\\&=-\frac{p_\vartheta}{\sin\phi}\hat{e}_\phi+p_\phi\hat{e}_\vartheta.\end{aligned} \tag{5.144}$$

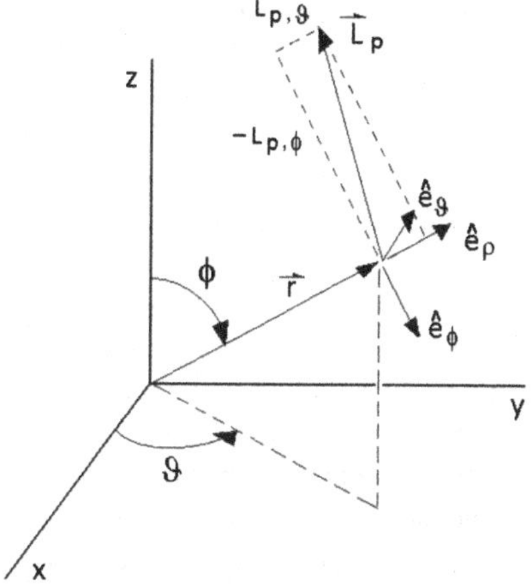

Fig. 5.3 Components of the angular momentum $\boldsymbol{L}_{\mathrm{p}}$. The position vector $\boldsymbol{r}$ locates the point particle

In Fig. 5.3 we have drawn the components of the total angular momentum $\boldsymbol{L}_{\mathrm{p}}$ and its components in the directions $\hat{e}_{\vartheta}$ and $\hat{e}_{\phi}$. From (5.144) the square of the angular momentum $\boldsymbol{L}_{\mathrm{p}}$ is

$$\begin{aligned} L_{\mathrm{p}}^2 &= p_{\phi}^2 + \frac{p_{\vartheta}^2}{\sin^2\phi} \\ &= p_{\phi}^2 + \frac{\alpha_2^2}{\sin^2\phi} = \alpha_3^2. \end{aligned} \tag{5.145}$$

The magnitude of the total angular momentum is then equal to the constant α_3.

From the separated Eqs. (5.140), (5.142), and (5.143) we have the set of action variables

$$J_{\vartheta}(\alpha) = \oint p_{\vartheta}\mathrm{d}\vartheta = \oint \alpha_2 \mathrm{d}\vartheta \tag{5.146}$$

$$J_{\phi}(\alpha) = \oint p_{\phi}\mathrm{d}\phi = \oint \mathrm{d}\phi\sqrt{\alpha_3^2 - \frac{\alpha_2^2}{\sin^2\phi}} \tag{5.147}$$

$$J_{\mathrm{r}}(\alpha) = \oint p_{\mathrm{r}}\mathrm{d}r = \oint \mathrm{d}r\frac{1}{r}\sqrt{2m\mathcal{E}r^2 + 2mKr - \alpha_3^2}. \tag{5.148}$$

To evaluate these integrals we must specify the ranges of integration of each of the coordinates (r, θ, ϕ) for a complete orbit of the point particle. We can find these ranges from the algebraic equations (5.140), (5.142), and (5.143) as the values of the coordinates for which the corresponding momenta vanish.

There are no limits on the azimuthal angle ϑ because from (5.140) the momentum p_{ϑ} is a constant of the motion. In the integral (5.146) the angle ϑ then takes on all values in the range $0 \rightarrow 2\pi$.

From (5.142) we find the limits on the polar angle ϕ of the orbit as the values of ϕ, for which the canonical momentum p_ϕ vanishes. With $p_\phi = 0$ (5.142) becomes

$$\sin^2 \phi = \alpha_2^2/\alpha_3^2. \tag{5.149}$$

Therefore, if the canonical momentum $p_\phi \neq 0$, the angle ϕ must have the values $\phi = \pm \sin^{-1} \alpha_2/\alpha_3$ at the limiting points and, with $p_\phi \neq 0$, Eq. (5.145) requires that $|\alpha_3| > |\alpha_2|$. If we designate $\phi_0 = \sin^{-1}(\alpha_2/\alpha_3)$, the limiting values of the polar angle may be $\pm\phi_0$ or may be separated by π, since $\sin(\phi_0 + \pi) = -\sin\phi_0$. But the polar angle is only defined in the range $0 \leq \phi \leq \pi$. There is, therefore, no angle $-\phi_0$, nor is there an angle $\phi_0 = \phi_0 + \pi$, accessible to the particle. The canonical momentum p_ϕ must then be zero and the motion of the point particle is in a plane. It is simplest to choose this plane to be that for which $\phi = \pi/2$. Then $\alpha_3 = \alpha_2$ and we have $J_\phi(\alpha) = 0$ for the general Kepler problem.

Since $\mathcal{E} < 0$ for bound motion of the particle, setting the canonical momentum $p_r = \mathrm{d}F_r/\mathrm{d}r = 0$ in (5.143) we have

$$m|\mathcal{E}|r^2 - mKr + \frac{1}{2}\alpha_2^2 = 0.$$

The maximum and minimum distances of the orbit from the origin are then

$$r_{\text{max/min}} = \frac{1}{2m|\mathcal{E}|}\left(mK \pm \sqrt{m^2K^2 - 2m|\mathcal{E}|\alpha_2^2}\right) \tag{5.150}$$

For real values of these radial distances we require that the discriminant $m^2K^2 - 2m|\mathcal{E}|\alpha_2^2 > 0$. In a complete orbit then the radial distance varies from $r_{\text{min}} \to r_{\text{max}}$ and then returns from $r_{\text{max}} \to r_{\text{min}}$. In Fig. 5.4 we have plotted the *effective potential*

$$V_{\text{eff}} = \frac{\alpha_3^2}{r^2} - \frac{K}{r} \tag{5.151}$$

as a function of the radial distance r and labeled the distances r_{max} and r_{min}.

With $\phi = \pi/2$, the action variable J_r becomes

$$J_r(\alpha) = \oint \frac{\mathrm{d}r}{r}\sqrt{2m\mathcal{E}r^2 + 2mKr - \alpha_2^2} \tag{5.152}$$

In the part of the cycle from r_{min} to r_{max} the differential $\mathrm{d}r > 0$ and from r_{max} to r_{min} the differential $\mathrm{d}r < 0$. The integral in (5.152) is then

$$J_r(\alpha) = 2\int_{r_{\text{min}}}^{r_{\text{max}}} \frac{\mathrm{d}r}{r}\sqrt{2m\mathcal{E}r^2 + 2mKr - \alpha_2^2} \tag{5.153}$$

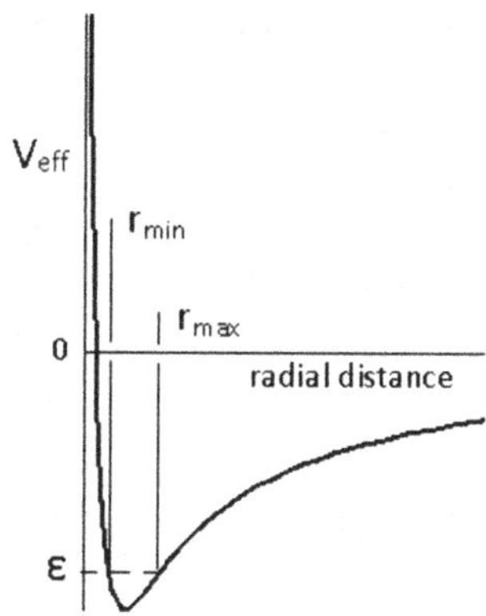

Fig. 5.4 Effective potential for the Kepler problem. The long range attractive potential is from the central force. The repulsive contribution is from the angular velocity

We can perform the indefinite integration in (5.153), which is

$$F_{\mathrm{r}}(r,\alpha) = \int \frac{\mathrm{d}r}{r}\sqrt{2m\mathcal{E}r^2+2mKr-\alpha_2^2}, \tag{5.154}$$

without difficulty, since the integrals required for final evaluation are tabulated [[35], GR 2.267, 2.266 and 2.261]. The result is the radial contribution to the generator

$$\begin{aligned} F_{\mathrm{r}}(r,\alpha) = & \sqrt{2m\mathcal{E}r^2+2mKr-\alpha_2^2} \\ & -\frac{mK}{\sqrt{-2m\mathcal{E}}}\sin^{-1}\frac{4m\mathcal{E}r+2mK}{\sqrt{(2mK)^2+8\alpha_2^2 m\mathcal{E}}} \\ & -a_2\sin^{-1}\frac{2mKr-2\alpha_2^2}{r\sqrt{(2mK)^2+8\alpha_2^2 m\mathcal{E}}}, \end{aligned} \tag{5.155}$$

where $\mathcal{E} < 0$. From (5.153) and (5.155) the action $J_{\mathrm{r}}(\alpha)$ is then

$$J_{\mathrm{r}}(\alpha) = 2\ F_{\mathrm{r}}(r,\alpha)]_{\mathrm{r_{min}}}^{\mathrm{r_{max}}}\,. \tag{5.156}$$

From (5.150) we see that the first term in (5.155) vanishes at $r = r_{\mathrm{max/min}}$. The argument of the first $\sin^{-1}$ in (5.155) at the limits $r = r_{\mathrm{max/min}}$ is

$$\frac{-4m\,|\mathcal{E}|\,r_{\mathrm{max/min}}+2mK}{\sqrt{(2mK)^2-8\alpha_2^2 m\,|\mathcal{E}|}} = \mp 1,$$

and the argument of the second $\sin^{-1}$ in (5.155) at the limits $r = r_{\mathrm{max/min}}$ is

$$\frac{2mKr_{\mathrm{max/min}}-2\alpha_2^2}{r_{\mathrm{max/min}}\sqrt{(2mK)^2-8\alpha_2^2 m\,|\mathcal{E}|}} = \pm 1.$$

Since $\sin^{-1}(\pm 1) = \pm\pi/2$, the action $J_{\mathrm{r}}(\alpha)$ in (5.156) is

$$J_{\mathrm{r}}(\alpha) = \pi K \sqrt{\frac{2m}{-\mathcal{E}}} - 2\pi\alpha_2. \tag{5.157}$$

The generator $F_{\vartheta}(\vartheta, \alpha)$ is

$$F_{\vartheta}(\vartheta, \alpha) = \int \alpha_2 \, \mathrm{d}\vartheta = \alpha_2 \vartheta. \tag{5.158}$$

The integration to obtain the action $J_{\vartheta}(\alpha)$ in (5.146) yields

$$J_{\vartheta}(\alpha) = 2\pi\alpha_2. \tag{5.159}$$

Using (5.159) the action $J_{\mathrm{r}}(\alpha)$ in (5.157) becomes

$$J_{\mathrm{r}}(\alpha) = \pi K \sqrt{\frac{2m}{-\mathcal{E}}} - J_{\vartheta}(\alpha). \tag{5.160}$$

Then the final Hamiltonian as a function of the action variables is

$$\mathcal{H}(J) = \mathcal{E}(J) = -\pi^2 K^2 \frac{2m}{(J_{\mathrm{r}} + J_{\vartheta})^2}. \tag{5.161}$$

From (5.123) we have

$$\begin{aligned} \dot{\omega}_{\mathrm{r}} &= \frac{\partial \mathcal{H}(J)}{\partial J_{\mathrm{r}}} = 2\pi^2 K^2 \frac{2m}{(J_{\mathrm{r}} + J_{\vartheta})^3} \\ &= -2\frac{\mathcal{E}(J)}{(J_{\mathrm{r}} + J_{\vartheta})} \end{aligned} \tag{5.162}$$

and

$$\begin{aligned} \dot{\omega}_{\vartheta} &= \frac{\partial \mathcal{H}(J)}{\partial J_{\vartheta}} = 2\pi^2 K^2 \frac{2m}{(J_{\mathrm{r}} + J_{\vartheta})^3} \\ &= -2\frac{\mathcal{E}(J)}{(J_{\mathrm{r}} + J_{\vartheta})} \end{aligned} \tag{5.163}$$

We then have a solution to the Kepler problem in action-angle variables. That the frequencies of the angle variables $\nu_{\mathrm{r}} = \dot{\omega}_{\mathrm{r}}$ and $\nu_{\vartheta} = \dot{\omega}_{\vartheta}$ are identical tells us that the orbit is a closed figure. The solution in action-angle variables, however, does not give us the geometrical picture of the orbit. That requires the relationship between p_{r} and ϑ. To obtain the orbit we must turn to the generator.

Until this point in our treatment we have not required identification of the generator as either F_1 or F_2. The choice depends on whether we identify the set of constants α

as final coordinates Q or final canonical momenta P. For the sake of variety, since the choice is completely arbitrary, we now choose the generator to be F_2 and the constants α to be the final canonical momenta P. Specifically the energy $\mathcal{E}$ becomes a final canonical momentum, which we shall designate as $P_{\mathcal{E}}$. And the constant α_2 is the final canonical momentum P_{ϑ}. We then have $F_{\mathrm{r}}(r, P_{\mathcal{E}}, P_{\vartheta})$ in Eq. (5.155) and $F_{\vartheta}(\vartheta, P_{\mathcal{E}}, P_{\vartheta})$ in (5.158).

Differentiating (5.155) involves extensive algebra. But if we hold the integration in the component of the generator F_{r} then, with (5.154) and (5.158), we have

$$\begin{aligned} F_2(q, P) &= F_{\mathrm{t}}(P_{\mathcal{E}}, t) + F_{\mathrm{r}}(r, P_{\mathcal{E}}, P_{\vartheta}) + F_{\vartheta}(\vartheta, P_{\vartheta}) \\ &= -P_{\mathcal{E}} t \pm \int \frac{\mathrm{d}r}{r} \sqrt{2mP_{\mathcal{E}} r^2 + 2mKr - P_{\vartheta}^2} + P_{\vartheta}\vartheta. \end{aligned} \tag{5.164}$$

From the generator $F_2(q, P)$ the final coordinates are $Q_{\mathrm{i}} = \partial F_2(q, P)/\partial P_{\mathrm{i}}$ (see (5.120)). Our expressions for the final coordinates, which we shall designate as $Q_{\mathcal{E}}$ and Q_{ϑ}, we then avoid extensive algebra. For $Q_{\mathcal{E}}$

$$Q_{\mathcal{E}} = \frac{\partial F_2(q, P)}{\partial P_{\mathcal{E}}} = -t \pm \int \mathrm{d}r \frac{mr}{\sqrt{2mP_{\mathcal{E}} r^2 + 2mKr - P_{\vartheta}^2}} \tag{5.165}$$

The integral in (5.165) is tabulated [[35], GR 2.261 and 2.264]. The result is

$$\begin{aligned} Q_{\mathcal{E}} &= -t \pm \frac{1}{2|\mathcal{E}|} \sqrt{2mKr - 2m|\mathcal{E}| r^2 - P_{\vartheta}^2} \\ &\mp \sqrt{\frac{mK^2}{8|\mathcal{E}|^3}} \sin^{-1}\left(\frac{2mK - 4m|\mathcal{E}| r}{\sqrt{(2mK)^2 + 8m|\mathcal{E}|\left(P_{\vartheta}^2\right)}}\right). \end{aligned} \tag{5.166}$$

We may then identify $Q_{\mathcal{E}} \propto -(t - \tau)$, provided we choose the positive sign on the square root in (5.164).

We can find the orbit from Q_{ϑ}, which, using the positive sign on the square root in (5.164), is

$$\begin{aligned} Q_{\vartheta} &= \frac{\partial F_2(q, P)}{\partial P_{\vartheta}} \\ &= -P_{\vartheta} \int \frac{\mathrm{d}r}{r} \frac{1}{\sqrt{2mP_1 r^2 + 2mKr - P_{\vartheta}^2}} + \vartheta. \end{aligned} \tag{5.167}$$

The final coordinate Q_{ϑ} is a constant of the motion, which we shall designate as ϑ_0 for simplicity. Then (5.167) becomes [[35], GR 2.266]

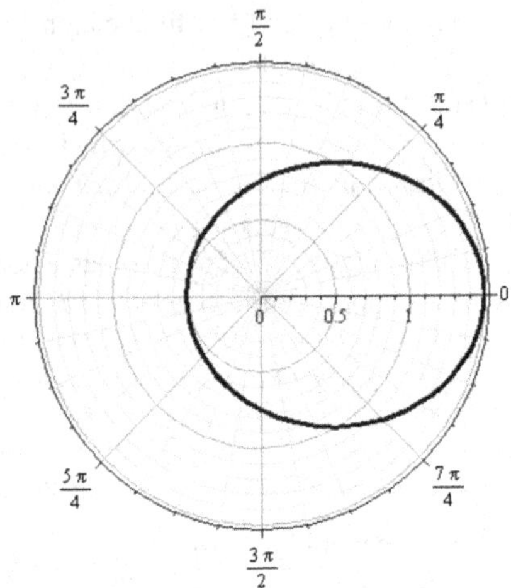

Fig. 5.5 The elliptical orbit for a mass moving in a plane under a central potential $V(r) = -K/r$

$$\begin{aligned}\vartheta - \vartheta_0 &= P_\vartheta \int \frac{\mathrm{d}r}{r} \frac{1}{\sqrt{2mP_1r^2 + 2mKr - P_\vartheta^2}} \\ &= \sin^{-1}\left(\frac{2mKr - 2P_\vartheta^2}{r\sqrt{(2mK)^2 - 8m\,|\mathcal{E}|\,P_\vartheta^2}}\right). \end{aligned} \tag{5.168}$$

Solving (5.168) for r we have

$$r = \frac{B}{A - \sin(\vartheta - \vartheta_0)}, \tag{5.169}$$

where

$$A = 2mK/\sqrt{(2mK)^2 - 8m\,|\mathcal{E}|\,P_\vartheta^2},$$

and

$$B = 2P_\vartheta^2/\sqrt{(2mK)^2 - 8m\,|\mathcal{E}|\,P_\vartheta^2}.$$

Equation (5.169) is the polar form of an ellipse centered on one of the foci. The angle ϑ_0 determines the orientation of the axes. In Fig. 5.5 we have plotted the orbit with $\vartheta_0 = -\pi/2$, $A = 2$, and $B = 1.5$. This is an ellipse with an eccentricity of 0.5.

Our result in Fig. 5.5 is the familiar orbit for the Kepler problem, which is an ellipse with one focus on the center of force. Our path to this result through the action and angle variables has been considerably less involved than the standard path through the canonical equations. It is simply a more elegant procedure. We have also found a straightforward argument for elimination of the action J_ϕ based on the extent of the range taken on by the polar coordinate ϕ. Our previous argument that

ϕ =constant= $\pi/2$ was based on symmetry of the force potential. The symmetry argument is no less acceptable. But there is a geometric finality to our treatment here.

5.8 Poisson Brackets

Some quantities are conserved during canonical transformation. In this section we will develop a general method to identify these conserved quantities working in terms of the extended method as the most general approach. We will also use Greek subscripts and the Einstein summation convention in order to keep our equations as simple as possible. We always sum on repeated Greek indices.

The time rate of change of a general function of the generalized coordinates, canonical momenta, and the time $f(q, p, t)$ is

$$\begin{aligned} \frac{\mathrm{d}f}{\mathrm{d}t} &= \frac{\partial f}{\partial q_\mu}\frac{\mathrm{d}q_\mu}{\mathrm{d}t} + \frac{\partial f}{\partial p_\upsilon}\frac{\mathrm{d}p_\upsilon}{\mathrm{d}t} + \frac{\partial f}{\partial t} \\ &= \frac{\partial f}{\partial q_\mu}\frac{\partial \mathcal{H}}{\partial p_\mu} - \frac{\partial f}{\partial p_\upsilon}\frac{\partial \mathcal{H}}{\partial q_\upsilon} + \frac{\partial f}{\partial t}. \end{aligned} \tag{5.170}$$

If we define the *Poisson Bracket*[7] (*PB*) of two general functions $f(q, p, t)$ and $g(q, p, t)$ as [[83], pp. 290–291, [125] Whittaker, pp. 299-301, [34], pp. 391–425][8]

$$\boxed{(f, g) = \left(\partial f/\partial q_\mu\right)\left(\partial g/\partial p_\mu\right) - \left(\partial f/\partial p_\mu\right)\left(\partial g/\partial q_\mu\right)} \tag{5.171}$$

Equation (5.170) can be written as

$$\frac{df}{dt} = (f, \mathcal{H}) + \frac{\partial f}{\partial t}. \tag{5.172}$$

The PB then provides a convenient form for the time rate of change of a dynamically varying function. As we shall now see, because the PB is preserved in canonical transformation, (5.172) is more than just a convenient formulation.

In the extended method the canonical integrals at the beginning and end of the transformation are (see (5.57), (5.58) and (5.60))

[7]Siméon-Denis Poisson (1781–1840), was a French mathematician, geometer and physicist. He was anti-aristocratic based on experiences of his father, and brought up in the stern era of the First Republic. When he was made a baron after the restoration of the Empire, never used the title nor accepted the diploma.

[8]The notation for the PB is not uniform. Goldstein and Dirac use a square bracket $[f, g]$. Whittaker and Morse and Feshbach use the round bracket (f, g) as we have and Morse and Feshbach change the order of the partial derivatives. The bracket $\{f, g\}$ is also sometimes used to designate PB.

$$S = \int_0^t p_\nu \dot{q}_\nu \mathrm{d}t \tag{5.173}$$

and

$$S = \int_0^t P_\sigma \dot{Q}_\sigma \mathrm{d}t. \tag{5.174}$$

As we realize, the integrands in (5.173) and (5.174) may differ by the differential of a function without affecting the variation, which produces the canonical equations. That is

$$\mathrm{d}F = p_\nu \mathrm{d}q_\nu - P_\sigma \mathrm{d}Q_\sigma. \tag{5.175}$$

Because $\mathrm{d}F$ is an exact differential $p_\nu = \partial F/\partial q_\nu$ and $P_\sigma = \partial F/\partial Q_\sigma$, and because the order of partial differentiation is immaterial,

$$\frac{\partial p_\nu}{\partial Q_\sigma} = -\frac{\partial P_\sigma}{\partial q_\nu} \tag{5.176}$$

Written separately in terms of initial and final canonical variables, the PBs of two functions f and g are

$$(f, g)_{\mathrm{q,p}} = \frac{\partial f}{\partial q_\mu}\frac{\partial g}{\partial p_\mu} - \frac{\partial f}{\partial p_\mu}\frac{\partial g}{\partial q_\mu}$$

and

$$(f, g)_{\mathrm{Q,P}} = \frac{\partial f}{\partial Q_\mu}\frac{\partial g}{\partial P_\mu} - \frac{\partial f}{\partial P_\mu}\frac{\partial g}{\partial Q_\mu}.$$

We now wish to show that provided (q, p) and (Q, P) are linked by a canonical transformation these two PBs are identical. That is

$$(f, g)_{\mathrm{q,p}} = (f, g)_{\mathrm{Q,P}}. \tag{5.177}$$

We shall carry out the demonstration in a straightforward manner. The partial derivatives of $f(Q, P)$ with respect to q and p are

$$\frac{\partial f}{\partial q_\mu} = \frac{\partial f}{\partial Q_\nu}\frac{\partial Q_\nu}{\partial q_\mu} + \frac{\partial f}{\partial P_\nu}\frac{\partial P_\nu}{\partial q_\mu}, \tag{5.178}$$

and

$$\frac{\partial f}{\partial p_\mu} = \frac{\partial f}{\partial P_\sigma}\frac{\partial P_\sigma}{\partial p_\mu} + \frac{\partial f}{\partial Q_\sigma}\frac{\partial Q_\sigma}{\partial p_\mu}. \tag{5.179}$$

Equivalent equations result for $g(Q, P)$. With (5.178), (5.179), and the equivalent for $g(Q, P)$, the PB $(f, g)_{\mathrm{q,p}}$ becomes

$$
\begin{aligned}
(f, g)_{\mathrm{q,p}} &= \left(\frac{\partial f}{\partial Q_\nu}\frac{\partial g}{\partial P_\sigma} - \frac{\partial f}{\partial P_\sigma}\frac{\partial g}{\partial Q_\nu}\right)\frac{\partial P_\sigma}{\partial p_\mu}\frac{\partial Q_\nu}{\partial q_\mu} + \left(\frac{\partial f}{\partial Q_\nu}\frac{\partial g}{\partial Q_\sigma} - \frac{\partial f}{\partial Q_\sigma}\frac{\partial g}{\partial Q_\nu}\right)\frac{\partial Q_\nu}{\partial q_\mu}\frac{\partial Q_\sigma}{\partial p_\mu} \\
&+ \left(\frac{\partial f}{\partial P_\nu}\frac{\partial g}{\partial P_\sigma} - \frac{\partial f}{\partial P_\sigma}\frac{\partial g}{\partial P_\nu}\right)\frac{\partial P_\nu}{\partial q_\mu}\frac{\partial P_\sigma}{\partial p_\mu} + \left(\frac{\partial f}{\partial P_\nu}\frac{\partial g}{\partial Q_\sigma} - \frac{\partial f}{\partial Q_\sigma}\frac{\partial g}{\partial P_\nu}\right)\frac{\partial P_\nu}{\partial q_\mu}\frac{\partial Q_\sigma}{\partial p_\mu}
\end{aligned} \tag{5.180}
$$

From (5.175) we see that the function F depends on the sets of independent variables q_ν and Q_σ. Then $\partial Q_\sigma/\partial q_\nu = 0$. And from $\mathrm{d}F$ we have the equivalence of the cross partial derivatives

$$\frac{\partial p_\nu}{\partial Q_\sigma} = -\frac{\partial P_\sigma}{\partial q_\nu}$$

Then

$$\frac{\partial P_\nu}{\partial q_\mu}\frac{\partial P_\sigma}{\partial p_\mu} = -\frac{\partial p_\mu}{\partial Q_\nu}\frac{\partial P_\sigma}{\partial p_\mu} = -\frac{\partial P_\sigma}{\partial Q_\nu} = 0$$

And

$$\frac{\partial P_\sigma}{\partial q_\mu}\frac{\partial Q_\nu}{\partial p_\mu} = -\frac{\partial p_\mu}{\partial Q_\sigma}\frac{\partial Q_\nu}{\partial p_\mu} = -\delta_{\sigma\nu}$$

Then (5.180) becomes

$$
\begin{aligned}
(f, g)_{\mathrm{q,p}} &= \frac{\partial f}{\partial Q_\nu}\frac{\partial g}{\partial P_\nu} - \frac{\partial f}{\partial P_\nu}\frac{\partial g}{\partial Q_\nu} \\
&= (f, g)_{\mathrm{Q,P}} \,.
\end{aligned} \tag{5.181}
$$

Therefore, from (5.172) we see that for any conservative system a function $f(q, p)$ for which

$$\frac{\mathrm{d}f(q, p)}{\mathrm{d}t} = (f, \mathcal{H}) = 0$$

initially will remain constant during the motion, since that motion is describable in terms of a canonical transformation. We call these constants *integrals of the motion* I_{j}. They are functions of generalized coordinates and momenta that are constants for which

$$\left(I_{\mathrm{j}}, \mathcal{H}\right) = 0. \tag{5.182}$$

We then say that the integral of the motion I_{j} *Poisson commutes*, or is in *involution*, with the Hamiltonian.

We may also write the canonical equations of Hamilton in terms of PBs as

$$\dot{q}_\mu = \left(q_\mu, \mathcal{H}\right) \text{ and } \dot{p}_\mu = \left(p_\mu, \mathcal{H}\right), \tag{5.183}$$

which is a form that will be unchanged during the system motion, which is the basis of the canonical transformation.

5.9 And the Quantum Theory

5.9.1 Heisenberg Indeterminacy

If we construct the PB of a coordinate and the corresponding canonical momentum we find that

$$\begin{aligned}\left(q_\alpha, p_\beta\right) &= \frac{\partial q_\alpha}{\partial q_\mu}\frac{\partial p_\beta}{\partial p_\mu} - \frac{\partial q_\alpha}{\partial p_\upsilon}\frac{\partial p_\beta}{\partial q_\upsilon} \\ &= \delta_{\alpha\mu}\delta_{\beta\mu} = \delta_{\alpha\beta},\end{aligned} \tag{5.184}$$

Paul A.M. Dirac identified something that resembled this in a paper on the emerging quantum theory by Werner Heisenberg, that Dirac had been asked to read by his advisor Ralph Fowler [[26], pp. 83–87]. Dirac later used this relationship as the bridge between quantum mechanics and classical mechanics. Specifically Dirac showed that the Poisson Bracket in classical mechanics has a quantum analog in the commutator [[18], p. 87]. That is

$$\boxed{(A, B) \Rightarrow (1/i\hbar)\,(AB - BA)\,,} \tag{5.185}$$

where $\hbar$ is Planck's constant of action divided by 2π. On the left hand side of the arrow $\Longrightarrow$ in (5.185) is the PB of the classical dynamical quantities A and B and on the right hand side is the commutator $(AB - BA)$ of the quantum mechanical operators representing these quantities.

In the quantum theory we define the average of the measured value of a quantity represented by the operator A as $\langle A\rangle$ and the square of the indeterminacy[9] in the value as

$$(\Delta A)^2 = \left\langle (A - \langle A\rangle)^2\right\rangle. \tag{5.186}$$

We can then show that if

$$AB - BA = iC,$$

then

$$\Delta A \Delta B \geq \frac{1}{2}\left|\langle C\rangle\right|. \tag{5.187}$$

[see e.g. [78], pp. 218–219] The quantum mechanical analog (5.185) of (5.184) is

$$qp - pq = i\hbar\mathbf{1}. \tag{5.188}$$

[9] This is sometimes called the error or uncertainty. But that suggests that there is a particular value of A before the measurement is made.

Then (5.187) results in

$$\Delta q \Delta p \geq \frac{1}{2}\hbar, \tag{5.189}$$

which is the Heisenberg Indeterminacy Principle.[10]

From (5.172) the rate of change of a quantum mechanical operator, which does not depend explicitly on the time is

$$i\hbar\frac{\mathrm{d}}{\mathrm{d}t}G = (\mathcal{H}G - G\mathcal{H})\,. \tag{5.190}$$

This is the general form taken by the equations of quantum mechanics in the *Heisenberg picture* in which the time dependence is carried by the basis vectors. The Poisson Brackets then reveal a structure of mechanics which persists in an analogous form in the quantum theory.

5.9.2 The Schrödinger Equation

The relationship of the quantum theory to the Hamilton–Jacobi approach is deeper than the analog of the PB and the quantum mechanical commutator. To see this we shall look briefly at the development by Erwin Schrödinger of his celebrated wave equation.

The equations of the quantum theory may be expressed in two equivalent forms. These are termed the Heisenberg and the Schrödinger pictures. They are completely equivalent, as Schrödinger showed. Whether one or the other picture is more convenient depends upon the problem we are considering. If we choose the basis vectors to be time independent we are treating the problem in the Heisenberg picture. Then the time dependence is carried by the operators and our equation of motion for the system is (5.190). This is particularly convenient in condensed matter (solid state) physics. If we choose the basis vectors to be time dependent the operators are time independent and the equation of motion becomes what is called the Schrödinger Equation. In the Schrödinger picture the basis vectors are called the wave functions for the system.

Schrödinger developed what we now know as the Schrödinger Equation during a two and a half week vacation over Christmas, 1925, at a villa in Arosa in the Swiss Alps. For the details surrounding this vacation we refer the reader to Walter Moore's book [82].

For us it is important to note that he only brought along Louis de Broglie's thesis as his source for the physics. Schrödinger, however, left no record of his thoughts as he worked on the wave theory. So we can only speculate based on any inspiration he may have received from de Broglie's thesis and what we can deduce from his

[10] This is often termed the Heisenberg Uncertainty Principle. Indeterminacy is closer to the original German meaning.

first publication on the wave equation, which appeared a few weeks after the alpine vacation (in January of 1926) [105].

De Broglie's thesis is a veritable survey of aspects of physics that point in some way to a wave and particle description of motion. The aspects of physics he considered important for his hypothesis that a wave that could be associated with a moving particle include Einstein's special theory of relativity and concept of the photon, Maupertuis' principle, Hamilton's principle of least action, and the Hamilton–Jacobi approach [19]. Although we may speculate that Schrödinger simply accepted the de Broglie hypothesis and put together his wave equation in the same simple manner we see in courses, there is little reason to believe this was indeed the case. We must assume that he actually read de Broglie's thesis. And if he did he saw the importance of the Hamilton–Jacobi approach to de Broglie.

Schrödinger begins his first paper on *Quantisierung als Eigenwertproblem* (Quantization as an Eigenvalue Problem) with the Hamilton partial differential equation (Schrödinger's designation)

$$\mathcal{H}\left(q, \frac{\partial S}{\partial q}\right) = \mathcal{E} \tag{5.191}$$

and then seeks a solution that is of the form of a sum of functions each dependent only on a specific independent variable q. To accomplish this he requires that the principal function S be of the form

$$S = K \ln \psi \tag{5.192}$$

where K has the dimensions of action and ψ is a product of functions of the individual variables. With (5.192) Eq. (5.191) becomes

$$\mathcal{H}\left(q, \frac{K}{\psi}\frac{\partial \psi}{\partial q}\right) = \mathcal{E}. \tag{5.193}$$

Schrödinger chooses to write this for the Kepler problem, which, in rectangular coordinates, is

$$\left(\frac{\partial \psi}{\partial x}\right)^2 + \left(\frac{\partial \psi}{\partial x}\right)^2 + \left(\frac{\partial \psi}{\partial x}\right)^2 - \frac{2m}{K^2}\left(\mathcal{E}+\frac{k}{r}\right)\psi^2 = 0. \tag{5.194}$$

He then asks for finite-valued functions ψ, which are twice differentiable and result in an extremum of the integral of the left hand side of (5.194) over all space. This results immediately in the requirement that ψ must satisfy

$$\left(\frac{\partial^2 \psi}{\partial x^2}\right) + \left(\frac{\partial^2 \psi}{\partial x^2}\right) + \left(\frac{\partial^2 \psi}{\partial x^2}\right) + \frac{2m}{K^2}\left(\mathcal{E}+\frac{k}{r}\right)\psi = 0, \tag{5.195}$$

which is what we now know as Schrödinger's Equation for the Kepler problem. With an appropriate choice of k (5.195) is the equation for the wave function ψ of an electron moving about a proton.

Schrödinger chose spherical coordinates in which to represent his problem and wrote ψ, as he indicated that he would, as a product of functions for each of the coordinates (ϑ, ϕ, r). With this choice of ψ Eq. (5.195) separates into three equations as our solution for the generator separated. The angular solutions were already well known to Schrödinger (the spherical harmonics). The radial equation presents some difficulty and in a footnote he thanks his friend and colleague Hermann Weyl for an introduction to the handling of this equation. The eigenvalues for the energy $\mathcal{E}$ emerged from the solution of the radial equation, which already contained the eigenvalues from the azimuthal (ϑ) equation. He obtained an infinite set of eigenvalues for the energy $\mathcal{E}$ when $\mathcal{E} > 0$ and a finite set when $\mathcal{E} < 0$. The latter agreed precisely with those of the Bohr atomic model provided $K = h/2\pi$. Schrödinger's result contained, as well, the polar (ϕ) and azimuthal quantum numbers. In terms of theoretical physics, this was a triumph. He had obtained the principal equation of what could now be called the quantum theory.

We note that, although we may recognize the quantum theory as more fundamental than the classical theory of Analytical Mechanics, Schrödinger had coaxed the principal equation of the new quantum theory from the principal equation of classical Analytical Mechanics.

Although de Broglie was certainly not referring to Schrödinger's Equation, it seems appropriate to close this section with a quote from de Broglie's thesis.

> This whole beautiful structure can be extracted from a single principle, that of Maupertuis, and later in another form as Hamilton's Principle of least action, of which the mathematical elegance is simply imposing.

5.10 Summary

In this chapter we discussed the reduction of Hamilton's two partial differential equations for the Principal Function to a set of ordinary differential equations. Our approach essentially followed Jacobi's critique of Hamilton's method. In this we have not attempted to present a method for applying Jacobi's Theorem. Instead we have based our presentation on the central role of the canonical transformation as the link between the initial and final states of the system and the role of the generating function or generator in that transformation. The fact that the generating function satisfies a partial differential equation of the same form as the Hamilton–Jacobi equation is a mathematical result. But the generating function is not Hamilton's Principal Function. It is a subset of the functions that satisfy the Hamilton–Jacobi equation. Therein lies the basis for a great simplification, since we are then free to choose the final coordinates and canonical momenta to be constants.

The approach, as we have shown, remains straightforward provided the Hamiltonian is separable. And this is the case for many applications. If the Hamiltonian is separable we are left with a small number of integrals over single variables. In many interesting cases these are tabulated.

We found a further simplification, particularly for complex multivariable systems through the introduction of action and angle variables. The mathematical simplification was through the introduction of closed path (definite) integrals in place of the indefinite integrals required for the computation of the generator. Use of the action and angle variables, as we found, is particularly advantageous if we are interested in the periodic behavior of certain aspects of the system motion that may not be easily identified otherwise.

We ended the chapter with a discussion of the connection between the Hamilton–Jacobi approach and the quantum theory. Dirac discovered the analog between the Poisson Brackets and the quantum mechanical commutators. This analog led to the Heisenberg Indeterminacy Principle and the Heisenberg rate equation for an operator. We then followed Schrödinger's development of his celebrated equation from the Hamilton–Jacobi Equation.

In summary we have presented the Hamilton–Jacobi approach as a simplification. It is also the most elegant formulation of Analytical Mechanics. The remainder of this text will grow out of this elegant formulation.

Some authors present the results of our development in this chapter as a method that can be followed [cf. [34], pp. 447-449]. We have elected not to do that, since it obscures the elegance of the Hamilton–Jacobi approach to mechanical problems. To appreciate that elegance the reader must understand that we have replaced the original partial differential equation of Hamilton for the Principal Function with an equation for the generator of a canonical transformation. And that we have then chosen a method, which is applicable whether or not the Hamiltonian depends on time, and which results in a canonical transformation to a single final phase point. Provided the Hamiltonian is separable, the actual mathematical solution is not difficult.

5.11 Exercises

5.1. Apply the Legendre transformation to obtain generating functions of $F_2(p, P, t)$ and $F_3(p, Q, t)$.

5.2. Consider a conservative system and suppose that you have solved the partial differential equation

$$\mathcal{H}\left(q, \frac{\partial F}{\partial q}\right) = \mathcal{E}$$

for the function $F(q, a, \mathcal{E})$. Now suppose that you choose to form the link to the final configuration of the system through the final canonical momenta P. That is, you choose $P_1 = \mathcal{E}$. Follow the procedure we used in the chapter for $Q_1 = \mathcal{E}$ to discuss the steps toward the final solution for the generator.

5.3. In the chapter we considered the simple harmonic oscillator as an example for which we could find a generator and a final solution. There we found equations relating the initial coordinates and momenta (q, p) to the final coordinates $(Q = \mathcal{E})$ and momenta P based on a generator constructed based on a simple choice of the final coordinate. These equations we found to be

$$q = \mp\sqrt{2\mathcal{E}/k}\sin\sqrt{\frac{k}{m}}P$$

and

$$p = \pm\sqrt{2m\mathcal{E}}\cos\sqrt{\frac{k}{m}}P.$$

The physics requires that the final momentum is related to the final coordinate by the canonical equations. Use these to obtain the final momentum as a function of time and, then, the initial coordinate and momentum as functions of the time.

5.4. From the action-angle solution we obtained for the harmonic oscillator in the text

$$J_{\mathrm{q}}(\mathcal{E}) = \frac{2\pi\mathcal{E}}{\omega_0}$$

and

$$\omega_{\mathrm{q}} = \frac{\omega_0}{2\pi}t + \delta_{\mathrm{q}},$$

or

$$2\pi\omega_{\mathrm{q}} = \omega_0(t - \tau),$$

obtain the standard (q, p) description of the motion of the harmonic oscillator.

5.5. Show that the Poisson Bracket is unchanged by canonical transformation. That is, show that

$$\begin{aligned}(f, g)_{\mathrm{q,p}} &= \frac{\partial f}{\partial q_\mu}\frac{\partial g}{\partial p_\mu} - \frac{\partial f}{\partial p_\mu}\frac{\partial g}{\partial q_\mu} \\ &= (f, g)_{\mathrm{Q,P}} = \frac{\partial f}{\partial Q_\mu}\frac{\partial g}{\partial P_\mu} - \frac{\partial f}{\partial P_\mu}\frac{\partial g}{\partial Q_\mu}.\end{aligned}$$

Choose the generator of the canonical transformation to be dependent on (q, Q), as in the text.

5.6. Consider the situation we considered for the time dependent Hamiltonian in which we defined new coordinates

$$q_{n+1} = t$$
$$p_{n+1} = -\mathcal{H}(q, p, t).$$

Show that classically

$$(t, \mathcal{H}) = \mathbf{1}.$$

Note that it then follows that quantum mechanically

$$(\mathcal{H}t - t\mathcal{H}) = -i\hbar\mathbf{1}$$

and that, therefore, there is an indeterminacy relation

$$\Delta\mathcal{E}\,\Delta t \geq \frac{1}{2}\hbar.$$

Quantum mechanically energy levels in an atom are measured by the light emitted by the atom in transitions between states. The Planck–Einstein relation $\mathcal{E} = h\nu$, where ν is the frequency of the light quantum (photon), relates the characteristic spectrum of an element to the energies of the atom. We may consider that the indeterminacy of the time of transition from the state is the lifetime of the state. That is the atom may decay from a state at any time up to approximately the lifetime. What is then $\Delta\mathcal{E}$?

5.7. Consider a mass m moving without friction on a wire. The wire makes an angle β with the vertical and is free to rotate about the vertical axis, also without friction. The situation is shown here.

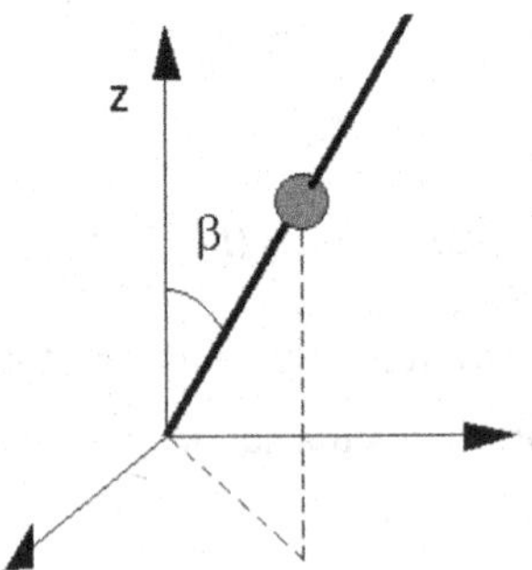

Bead on frictionless wire held at an angle β with the vertical. The wire rotates without friction about the vertical axis.

Consider that at the initial time the mass is located at a distance z_0 above the reference plane and has an angular momentum of $\ell = mr_0^2\dot{\vartheta}_0$ where r_0 and $\dot{\vartheta}_0$ are the initial radial distance from the axis and the initial angular velocity. There is no initial radial velocity.

Formulate and study the problem in the Hamilton–Jacobi formulation. Obtain the phase plot $p(r)$ versus r for the motion.

5.8. Consider a marble in a fishbowl. We shall assume that the marble slides without friction so we can neglect rotation. We have illustrated the situation in the figure below. We shall use cylindrical coordinates because then it is easiest to specify the surface of the fishbowl.

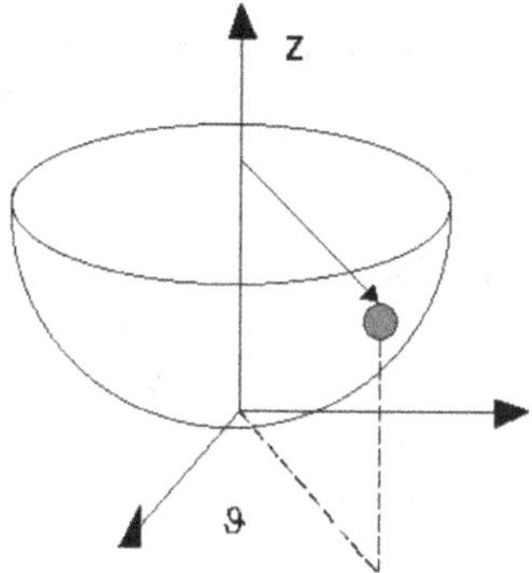

Marble in a fishbowl.

Pursue the problem employing the Hamilton–Jacobi approach. Find the phase plot(s).

5.9. Consider the two masses connected as shown here.

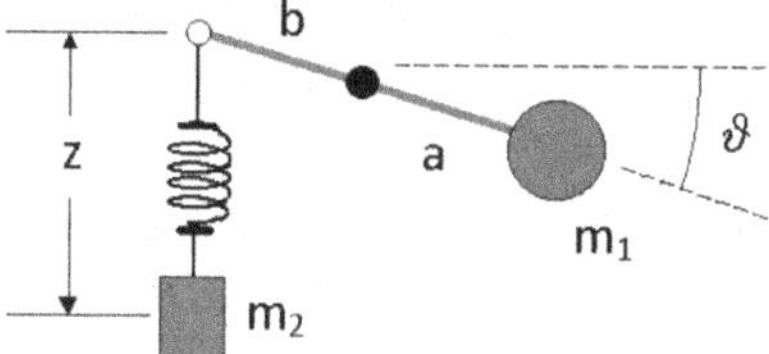

Rotating and suspended masses.

Obtain the Lagrangian for unequal masses and then simplify for equal masses and $a = b$.

Show that the motion cannot be easily studied using the Hamilton–Jacobi approach. So you should turn to the canonical equations.

Linearize the canonical equations for small vibrations and find the (eigen) frequencies of vibration.

[Answers:

$$\omega = \frac{1}{a\sqrt{m}}\sqrt{K_1 + \sqrt{K_2^2}}, \frac{1}{a\sqrt{m}}\sqrt{K_1 - \sqrt{K_2^2}}$$

with

$$K_1 = k' + a^2 k$$

and

$$K_2^2 = a^4 k^2 + k'^2 - a^2 k k' \big] .$$

5.10. Consider a charged point particle of mass m and charge Q moving in a magnetic field of induction $\boldsymbol{B} = \hat{e}_z B$. Assume a motion in the direction of $\hat{e}_z$ as well as in the (x, y) −plane. In the text we have shown that the Hamiltonian for the charged particle in the electromagnetic field is generally

$$\mathcal{H} = (1/2m)\left(p_\mu - QA_\mu\right)^2 + Q\varphi,$$

summing on Greek indices. In our situation there is no electric field. Therefore $\varphi = 0$.

Show that the vector potential $\boldsymbol{A} = -\hat{e}_x y B$ produces the required magnetic field induction. Obtain the trajectory of the charged particle using the Hamilton–Jacobi approach.

5.11. As an example in the text we considered the motion of a charged point particle of mass m and charge Q moving in a constant magnetic field of induction $\boldsymbol{B} = \hat{e}_z B$ using cylindrical coordinates. The vector potential is then

$$\boldsymbol{A} = \frac{1}{2} B r \hat{e}_\vartheta .$$

Treat this problem using the Hamilton–Jacobi approach.

[Note that a positive charge has a negative angular momentum, i.e. rotates clockwise.]

5.12. Consider the motion of a charged particle in a region of space in which there is a uniform magnetic field with induction $\boldsymbol{B} = \hat{e}_z B$ and a uniform electric field $\boldsymbol{E} = \hat{e}_y E$.

For a static magnetic field with induction $\boldsymbol{B} = \hat{e}_z B$ the vector potential is $\boldsymbol{A} = -\hat{e}_x y B$. And for an electric field $\boldsymbol{E} = \hat{e}_y E$ the electrostatic potential is

$$\varphi = -Ey.$$

Find the orbit of the charged particle using the Hamilton–Jacobi approach.

Chapter 6
Complex Systems

> A more complete study of the movements of the world will oblige us, little by little, to turn it upside down; in other words, to discover that if things hold and hold together, it is only by reason of complexity, from above.
>
> Pierre Teilhard de Chardin

6.1 Introduction

Historically the development of classical mechanics is related to our interest in astronomy and astrophysics. Newton was concerned that the planets and the stars, if left to themselves, would be unstable. He, therefore, was convinced that God intervened to make adjustments [[91], pp. 216–217]. We do not harbor the fears Newton had. But our interest in interacting n body systems is based, at least initially, on our interest in the solar system. McCuskey, for example, devotes a chapter of his book *Introduction to Celestial Mechanics* to a detailed treatment of particularly the three body problem, but with generalization to the n body problem. As McCuskey points out, neither the three, nor the general n body problem can be solved in closed form [[81], pp. 92–127].

We have, however, been able to approximate the system of two planets and the sun by considering that the interaction of each planet with the sun is much greater than the interaction between the two planets. We may then first solve the problem of the motion of the two noninteracting planets around the sun and then add the interaction between the two planets as a perturbation. This perturbation approach to the integrable system of the noninteracting planets was studied extensively in the 19th century. However, no one was able to prove that these perturbations converged, even though a substantial prize was offered by King Oscar II of Sweden and Norway for the proof [[120], [41], p. 423].

C.S. Helrich, *Analytical Mechanics*, Undergraduate Lecture Notes in Physics, DOI 10.1007/978-3-319-44491-8_6

In this chapter we have used the two dimensional oscillator, the Kepler problem, and the Hénon–Heiles system as examples for the approaches normally used in the study of complex systems. Of these only the Hénon–Heiles system is truly complex. The other systems were chosen to provide examples of the general approaches. Because complex systems do not yield to analytical solutions we have no exercises at the end of this chapter. If the students have access to appropriate software an instructor may choose to use the Hénon–Heiles system as a basis for student exercises.

We shall begin our study of complex systems by first introducing the concept of an integrable system.

6.2 Integrable Systems

The concept of a *complete integrable system* originated in the 19th century following the work of Hamilton and Jacobi. A system with a Hamiltonian of $2n$ dimensions, that is with n generalized coordinates and n conjugate momenta, is said to be integrable if there are n integrals of the motion. These integrals must all be independent and Poisson commute (see Sect. 5.8) with the Hamiltonian as in (5.182). There are then as many constants of the motion as there are degrees of freedom in an integrable system [[102], [48]]. Robert Hilborn points out that this definition is, perhaps, unfortunate, since it leads us to think that integrable systems are those that can be solved in the sense that the canonical equations can be integrated [[48], p. 281]. That is not the case. However the Liouville–Arnold theorem ensures that for an integrable system there exists a canonical transformation to the action angle variables. This canonical transformation also results in a Hamiltonian and all of the integrals (also termed 'Hamiltonians' in these systems) which are functions only of the action variables.

The original system, for which we are seeking a solution, will normally have highly nonlinear canonical equations. In practical terms the numerical integration of these equations may not be impossible with the computers available in the 21st century. But nonlinearities limit the time step sizes and the total time intervals that can be treated.[1] Our numerical solution does not, therefore, answer questions regarding the long term stability of these nonlinear systems. And numerical solutions alone do not improve our understanding of the structure of classical mechanics. We shall, therefore, return to the solution of integrable systems in terms of action and angle variables, the existence of which is assured by the Liouville–Arnold theorem. We have already carried out a solution to the Kepler problem in terms of action and angle variables, which we may use as an example.

In our treatment of the Kepler problem (Example 5.7.2) there are two tori $(J_\vartheta, \omega_\vartheta)$ and $(J_\text{r}, \omega_\text{r})$ in which the action variables J_ϑ and J_r are constants and the angle variables are linear functions of the time. The intersection of the two tori $(J_\vartheta, \omega_\vartheta)$ and $(J_\text{r}, \omega_\text{r})$ is a two dimensional curve, which is the trajectory of the Kepler system in the action angle variables. We can then, at least mentally, project this trajectory onto

[1]This is a well known problem in molecular dynamics.

the basis of the canonical variables (r, p_r). We presented the result of this projection in Fig. 5.2, although we obtained this plot in a simpler fashion.

The Kepler problem can, of course, be treated rather simply in the original canonical coordinates $(r, \vartheta, p_r, p_\vartheta)$ with the Hamiltonian and the angular momentum as integrals of the motion. Figure 5.2 is then a plot of (5.143). And we understand Fig. 5.2 in terms of a (distorted) harmonic motion of a planet around the equilibrium distance from the sun. The frequencies ω_r and ω_ϑ are also identical (see (5.162) and (5.163)) so that the orbit is closed. Our projection onto the the basis (r, p_r) in Fig. 5.2 is also a closed curve.

If we modify the dependence of the potential on r so that the actions J_r and J_ϑ are no longer linearly related as in (5.160) the angle frequencies will no longer be equal (see (5.162) and (5.163)) and the orbit in (r, p_r) coordinates will become open. For more complex systems we may expect open orbits to be common.

For a general integrable system with n generalized coordinates our description of the motion will be in terms of n action and n angle variables. Each combination of an action and a corresponding angle (J_i, ω_i) is a torus in the higher dimensional space consisting of the $2n$ variables $(J_1, J_2, \cdots, J_n, \omega_1, \omega_2, \cdots, \omega_n)$. Such general statements, however, seldom provide an intuitive picture of the motion that can be more readily understood in terms of the details that we may want to investigate. *Henri Poincaré* suggested that we can obtain a more detailed picture of the motion of complex integrable systems if we consider a surface of one dimension less than that of the space required to describe the state of the system. The trajectory will then cross this surface at a set of points. If the system trajectory lies on a torus the points at which the trajectory intersects the surface will form an image of the torus on the surface. This is called a Poincaré section.

To obtain the surface for the Poincaré section we normally begin with a numerical solution of the canonical equations for the system. We then select one of the phase variables to have a particular constant value for the Poincaré section. From the numerical solution we then obtain the values of the other canonical variables when the selected variable takes on that constant value to within limits we select. We then have the values attained by all of the remaining canonical variables as the trajectory crosses the Poincaré plane. If, during the time we selected for the numerical solution, the system trajectory crosses the Poincaré plane N_P times there will be N_P points for our representation of the Poincaré section. We may then project this N_P point representation of the Poincaré section onto a subspace of two of the remaining canonical variables to obtain a representation of the Poincaré section that provides the insight into the motion that we desire. Because the choice of the variable that is to be held constant in the Poincaré section and the value of that constant are arbitrary the number of possible Poincaré sections we may construct for a system is infinite [[25], p. 64].

The Poincaré section does not provide a simpler picture of the system motion. It provides insight into the character of the motion. For example, if the Poincaré section is, indeed, a single curve then we know that the original system is integrable. The Liouville–Arnold theorem guarantees that the integrable system will produce such a curve resulting from the intersection of two of the tori in the space of action and angle variables.

The Poincaré section also provides insight into the stability of the motion of the complex system. If the Poincaré section breaks up under certain conditions we know that the system has transitioned from integrability to chaos.

The two dimensional oscillator is an integrable system that is simple enough to be solved analytically and yet, with four canonical variables, can still serve as an example for construction of a Poincaré section.

Example 6.2.1 The Hamiltonian for the two dimensional harmonic oscillator is

$$\mathcal{H} = \frac{1}{2}\left(\frac{1}{m}p_x^2 + k_x x^2\right) + \frac{1}{2}\left(\frac{1}{m}p_y^2 + k_y y^2\right),$$

which is separable. If we identify the constants $\mathcal{E}_x$ and $\mathcal{E}_y$ as

$$2\mathcal{E}_x = \frac{1}{m}p_x^2 + k_x x^2 \text{ and } 2\mathcal{E}_y = \frac{1}{m}p_y^2 + k_y y^2$$

the Hamiltonian becomes $\mathcal{H} = \mathcal{E}_x + \mathcal{E}_y$. The momenta are

$$p_x = \pm\sqrt{2m\mathcal{E}_x - mk_x x^2} \text{ and } p_y = \pm\sqrt{2m\mathcal{E}_y - mk_y y^2}$$

And the limits of motion $x = \pm\sqrt{2\mathcal{E}_x/k_x}$ and $y = \pm\sqrt{2\mathcal{E}_y/k_y}$ are found by setting $p_{x,y} = 0$. The actions are then

$$\begin{aligned} J_x &= \int_{-\sqrt{2|\mathcal{E}_x|/k_x}}^{\sqrt{2|\mathcal{E}_x|/k_x}} \sqrt{2m\mathcal{E}_x - mk_x x}\,dx \\ &= \frac{\pi}{\omega_{0,x}}\mathcal{E}_x, \end{aligned}$$

and

$$J_y = \frac{\pi}{\omega_{0,y}}\mathcal{E}_y$$

where $\omega_{0,i} = \sqrt{k_i/m}$ for $i = x, y$. From the Hamiltonian

$$\mathcal{H} = \frac{1}{\pi}\left(\omega_{0,x} J_x + \omega_{0,y} J_y\right).$$

we have the angle frequencies as (see (5.121))

$$\nu_i = \frac{\partial \mathcal{H}}{\partial J_i} = \frac{\omega_{0,i}}{\pi}.$$

The ratio of the frequencies for this two dimensional oscillator is

$$\frac{\nu_x}{\nu_y} = \frac{\omega_{0,x}}{\omega_{0,y}} = \sqrt{\frac{k_x}{k_y}},$$

which may be integer, rational, or irrational. The angle variables are

$$\omega_x = \frac{\omega_{0,x}}{\pi} t$$

and

$$\omega_y = \frac{\omega_{0,y}}{\pi} t,$$

neglecting the additive constants.

There are four variables for the double oscillator. These are $(x, y, p_x p_y)$ or $(J_x, J_y, \omega_x, \omega_y)$. The phase space is then four dimensional and phase plots cannot be drawn. We may, however, simply plot two or three variables by projecting the phase plot onto two or three dimensions. We show the two dimensional projection of the double oscillator with $k_y = 2.5k_x$ onto the space of (x, y) in Fig. 6.1. We obtained this plot from a numerical solution of the canonical equations using Maple for the solution and graphics.

And in Fig. 6.2 we show the three dimensional projection of this double oscillator onto the space of (x, y, p_x). In this plot we can see the two dimensional phase space (x, p_x) projection of the motion without difficulty.

We can investigate the affect of changing the frequency ratios ω_x and ω_y from rational to irrational values by choosing values for the spring constants k_x and k_y. If we select values of the constants k_x and k_y such that the square root of their ratio is very close to an integer value we get a projected phase plot that is very close to a single trajectory for the same time interval as we used in Figs. 6.1 and 6.2. In the case for which $k_y = 4.01k_x$, i.e. the case for which the ratio of the frequencies is very close to $1/2$, we have the projection of the double oscillator onto the space of (x, y, p_x) in Fig. 6.3.

If we increase the ratio to $k_y = 4.1k_x$ we have the projection of the double oscillator onto the space of (x, y, p_x) in Fig. 6.4.

We may also obtain a Poincaré section by calculating, for example, the values of y, p_x, and p_y when $x = 0$ and then projecting this result onto the space (y, p_y). The result for $k_y = 2.5k_x$ we show in Fig. 6.5.

If we had chosen the spring constants to be in a ratio of precisely 4 the plot in Fig. 6.3 would have been a single line. Our choice to make it appear as a ribbon provides some sense of the form of the trajectory. In Fig. 6.4 we increased the ratio of the spring constants to $k_y = 4.1k_x$ which changed the frequencies sufficiently to make the trajectory differences visible. But we have still not reduced the representation to a point at which we can easily recognize a single oscillator. This requires us to obtain a plot of p_x versus x or p_y versus y. It is almost obvious that the rotation of Fig. 6.4

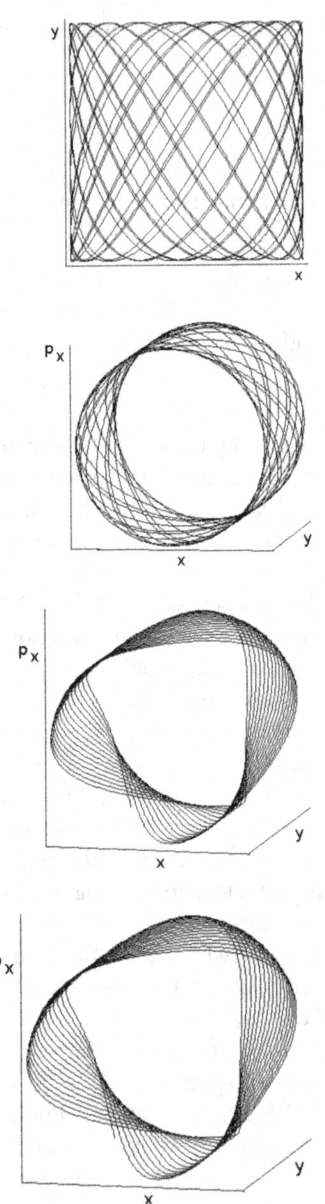

Fig. 6.1 Projection of the phase plot for a double oscillator with $k_y = 2.5k_x$ onto the space (x, y)

Fig. 6.2 Projection of the phase plot of the double oscillator with $k_y = 2.5k_x$ onto the space (x, y, p_x)

Fig. 6.3 Projection of the phase plot of the double oscillator with $k_y = 4.01k_x$ onto the space (x, y, p_x)

Fig. 6.4 Projection of the phase plot of the double oscillator with $k_y = 4.1k_x$ onto the space (x, y, p_x)

will reveal a plot of p_x versus x that is an ellipse (a circle in stretched coordinates). So in Fig. 6.5 we have done this by first obtaining a three dimensional representation of the trajectory in terms of y, p_x, and p_y and then finding the points for which the value of x lies between small limits centered on $x = 0$. This produces the Poincaré section in Fig. 6.5.

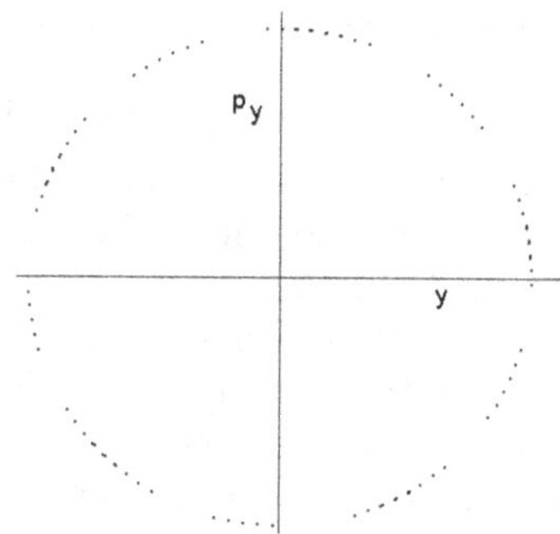

Fig. 6.5 Poincaré section of the double oscillator with $k_y = 2.5k_x$. This results from a determination of the variables (y, p_x, p_y) when $x = 0$ and then projecting the result onto the plane (y, p_y)

Our use of projections and finally a Poincaré section has shown us the relatively simple structure that we realized was present in the double oscillator. Of course in our usual application of these techniques we may not be aware of the detailed structure of the system we are studying and the Poincaré section might reveal a structure that may not be at all obvious.

6.3 The Hénon–Heiles System

6.3.1 Integrals of the Motion

As an example of an interesting and somewhat complicated system we have chosen to consider the system designed by Michel Hénon and Carl Heiles [43]. For the year during which this work was conducted Hénon was a visiting professor and Heiles a graduate student at Princeton University.

Ostensibly this system models the motion of a star inside a galaxy. The primary interest of Hénon and Heiles was, however, to investigate the existence of a third integral of galactic motion.

A conservative and axially symmetric system has constant energy and angular momentum about the axis of symmetry. These are the first and second integrals of the motion. The constant energy integral is denoted as I_1 and constant angular momentum is I_2. The third integral is then I_3.

Hénon and Heiles began with a time independent gravitational potential within the galaxy and chose cylindrical coordinates for the description (r, ϑ, z). The phase space then has six dimensions $(r, \vartheta, z, p_r, p_\vartheta, p_z)$. Because they did not consider the mass of the star being studied, Hénon and Heiles used velocities rather than momenta as phase coordinates. We shall consider unit mass in our numerical work to bring our results into line with those of Hénon and Heiles.

The trajectory in phase space, which is a one-dimensional line, must result from five independent integrals of the motion $I_j\,(r, \vartheta, z, p_r, p_\vartheta, p_z) = C_j$, where $j = 1 \ldots 5$ and C_j are constants. Two of these constant integrals we have already identified. They are isolating integrals, as contrasted to nonisolating integrals. From a physical point of view nonisolating integrals have no significance. Prior to 1958 the third integral had been considered to be nonisolating. Then George Contopoulos and Alexander Ollongren published the results of numerical computations on galactic orbits, which indicated that there may be 3 isolating integrals, rather than just 2 [[7], [8], [9], [10], [97]].

Hénon and Heiles again approached the problem numerically. But, as they pointed out, in order to have more freedom of experimentation they neglected the astronomical origin of the problem. Instead they considered the more general question of whether an axisymmetric potential admits a third isolating integral of the motion. Therefore the potential they constructed did not necessarily represent an actual galactic potential. Their analysis of the question was based on studies of the orbits in phase space as the system energy was varied.

Hénon and Heiles were not able to conclude that there always existed or that there was no isolating third integral. They could conclude, however, that a third isolating integral always exists for low energies. And at energies above a critical value there are an infinite number of separate regions in phase space in which a third isolating integral seems to exist. Between these regions is the *ergodic region*[2] in which the third integral is nonisolating. And as the energy is further increased the ergodic region rapidly fills the entire phase space.

Our intention is to first explore the phase space structure of the Hénon–Heiles system, as a system that is interesting in its own right, without reference to the question Hénon and Heiles pursued. We will then consider the Poincaré section on which Hénon and Heiles based their analysis of the third integral of the motion. Our approach will be direct integration of the canonical equations using a *Runge–Kutta* algorithm. We carried out all computations on Maple.

6.3.2 Equations of Motion

Hénon and Heiles began with a gravitational potential V_{g} which was independent of azimuthal angle ϑ. The first two integrals of the motion were (with $m = 1$)

$$I_1 = V_{\mathrm{g}}(r, z) + \frac{1}{2}\left(\dot{r}^2 + r^2\dot{\vartheta}^2 + \dot{z}^2\right)$$

and

$$I_2 = r^2\dot{\vartheta}.$$

Introducing

$$V(r, z) = V_{\mathrm{g}}(r, z) + \frac{C_2^2}{r^2},$$

[2]The trajectory fills the whole of the ergodic regions and is not confined to a single surface. The ergodic hypothesis arises in statistical mechanics [[44], p170].

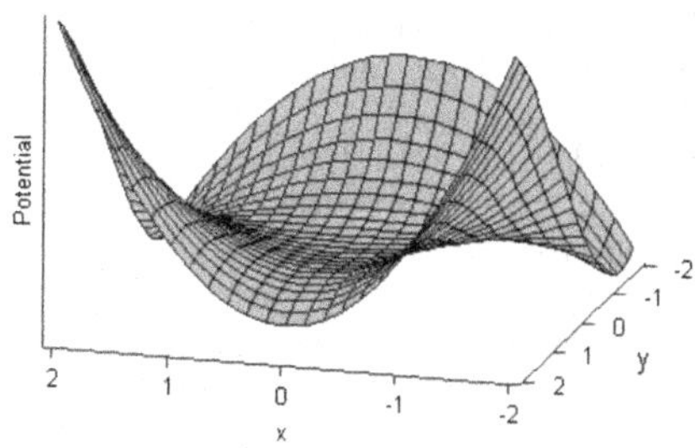

Fig. 6.6 Hénon–Heiles Potential. This is a two-dimensional potential with a very shallow basin at the origin

where $C_2^2 = I_2^2/2$, the problem was then completely equivalent to the motion of a particle in a plane. So Hénon and Heiles adopted a new formulation by substituting x and y for r and z. They then (after a number of trials) simplified the effective potential to

$$V(x, y) = \frac{1}{2}\left(x^2 + y^2\right) + x^2 y - \frac{1}{3} y^3. \tag{6.1}$$

Their bases for this simplification were

1. this potential is analytically simple
2. it is sufficiently complicated to yield trajectories that are far from simple.
   They believed that nothing would be fundamentally changed by adding higher order terms [cf. [60]].

The Hamiltonian for the Hénon–Heiles system is (with $m = 1$)

$$\mathcal{H}\left(x, y, p_{\mathrm{x}}, p_{\mathrm{y}}\right) = \frac{1}{2}\left(p_{\mathrm{x}}^2 + p_{\mathrm{y}}^2\right) + \frac{1}{2}\left(x^2 + y^2\right) + x^2 y - \frac{1}{3} y^3. \tag{6.2}$$

The canonical equations are then

$$p_{\mathrm{x}} = \dot{x},\ \ p_{\mathrm{y}} = \dot{y}, \tag{6.3}$$

$$\dot{p}_{\mathrm{x}} = -x - 2xy \tag{6.4}$$

and

$$p_{\mathrm{y}} = -y - x^2 + y^2. \tag{6.5}$$

All trajectories we discuss below were obtained from a simultaneous numerical integration of these canonical equations.

In Fig. 6.6 we have plotted the potential (6.1). This potential has a slight depression around the origin. As long as the energy is kept low we may expect bound motion in this potential energy depression. The coordinate x takes on positive and negative values. The original coordinate for the star was r, which could take on only positive values. So we cannot easily interpret our results in terms of the motion of a star.

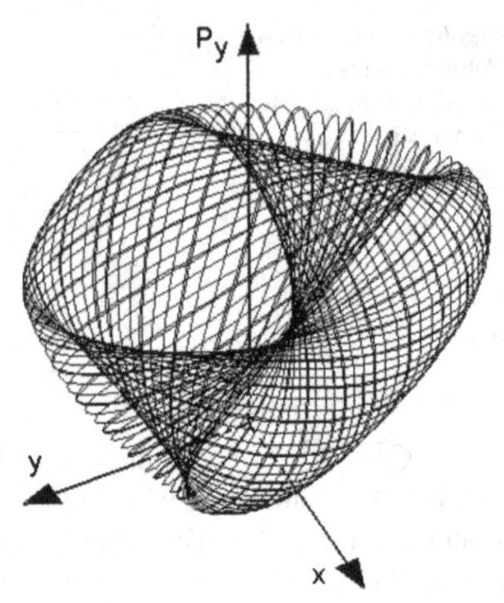

Fig. 6.7 Hénon–Heiles phase plot. Initial conditions were ($x = 0$, $y = -0.1475$, $p_x = 0.3101$, $p_y = 0$). The presence of a torus is evident from the pattern of the trajectory

6.3.3 Trajectories

If we initially locate the (unit) mass slightly up the potential hill in the negative $y-$direction with $x = 0$ and impart a small momentum in the $x-$direction ($y\,(0) = -0.1475$ and $p_x\,(0) = 0.3101$ with energy $= 0.06003$) the mass executes the bound motion we have plotted in Fig. 6.7. This corresponds to the motion of the double oscillator represented in Figs. 6.2 and 6.4. In Fig. 6.7 we chose to plot the momentum p_y against x and y with an orientation that shows that the trajectory lies on a surface. The complete phase plot has four dimensions $\left(x, y, p_x, p_y\right)$. So we have here a projection of the complete phase plot.

If we project the trajectories in Fig. 6.7 onto the (x, y) $-$plane we have the plot in Fig. 6.8, which corresponds to the plot of the double oscillator motion in Fig. 6.1. If we change the initial conditions slightly, the projection of the trajectory onto the (x, y) $-$plane takes on a different appearance. We have shown an example in Fig. 6.9.

Although very beautiful in its symmetry, the length of time over which we plotted the trajectories in Figs. 6.7 and 6.8 obscures the details. In Fig. 6.10 we allowed the motion to first settle and then plotted the trajectory over a relatively short time interval. In Fig. 6.10 the beginning point is indicated by a closed circle and the final point by an arrow. We picked the length of time to include a clear indication that the orbit is not closed. The open character of the trajectory was noticed, as well, in the original publication [43]. The trajectories then are densely packed as we can see in the figures above.

In Fig. 6.11 we have plotted a trajectory with energy above the critical energy for the mass to escape from the slight energy depression near the origin of the potential. The initial point for the trajectory is indicated by a filled circle and the direction of motion at the instant before the mass escapes the depression is indicated by an arrow.

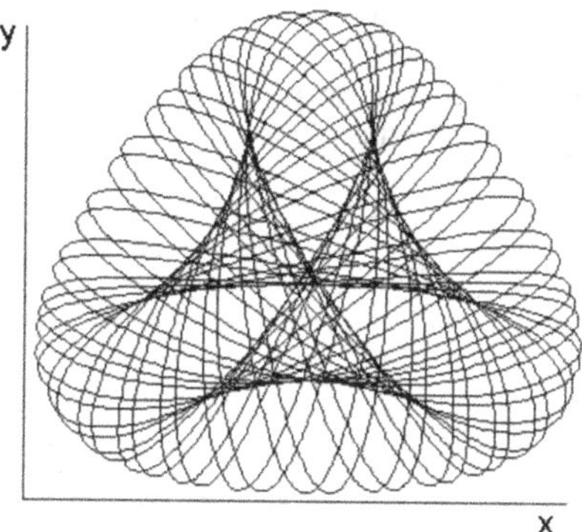

Fig. 6.8 Hénon–Heiles trajectory in (x, y) −plane

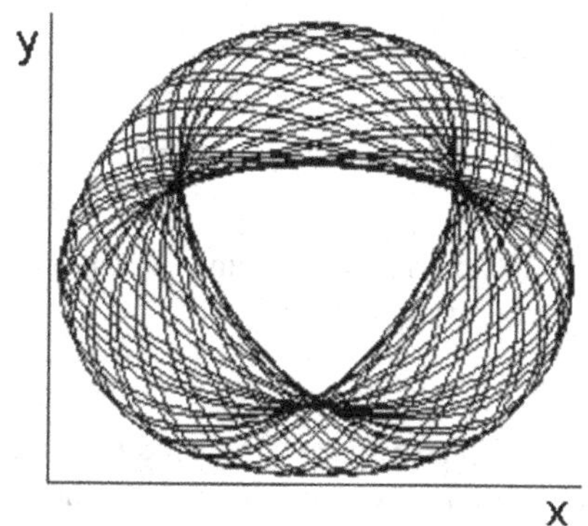

Fig. 6.9 Hénon–Heiles trajectory in (x, y) −plane

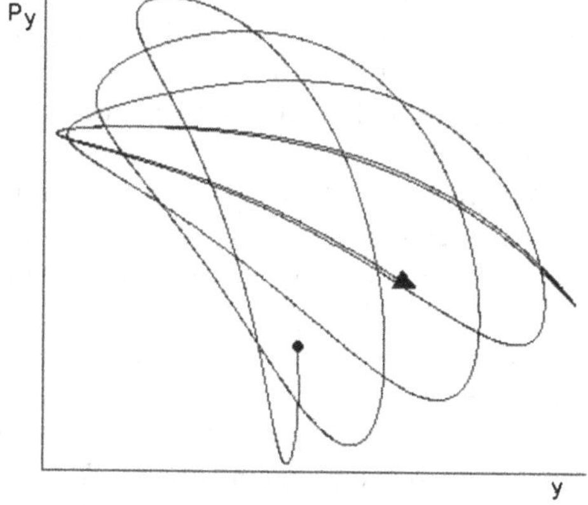

Fig. 6.10 Hénon–Heiles phase space trajectory in (y, p_y) −plane. The plot time begins at the filled circle after the system has settled and continues briefly to the arrow point. The orbit in phase space is not closed

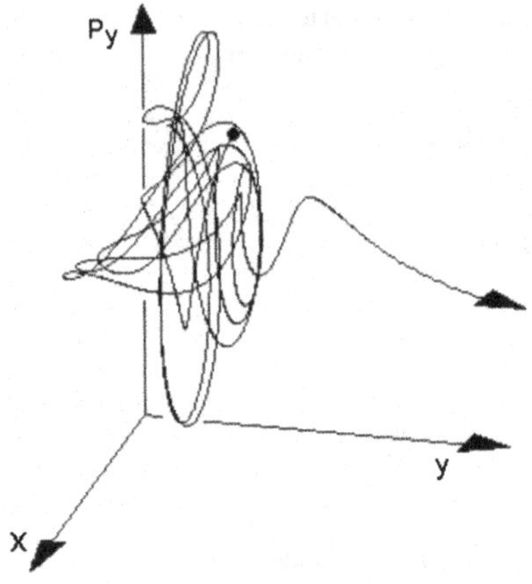

Fig. 6.11 Hénon–Heiles trajectory for an energy above the critical energy. The energy for this plot is $\mathcal{E} = 0.17406$

6.3.4 *Poincaré Sections*

The Hénon–Heiles system is a conservative Hamiltonian system. There is then a first integral of the motion

$$I_1 = \mathcal{H}\left(x, y, p_x, p_y\right) \tag{6.6}$$

and the phase space is reduced to three dimensions. For the potential (6.1) there is, however, no axial symmetry and, therefore, no second integral I_2. If a third integral exists there will be a reduction of the phase space to two dimensions. A Poincaré section will reduce this phase space to a single dimension, which is a line. Therefore the question of whether or not a third integral exists can be answered by considering the Poincaré section. If the Poincaré section is a line the third integral exists. But if the third integral does not exist the Poincaré section will be a scatter of points that eventually fill the space.

We may obtain a Poincaré section by limiting the coordinate x to values close to zero and recording the corresponding values of y and p_y as we did in Fig. 6.5 for the double oscillator. We show the resulting Poincaré section for the Hénon–Heiles system in Fig. 6.12. Here we see the lines from the projections of two toroidal surfaces that indicate the presence of a third integral I_3. The system energy in Fig. 6.12 was $\mathcal{E} = 0.119$.

If we increase the energy to $\mathcal{E} = 0.145$ we obtain the Poincaré section in Fig. 6.13. Here we notice the beginning of a breakup in the projection of the torus to the right. A further increase in energy results in the randomization of the points at which the trajectory crosses the plane indicating that the trajectory is no longer isolated to the tori observed in Figs. 6.12 and 6.13. We show this breakup of the trajectories in Fig. 6.14. The trajectories plotted in Fig. 6.14 are still deterministic in the sense that they result from the canonical equations. Compared to the plots in Figs. 6.12 and 6.13, however, the Poincaré plot has a random or chaotic appearance. The region has become ergodic and the third integral I_3 is no longer isolating.

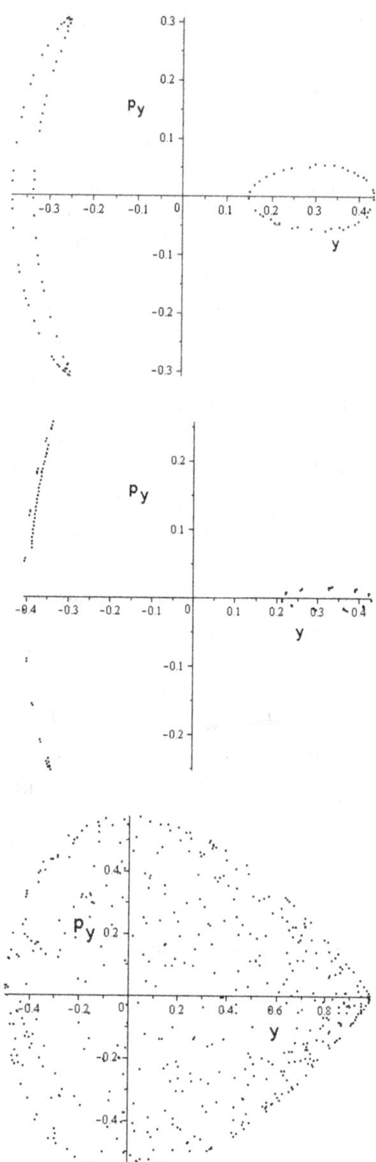

Fig. 6.12 Poincaré section for the Hénon–Heiles system. The presence of two tori on which the motion is confined are evident. The energy for this plot is $\mathcal{E} = 0.119$

Fig. 6.13 Poincaré section for the Hénon–Heiles system. The image of the torus on the right hand side of this plot is on the verge of breaking up. The energy for this plot is $\mathcal{E} = 0.145$

Fig. 6.14 Poincaré section for the Hénon–Heiles system. Both of the tori in the preceding Figs. 6.12 and 6.13 have broken up and the region has become ergodic. The energy for this plot is $\mathcal{E} = 0.17406$

6.4 Summary

Our intention in this chapter was to introduce the analysis of integrable complex systems. This analysis will always involve numerical solutions. And our analysis of the dynamical behavior of the system will be based on our comprehension of the manner in which the numerical results are presented. We can obtain insight into the character and the frequencies of the motion by turning to the action and angle variables. If we seek a more detailed picture of the motion we may turn to a numerical integration of the canonical equations and phase space representations. Then we can identify the surfaces on which the trajectories of the system lie using Poincaré sections.

We considered two systems of dramatically different levels of complexity. The first of these, the double oscillator, we could actually solve in terms of eigenvectors if we chose. And the solution in terms of action and angle variables was simple, providing an algebraic form for the ratio of frequencies. The final Poincaré section provided nothing that we could not have discovered algebraically. But the relationship between the projections of the phase plots and the Poincaré section is instructive.

The Hénon–Heiles system is very familiar in any study of complex systems. And we chose this system for the same reasons most authors have. It is a system that can be treated numerically without great difficulty and is a logical step in our understanding of the analysis of complex systems. We have, however, based our discussion on the problem posed initially by Hénon and Heiles, which was to study the possibility of a third integral of the motion. They began with a star in a galaxy. But the system they finally studied was no longer appropriate for a star. It was, however, appropriate for the question of interest to them.

In the first part of our study of the Hénon–Heiles system we simply presented the results of numerical studies of the bound motion and the escape of the system from the shallow potential well. Although the beautiful symmetries of the bound motion reveal that the motion is confined to tori for certain energy limits, they do not provide insight into the breakup of that symmetric motion and the transition to a more complicated motion as the energy increases. This was the original objective of the study and was ours here. For this study we turned, as did Hénon and Heiles, to a Poincaré section. We found that even for motion confined to the shallow potential well there was a transition from motion confined to tori to a motion in which the tori break up and the motion, although still deterministic, becomes chaotic.

Chapter 7
Chaos in Dynamical Systems

The phenomenon of chaos could have been discovered long, long ago. It wasn't, in part because this huge body of work on the dynamics of regular motion didn't lead in that direction. But if you just look, there it is.

Norman Packard

7.1 Introduction

At the end of our study of the Hénon–Heiles system we increased the energy to a value for which the system trajectory filled the entire phase space and was no longer confined to the tori we initially identified. This motion was still determined by the canonical equations. But the phase space had become ergodic and the motion of the system became very sensitive to the conditions at any instant. The motion of such systems is termed chaotic.

When we first became interested in chaotic dynamical systems, as a result primarily of the ability to integrate complex sets of differential equations on desktop computers, we found it relatively easy to identify the universal characteristics of a chaotic system. And we defined the concept of dynamical chaos based on those characteristics. That is no longer possible. Our understanding of chaos has shown the limitations of our previous ideas. One characteristic remains as a universal property of chaotic systems. There is a *sensitive dependence on initial conditions* (SDIC). This is, however, not limited to dynamical chaos. We must, finally, realize that there is no set of defining characteristics for dynamical chaos agreed upon among physicists and mathematicians [4].

Here we will investigate dynamical chaos by considering a single system as an example. We chose the Rössler (Otto Rössler) system as this example.

C.S. Helrich, *Analytical Mechanics*, Undergraduate Lecture Notes in Physics, DOI 10.1007/978-3-319-44491-8_7

We again have essentially no exercises for this chapter for the same reasons in the previous chapter. The examples we suggest are rather simple. If students have access to appropriate software the Rössler system is simple and fruitful.

7.2 The Rössler System

The Rössler system is a mathematical system of three equations, which is not obtainable from any set of canonical equations. It is, however, a system that represents, fairly simply, the salient properties of chaotic systems. As we study this system we must only realize that we cannot base the appreciation we may have for the numerical results on any mental picture of particle dynamics.

In using the Rössler system as a fruitful and mathematically beautiful system we will avoid any speculation regarding Rössler's thoughts or concerns beyond the development of this system.

7.2.1 Rössler Equations

The Rössler system is completely described by the set of first order differential equations

$$\frac{\mathrm{d}x}{\mathrm{d}t} = -y - z, \tag{7.1}$$

$$\frac{\mathrm{d}y}{\mathrm{d}t} = x + ay, \tag{7.2}$$

and

$$\frac{\mathrm{d}z}{\mathrm{d}t} = bx - cz + xz. \tag{7.3}$$

The quadratic term in (7.3) creates the interest in an otherwise linear system. Even though the system does not represent particle motion of any sort, we will refer to the solution of the Eqs. (7.1)–(7.3) as trajectories for the system in the phase space (x, y, z).

The equilibrium points for the system are obtained when $\mathrm{d}x/\mathrm{d}t = \mathrm{d}y/\mathrm{d}t = \mathrm{d}z/\mathrm{d}t = 0$. At these equilibrium points the system of equations (7.1)–(7.3) becomes

$$0 = -y - z \tag{7.4}$$

$$0 = x + ay \tag{7.5}$$

$$0 = bx - cz + xz. \tag{7.6}$$

The solutions to the set of Eqs. (7.4)–(7.6) are at the origin of coordinates $x = y = z = 0$ and the point

$$x = c - ab$$
$$y = b - \frac{c}{a}$$
$$z = \frac{c}{a} - b. \qquad (7.7)$$

(see exercises). We will use points very near these equilibrium points as initial conditions for our numerical studies of the Rössler system.

7.2.2 Numerical Solution

We solved the system of equations (7.1)–(7.3) using a *Runge–Kutta* algorithm. For our numerical calculations we chose the parameters of the system to be $a = 0.32, b = 0.3$, and $c = 4.5$, except where otherwise indicated. We conducted all calculations on Maple.

If we release the system from a point very close to the origin we obtain the time evolution of the coordinate $x(t)$ shown in Fig. 7.1 and the time evolution of the coordinate $z(t)$ shown in Fig. 7.2. Aside from noting the aperiodic character of the

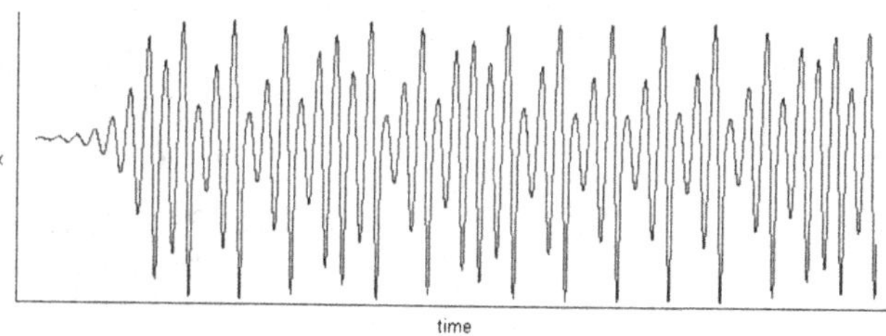

Fig. 7.1 Plot of $x(t)$ versus t for the Rössler system. The initial point was $x = 0.00001$, $y = 0.00001, z = 0.00001$

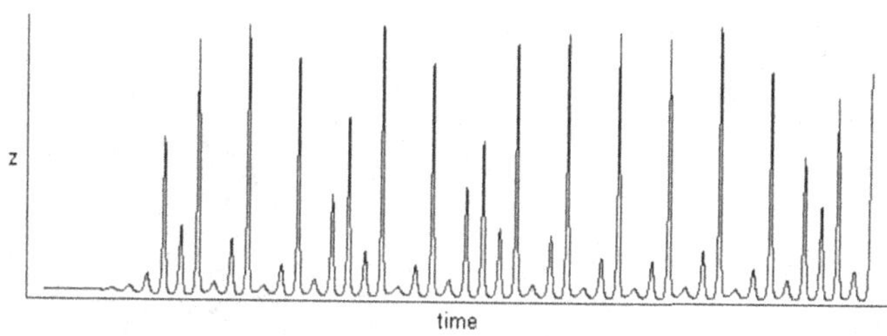

Fig. 7.2 Plot of $z(t)$ versus t for the Rössler system. The initial point was $x = 0.00001$, $y = 0.00001, z = 0.00001$

motion, which was once considered characteristic of dynamical chaos, and that the trajectory is confined we learn very little from these plots.

7.2.3 Rössler Attractor I

If we plot the three phase space coordinates (x, y, z) we obtain the plot in Fig. 7.3 that corresponds to that in Fig. 6.7 for the Hénon–Heiles system, except there is no momentum in the Rössler system.

The darkened origin in Fig. 7.3 results from the initial part of the trajectory, which we can see at the left of Figs. 7.1 and 7.2. The trajectory never returns to the neighborhood of this point, and remains on a surface that is becoming evident in Fig. 7.3.

If we release the system from a point very near the second equilibrium point (7.7) we obtain the plot in Fig. 7.4. In Fig. 7.4 we see that the trajectory remains on the same surface as that which we identified in Fig. 7.3. Although the system has

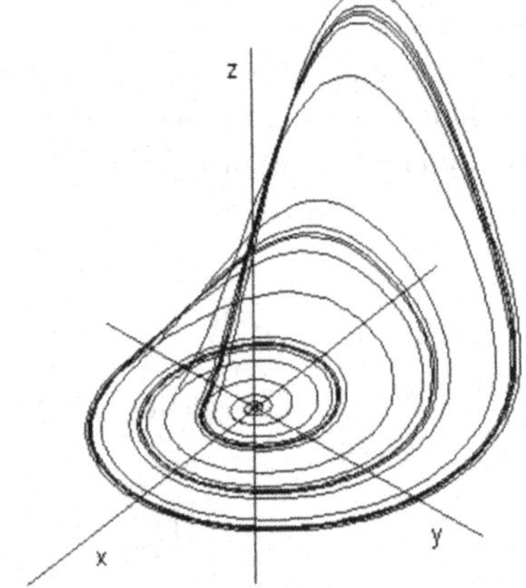

Fig. 7.3 Plot of (x, y, z) for the Rössler system released from the initial point $x = 0.00001$, $y = 0.00001$, $z = 0.00001$

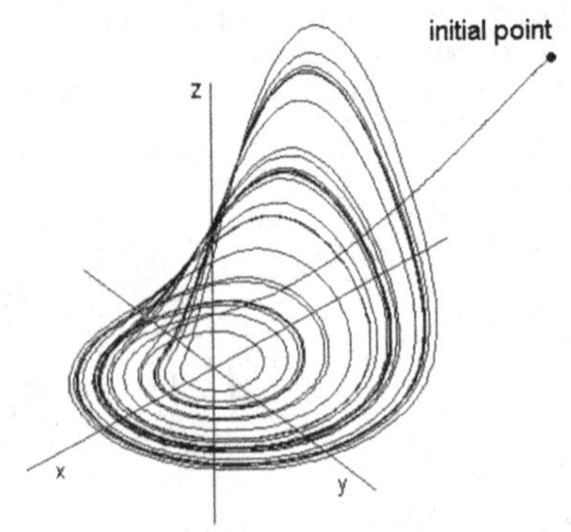

Fig. 7.4 Plot of (x, y, z) for the Rössler system released from the initial point $x = c - ab + 0.00001$, $y = b - c/a + 0.00001$, $z = c/a - b + 0.00001$

SDIC the trajectory is confined to a surface. The neighborhood of the initial point is excluded from the surface.

If we increase the length of time for the solution in Fig. 7.3 we obtain a more complete picture of the surface on which the trajectory of the Rössler system lies. We show this surface in Fig. 7.5. In Fig. 7.5 we have plotted only the final portion (950 out of 1000 points) of the data gathered from the numerical solution of the Rössler system. The surface we have found in Figs. 7.3, 7.4, and 7.5 is known as an *attractor* for the motion. The trajectory is attracted to and remains on the attractor.

If we vary the parameter c in the Rössler system we obtain some understanding of what is termed *bifurcation* in a chaotic system. A chaotic system may have multiple periodicities of the motion, which may appear as certain parameters of the system are altered. We studied this phenomenon in the Rössler system by holding the parameters a and b fixed and varying the parameter c. In Fig. 7.6 we have plotted the attractor in (x, y, z) coordinates for $a = 0.32, b = 0.3$, and (a) $c = 1.0$, (b) $c = 3.0$, (c) $c = 3.5$, (d) $c = 3.6$, (e) $c = 3.9$, and (f) $c = 8.0$. In each case we have plotted only the final portion of the trajectory so that the attractor alone is visible. The values of c chosen for the plots are characteristic for the ranges in which the patterns exist. In Fig. 7.6 we can see the development of the attractor characterized by a general increase in the number of bands in the trajectory. In panels (a), (b), and (c) the bands are 1, 2 and 4 respectively. There are separate bands in (d) and (f). In panel (e), however, the bands have begun to dissolve and we have a chaotic pattern emerging, although the trajectory remains on the attractor.

If we study the attractor in Fig. 7.4 or in Fig. 7.6f we note that where the trajectory appears to leave that part of the attractor approximately in the (x, y) −plane there is an apparent branching of the trajectory. But this is not the branching of a single trajectory. Rather it is a folding of surfaces onto, but not into, one another. This is indicative of an unfamiliar topology [see [25], pp. 94–97].

At the very least we see that there are striking differences between the tori we found supporting the trajectories of the Hénon–Heiles system (at lower energies) and the attractor we identified for the trajectories of the R össler system. We may,

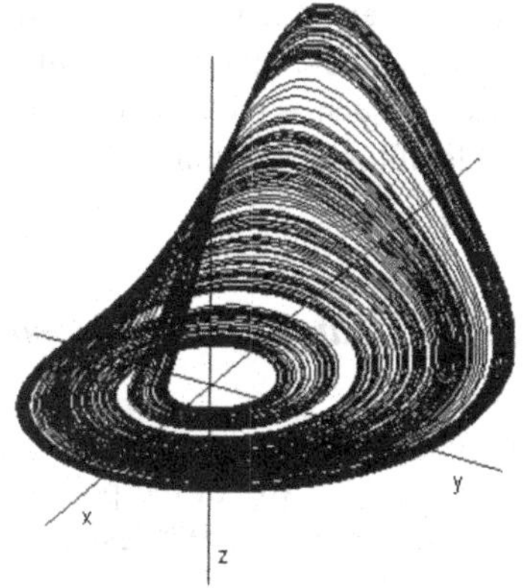

Fig. 7.5 Rössler attractor with longer integration time than in Figs. 7.3 and 7.4

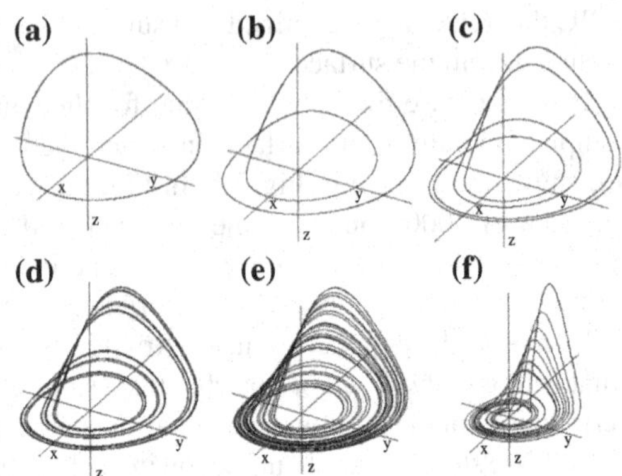

Fig. 7.6 Bifurcation in the Rössler system. Separate bands appear in the trajectory as the parameter c is varied, while a and b are held constant

therefore, expect a fundamental difference in the systems. The difference lies in the fact that the Hénon–Heiles system is a Hamiltonian system and the Rössler system is not. The Rössler system is what is termed a *dissipative* system.

7.3 Dissipation and Attractors

All systems do not have attractors. The systems to which we have devoted most of our study, for example, have been conservative systems for which the Hamiltonian is a constant. For these systems the trajectory lies on the surface $\mathcal{H} = \text{constant}$ for all time. The trajectories of integrable systems, such as the Hénon–Heiles system for low energies, lie on tori. These can be detected, as we did, and as Hénon and Heiles originally did, through Poincaré sections. Such systems, however, are not attracted to these surfaces. The trajectories simply lie on families of these surfaces. The initial conditions determine which of the families of surfaces are occupied by the system trajectory.

The fundamental difference between this behavior and that of a system such as the Rössler system is the fact that initial phase point does not necessarily lie on the surface eventually containing the trajectories. We saw this in Figs. 7.3 and 7.4. Systems such as the Rössler system have a distinct time directionality that is lacking in Hamiltonian systems.

7.3.1 Hamiltonian Systems

We consider a Hamiltonian system with N coordinates and N momenta $(q_1, \ldots, q_N, p_1, \ldots, p_N)$. Any of the points in this phase space may, at any time, be the state point of the system. We designate the density of the state points as τ and consider an infinitesimal volume in the phase space for which the ranges of the ith generalized coordinate and momentum are $q_{\mathrm{i}} \rightarrow q_{\mathrm{i}}+\mathrm{d}q_{\mathrm{i}}$ and $p_{\mathrm{i}} \rightarrow p_{\mathrm{i}}+\mathrm{d}p_{\mathrm{i}}$. These ranges define an infinitesimal volume in the $2N$ dimensional phase space

$$\begin{aligned} \mathrm{d}\Omega &= \mathrm{d}q_1 \cdots \mathrm{d}q_{\mathrm{N}} \mathrm{d}p_1 \cdots \mathrm{d}p_{\mathrm{N}} \\ &= \prod_{\mathrm{j}}^{\mathrm{N}} \mathrm{d}q_{\mathrm{j}} \mathrm{d}p_{\mathrm{j}}. \end{aligned} \tag{7.8}$$

If we fix the volume $\mathrm{d}\Omega$ we can compute the number of state points (trajectories) entering $\mathrm{d}\Omega$ across the differential area

$$\mathrm{d}S_{\mathrm{q,i}} = \mathrm{d}q_1 \cdots \mathrm{d}q_{\mathrm{i-1}} \mathrm{d}q_{\mathrm{i+1}} \cdots \mathrm{d}q_{\mathrm{N}} \mathrm{d}p_1 \cdots \mathrm{d}p_{\mathrm{N}} \tag{7.9}$$

normal to the generalized coordinate q_{i} as

$$\tau \dot{q}_{\mathrm{i}} \mathrm{d}S_{\mathrm{q,i}}$$

and those leaving diagonally across $\mathrm{d}\Omega$ as

$$\tau \dot{q}_{\mathrm{i}} \mathrm{d}S_{\mathrm{q,i}} + \frac{\partial}{\partial q_{\mathrm{i}}} \left(\tau \dot{q}_{\mathrm{i}}\right) \mathrm{d}q_{\mathrm{i}} \mathrm{d}S_{\mathrm{q,i}}$$

The net rate of increase of state points in $\mathrm{d}\Omega$ from motion of state points along q_{i} is then

$$-\frac{\partial}{\partial q_{\mathrm{i}}} \left(\tau \dot{q}_{\mathrm{i}}\right) \mathrm{d}q_{\mathrm{i}} \mathrm{d}S_{\mathrm{q,i}} = -\frac{\partial}{\partial q_{\mathrm{i}}} \left(\tau \dot{q}_{\mathrm{i}}\right) \mathrm{d}\Omega,$$

since $\mathrm{d}q_{\mathrm{i}} \mathrm{d}S_{\mathrm{q,i}} = \mathrm{d}\Omega$. Summing this result over all faces of the volume $\mathrm{d}\Omega$ we have the total rate of increase of state points in $\mathrm{d}\Omega$ as

$$-\sum_{\mathrm{i}} \left[\frac{\partial}{\partial q_{\mathrm{i}}} \left(\tau \dot{q}_{\mathrm{i}}\right) + \frac{\partial}{\partial p_{\mathrm{i}}} \left(\tau \dot{p}_{\mathrm{i}}\right)\right] \mathrm{d}\Omega. \tag{7.10}$$

The summation includes, of course, the canonical momenta on the same basis as the generalized coordinates. In the case of the canonical momentum p_{i} the facial area is $\mathrm{d}S_{\mathrm{p,i}}$ and $\mathrm{d}p_{\mathrm{i}} \mathrm{d}S_{\mathrm{q,i}} = \mathrm{d}\Omega$.

The result in Eq. (7.10) is equal to $(\partial \tau / \partial t) \mathrm{d}\Omega$. Therefore

$$\begin{aligned} \frac{\partial \tau}{\partial t} + \sum_{\mathrm{i}} \left[\dot{q}_{\mathrm{i}} \frac{\partial \tau}{\partial q_{\mathrm{i}}} + \dot{p}_{\mathrm{i}} \frac{\partial \tau}{\partial p_{\mathrm{i}}}\right] &= \tau \sum_{\mathrm{i}} \left[\frac{\partial}{\partial q_{\mathrm{i}}} \left(\dot{q}_{\mathrm{i}}\right) + \frac{\partial}{\partial p_{\mathrm{i}}} \left(\dot{p}_{\mathrm{i}}\right)\right] \\ &= \tau \sum_{\mathrm{i}} \left[\frac{\partial}{\partial q_{\mathrm{i}}} \left(\frac{\partial \mathcal{H}}{\partial p_{\mathrm{i}}}\right) + \frac{\partial}{\partial p_{\mathrm{i}}} \left(-\frac{\partial \mathcal{H}}{\partial q_{\mathrm{i}}}\right)\right]. \end{aligned} \tag{7.11}$$

Because the order of partial differentiation is immaterial, (7.11) becomes

$$\boxed{\partial\tau/\partial t + \sum_{\mathrm{i}} \left[\dot{q}_{\mathrm{i}}\left(\partial\tau/\partial q_{\mathrm{i}}\right) + \dot{p}_{\mathrm{i}}\left(\partial\tau/\partial p_{\mathrm{i}}\right)\right] = 0,} \tag{7.12}$$

Equation (7.12) is the mathematical expression of the *Liouville theorem*, which is fundamental in statistical mechanics [see [28], p. 13; [32], pp. 7–18].

We may also consider the rate of change of a volume in phase space

$$\Omega = \int_{\Omega} \mathrm{d}\Omega = \int_{\Omega} \mathrm{d}q_1 \cdots \mathrm{d}q_{\mathrm{N}} \mathrm{d}p_1 \cdots \mathrm{d}p_{\mathrm{N}}, \tag{7.13}$$

which is called the *extension in phase* [[32], p. 18]. The size of the volume will change as a result of the changes in the coordinates and momenta on the surface of the volume, which satisfy Hamilton's canonical equations. The change in the volume Ω in (7.13) is then

$$\frac{\mathrm{d}}{\mathrm{d}t}\int_{\Omega} \mathrm{d}\Omega = \oint_{\mathrm{S}_\Omega} \sum_{\mathrm{i}}^{\mathrm{N}} \left(\dot{q}_{\mathrm{i}}\mathrm{d}S_{\mathrm{q,i}} + \dot{p}_{\mathrm{i}}\mathrm{d}S_{\mathrm{p,i}}\right). \tag{7.14}$$

The integral on the right hand side of (7.14) is over the entire bounding surface of the volume S_Ω. Applying Gauss' theorem to (7.14) we obtain

$$\begin{aligned}\frac{\mathrm{d}}{\mathrm{d}t}\int_{\Omega} \mathrm{d}\Omega &= \int_{\Omega} \sum_{\mathrm{i}}^{\mathrm{N}} \left(\frac{\partial}{\partial q_{\mathrm{i}}}\dot{q}_{\mathrm{i}} + \frac{\partial}{\partial p_{\mathrm{i}}}\dot{p}_{\mathrm{i}}\right) \mathrm{d}\Omega \\ &= \int_{\Omega} \sum_{\mathrm{i}}^{\mathrm{N}} \left(\frac{\partial}{\partial q_{\mathrm{i}}}\frac{\partial \mathcal{H}}{\partial p_{\mathrm{i}}} - \frac{\partial}{\partial p_{\mathrm{i}}}\frac{\partial \mathcal{H}}{\partial q_{\mathrm{i}}}\right) \mathrm{d}\Omega \\ &= 0. \end{aligned} \tag{7.15}$$

The volume containing the system trajectory then does not change in time provided the system is a Hamiltonian system. This is the constancy of the extension in phase in statistical mechanics [see [28], pp. 11–13, [48], pp. 208-213, [32], pp. 7–11].

We have then a fundamental picture of the behavior in phase space of the trajectory of a Hamiltonian system. The volume containing the trajectory remains constant. But the volume may change shape. The content of this theorem is that the density of states τ for a Hamiltonian system behaves like an incompressible fluid in the phase space. There can be then no compression of the representative points (the trajectories) of a Hamiltonian system. Therefore there is no attractor for a Hamiltonian system.

7.3.2 Dissipative Systems

A physical dissipative system loses energy to the surroundings and is kept in motion by a driving force. A common example of such a system is the damped and driven pendulum. Defining $q = \vartheta$, $p = \mathrm{d}\vartheta/\mathrm{d}t$, and the driving force as $T \sin(\omega t)$, the equations of motion are

$$\frac{\mathrm{d}}{\mathrm{d}t}p = -\nu p - \sin(q) + T\sin(\omega t)$$
$$\frac{\mathrm{d}}{\mathrm{d}t}q = p.$$

For this system the first line of (7.15) is

$$\frac{\mathrm{d}}{\mathrm{d}t}\int_{\Omega}\mathrm{d}\Omega = \frac{\mathrm{d}}{\mathrm{d}t}\Omega(t) = -\nu\Omega(t). \tag{7.16}$$

Then

$$\Omega(t) = \exp(-\nu t)$$

and the volume of the system in phase space decreases exponentially in time. Dissipative systems will contract to attractors in phase space [cf. [98], p. 9].

For the Rössler system the rate of change of the volume in phase space is

$$\frac{\mathrm{d}}{\mathrm{d}t}\Omega = (a - c)\Omega + F_{\Omega}(x, y, z), \tag{7.17}$$

where

$$F_{\Omega}(x, y, z) = \int_{\Omega} x\mathrm{d}\Omega. \tag{7.18}$$

and $F_{\Omega}(x, y, z)$ is generally sensitive to the motion of the system in phase space (see exercises). The rate of change of the phase space volume is negative and the trajectory has an attractor only when $(a - c)\Omega + F_{\Omega}(x, y, z) < 0$. If at some time $(a - c)\Omega + F_{\Omega}(x, y, z) > 0$ then in the next instant there will be a growth in the volume of phase space (extension in phase) available to the Rössler system and the next phase point may result in a further increase in the value of Ω. The trajectory may then deviate completely from the confines of the attractor.

A Hamiltonian system is what is generally called a conservative system for which the total mechanical energy is constant. But in many systems, particularly in model systems, there may not be an identifiable energy. If the volume in phase space remains constant we still, however, refer to the system as conservative. Similarly we refer to a system whose volume is not constant as dissipative. This definition includes decreasing volumes, as in the case of the damped and driven oscillator, and systems for which the volume may decrease or increase, as the Rössler system. In the context of this general definition only dissipative systems may have attractors. But there is no claim that a dissipative system necessarily loses energy.

7.4 Rössler Attractor II

7.4.1 Dissipation

To understand the separation of the dissipative Rössler system into regions for which $\mathrm{d}\Omega/\mathrm{d}t$ is either positive or negative we choose $a = 0.32$, $b = 0.3$ and investigate the time dependence of the coordinates x and z that results from variation in the parameter c. In Fig. 7.7 we plot the time development of the coordinates x and z for $c = 0.737$ (panels (a) and (b)) and $c = 0.736$ (panels (c) and (d)). We carried out the calculations using a Runge–Kutta algorithm with discrete time steps and limited time durations. For the case in which $c = 0.737$ the volume occupied by the trajectory is confined for the time duration considered. With $c = 0.736$, however, we see the development of a rapid growth in x and a corresponding very rapid growth in z. This numerical experiment has at least indicated that for certain ranges of the parameters of the system there may be no attractor.

We have plotted the phase space trajectory of the system when $c = 0.737$ in Fig. 7.8. For this value of the parameter c the trajectory converges on the single band attractor that we discovered in Fig. 7.6a.

If we plot the phase space trajectory for the system with $c = 0.736$, as we have done in Fig. 7.9, we find the unbounded behavior indicated in Fig. 7.7 panels (c) and (d). The arrow indicates the direction of the trajectory for increasing time. Here there is no attractor.

In both of the cases we considered $a - c < 0$. So the determining factor is $F_{\Omega}(x, y, z)$ defined in (7.18). We can also see that $F_{\Omega}(x, y, z)$ determines the transition in $\mathrm{d}\Omega/\mathrm{d}t$ from negative to positive in Fig. 7.7 panels (c) and (d). There the sudden comparatively large change in x just before $t = 500$ drives the very large change in z. Once a threshold has been surpassed the system begins a rapid divergence.

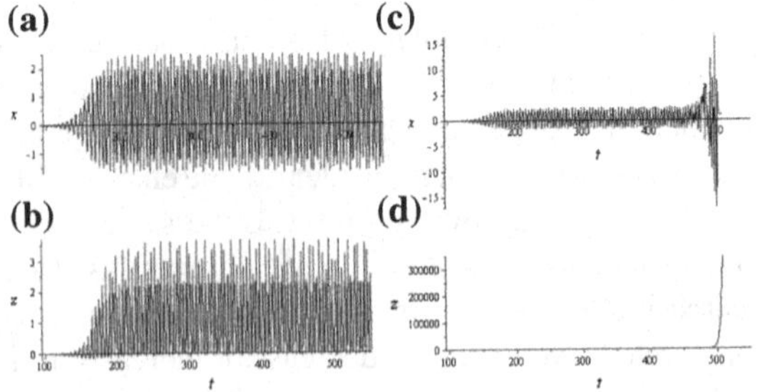

Fig. 7.7 Attraction and divergence of the Rössler system

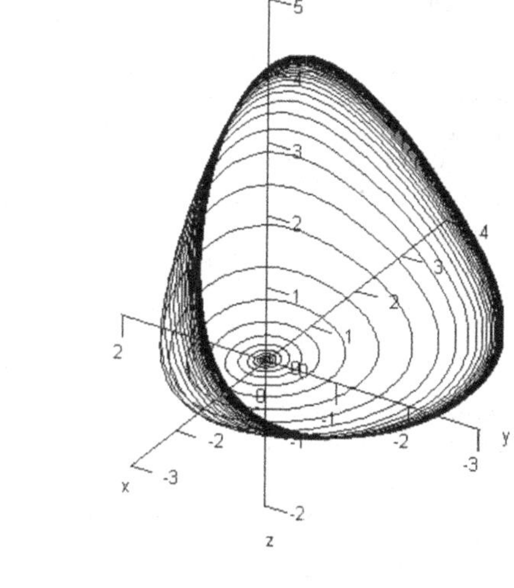

Fig. 7.8 The Rössler attractor for $a = 0.32$, $b = 0.3$ and $c = 0.737$. The initial point was $x = 0.00001$, $y = 0.00001$, $z = 0.00001$

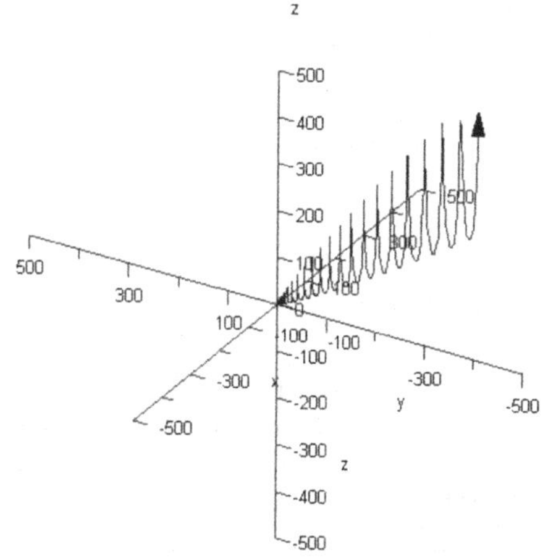

Fig. 7.9 The Rössler system for $a = 0.32$, $b = 0.3$ and $c = 0.736$. for this choice of parameters there is no attractor. The initial point was $x = 0.00001$, $y = 0.00001$, $z = 0.00001$

7.4.2 Poincaré Sections

In Fig. 7.10 we have plotted two Poincaré sections of the Rössler system using parameter values for which there is definitely an attractor. These Poincaré sections correspond to the attractor in Figs. 7.4 and 7.5. The sections differ only in the choice of which variable (x or y) is to be set to zero. Although these Poincaré sections may initially appear to be representations of a planar surface, if we consider closely the portion of the plot in panel (a) between the arrows we see evidence of the two surfaces we previously indicated were present in the attractor. The parts of the trajectory on these two surfaces become almost intermingled into a single surface, although the trajectory never crosses itself.

A Poincaré section of the attractor in Fig. 7.6 panel (f) with $a = 0.32$, $b = 0.3$, and $c = 8.0$ opens the two surfaces we identified in Fig. 7.10 panel (a). We show this Poincaré section in Fig. 7.11. In Fig. 7.11 we have indicated the point at which the

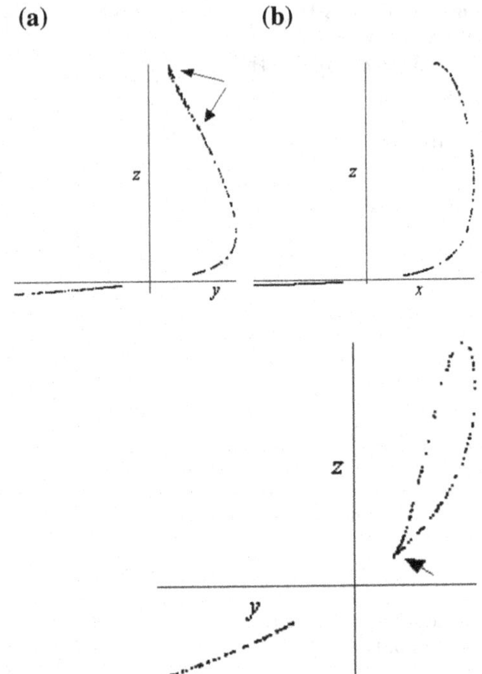

Fig. 7.10 Poincare plots for the Rössler system. **a** z versus y with $x = 0$, **b** z versus x with $y = 0$. Parameters in each plot are $a = 0.32$, $b = 0.3$, and $c = 4.5$

Fig. 7.11 Poincaré section of Rössler attractor with parameters $a = 0.32$, $b = 0.3$, and $c = 8.0$. The arrow indicates the point at which the two surfaces join

two surfaces almost join by an arrow. This folding of surfaces into almost a single surface is a characteristic of fractal geometry [cf. [25], pp. 94–97].

Another characteristic of fractal geometry is self similarity at different scales [cf. [61], pp. 4–5, [73]]. For the Rö ssler system we can see this if we read the data of our numerical solution of the equations on different time scales. In the top panel of Fig. 7.12 we have plotted the values of the coordinate z for each point of the numerical solution and in the bottom panel we have plotted the values for every fifth point of the numerical solution. We notice that, although a more detailed structure is revealed in the top panel in Fig. 7.12, the time self similarity is evident.

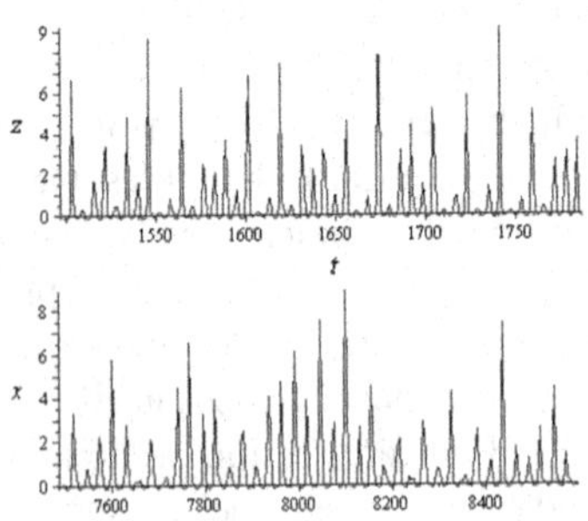

Fig. 7.12 Self similarity on different time scales in the Rössler system

The attractor for the Rössler system is not a two dimensional surface. The dimension of the attractor is not an integer but a fraction. Such a surface is termed a fractal and an attractor with a fractal dimension is traditionally called a "strange attractor".

7.5 Attractor Dimension

7.5.1 Definition of Dimension

We consider the problem of covering an arbitrary surface that we have drawn on a sheet of paper by dividing the surface into a number of small squares. If the side of the square has a length ε and if the surface has an area A, we can cover the surface approximately with

$$N_{\varepsilon} = \frac{A}{\varepsilon^2} \tag{7.19}$$

of these small squares. In the limit as $\varepsilon \to 0$ the squares become infinitesimal and we cover the surface completely. So we must consider the concept of covering in terms of this limit.

From (7.19) we have

$$\ln(N_{\varepsilon}) = 2\ln\left(\frac{1}{\varepsilon}\right) + \ln(A). \tag{7.20}$$

Or

$$2 = \frac{\ln(N_{\varepsilon})}{\ln\left(\frac{1}{\varepsilon}\right)} - \frac{\ln(A)}{\ln\left(\frac{1}{\varepsilon}\right)}. \tag{7.21}$$

Taking the limit of infinitesimal squares, $\lim_{\varepsilon\to 0} \ln(A) / \ln(1/\varepsilon) = 0$, therefore the dimension of our arbitrary surface is

$$2 = \lim_{\varepsilon\to 0} \frac{\ln(N_{\varepsilon})}{\ln\left(\frac{1}{\varepsilon}\right)}. \tag{7.22}$$

In the same fashion we may cover a line by using segments of length ε and a volume by using cubes with sides of length ε. And for general phase spaces we may speak in terms of volumes of higher dimension, which we may choose to call hypercubes. The concept of dimension, however, remains the same. A volume V_{n} in an n-dimensional (hyper) space is covered by N_{ε} infinitesimal hypercubes of volume ε^{n} and

$$n = \lim_{\varepsilon\to 0} \frac{\ln(N_{\varepsilon})}{\ln\left(\frac{1}{\varepsilon}\right)}. \tag{7.23}$$

If we consider a general surface, in a phase space, for example, we can define the dimension d of the surface as

$$d = \lim_{\varepsilon \to 0} \frac{\ln(N_\varepsilon)}{\ln\left(\frac{1}{\varepsilon}\right)}. \tag{7.24}$$

In particular we are interested in attractors on which trajectories of complex systems reside. Our investigations of these surfaces have been, and will always be, based on the numerical integration of the nonlinear system equations. Therefore our information about the attractor will be in the form of discrete points lying on the surface. The procedure we have outlined here for determining the dimension of a surface is known as *box counting*. We may apply this procedure readily to find the dimension of an attractor for which we have data in the form of discrete points. We require only an algorithm which defines the box size in terms of ε and determines, at each level ε the number of boxes that contain contain points at that level N_ε. This approach was first used by Andrey N. Kolmogorov in the study of dynamical systems in 1958 [[48], p. 343]. This is not the most efficient approach. But it is the approach we shall use here for the sake of transparency.

7.5.2 Fractal Dimension

The term fractal to designate a surface with noninteger dimension was introduced by Benoit Mandelbrot [73], but the concept of noninteger dimension had been recognized by mathematicians much earlier. We shall introduce the concept here using the Cantor set introduced by the German mathematician Georg Cantor (1845–1919).

The Cantor set is constructed beginning with a line of unit length. The first step divides this line into three equal lengths and removes the central section. What remains is two lines of length $1/3$. In step 2 each of these lines is divided into three equal lengths and the central section of each line removed. What remains is four lengths of $(1/3)^2$. At the Nth step there are 2^{N} line segments each of length $(1/3)^{\text{N}}$. We have presented a diagram of the Cantor set in Fig. 7.13. At the Nth step we then have a set consisting of 2^{N} elements, which we cover with segments (boxes) of length $\varepsilon = (1/3)^{\text{N}}$. From (7.24) the Cantor set has dimension

$$d_{\text{Cantor}} = \frac{\ln\left(2^{\text{N}}\right)}{\ln\left(3^{\text{N}}\right)} = \frac{\ln(2)}{\ln(3)} = 0.63093\ldots \tag{7.25}$$

We have then a set which, in the limit $N \to \infty$, approaches a collection of points (which still remain line segments) with noninteger dimension. The total length of the Cantor set can be found by subtracting the removed lengths from unity

Fig. 7.13 The Cantor set. In each step the central one third of each line segment is removed. The Cantor set is the limit of this process for an infinite number of steps

$$\begin{aligned}\ell_{\text{Cantor}} &= 1 - \frac{1}{3} - \frac{1}{3}\left(\frac{2}{3}\right) - \frac{1}{3}\left(\frac{2}{3}\right)^2 - \cdots \\ &= 1 - \frac{1}{3}\sum_{\text{k}=0}^{\infty}\left(\frac{2}{3}\right)^{\text{k}}. \end{aligned} \tag{7.26}$$

The geometrical series sum here evaluates to [see [12], p8]

$$\frac{1}{3}\sum_{\text{k}=0}^{\infty}\left(\frac{2}{3}\right)^{\text{k}} = \frac{1}{3}\left(1 - \frac{2}{3}\right)^{-1} = 1,$$

and $\ell_{\text{Cantor}} = 0$. The measure, or total length, of the Cantor set is zero. In the limit, then, the Cantor set resembles a collection of points, with zero total length. But the dimension of the Cantor set is not zero.

7.6 Rössler Box Counting

To obtain the dimension of the Rössler attractor we followed the box counting procedure outlined in Sect. 7.5.1. We chose the limits on the variables (x, y, z) from the extent of the attractor in Fig. 7.5 and began with a 5000 step numerical solution to the Rössler system with $a = 0.32$, $b = 0.3$, and $c = 4.5$. Each step in the solution was a point on the trajectory. Except for the initial points, as the trajectory settled on the attractor, each of these points was then a point on the attractor. At each step k in the algorithm we defined $\varepsilon \propto 1/k$, which divided the region of (x, y, z) space into a grid of three dimensional cubes. For each of these cubes we stepped through the trajectory points determining which of the cubes was occupied. For a particular k the number of occupied cubes was N_ε.

In Fig. 7.14 we have plotted $\ln(N_\varepsilon)$ versus $\ln(1/\varepsilon)$ for sixty steps in k, i.e. sixty values of the parameter ε. We carried out two separate calculations. In the first we included all 5000 points from the numerical solution. In the second we included only

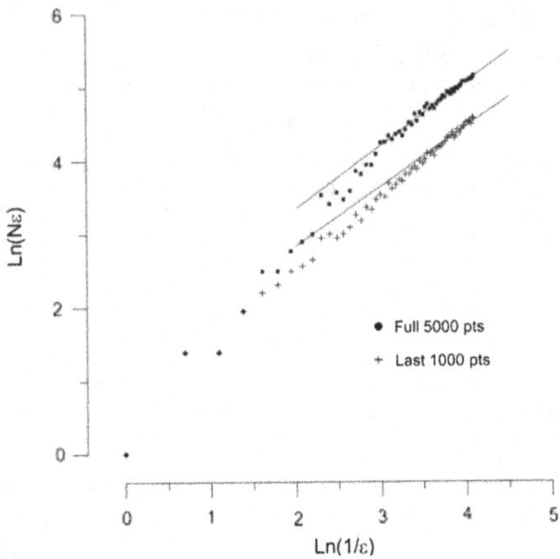

Fig. 7.14 Box Counting of Rössler system. Parameters were $a = 0.32$, $b = 0.3$, and $c = 4.5$. Numerical solution with 5000 pts. *Top plot* included full 5000 pts (fractal dim 0.86). *Bottom plot* included last 1000 pts (fractal dim 0.81)

the last 1000 points to eliminate any effects of the initial relaxation. In each case we fitted the last 15 points in the graph ($k = 45..60$) to the Eq. (7.24). This is the straight line in each plot. We then took these fits as approximations to the limit in (7.24). The slope of each line is then an approximation to the fractal dimension and the intercept is the value of $\ln(A)$ (see (7.20)). The fits are

$$\ln(N_\varepsilon) \approx 0.86 \ln\left(\frac{1}{\varepsilon}\right) + 1.62, \tag{7.27}$$

for the full 5000 points, and

$$\ln(N_\varepsilon) \approx 0.81 \ln\left(\frac{1}{\varepsilon}\right) + 1.22, \tag{7.28}$$

for the last 1000 points. From these results we see that, for the parameters considered, the Rössler attractor has a fractal dimension between 0.81 and 0.86. We then understand the Rössler system as a dissipative system whose attractor is a surface with fractal dimension. Such an attractor is called a *strange attractor*.

7.7 Summary

In this chapter we have considered the elements of dynamical chaos in the context of a single example: the Rössler system. Although our present understanding of the dynamics of complex systems allows for no set of defining characteristics for dynamical chaos, we were able to identify certain properties of the Rössler system that differed from what we had found for the Hénon–Heiles system. These properties are common to dynamical chaos, although they are not defining characteristics.

For our initial choice of parameters we found that the Rössler system contracted onto a surface that we identified as an attractor. By varying the third parameter c and holding the others constant we found a bifurcation in the motion, which we identified in the bands of the trajectory. For certain choices of c these bands were no longer distinguishable and the motion became what we termed chaotic.

We concentrated on the attractor for the Rössler system noting that this attractor had properties that distinguished it from the tori that held the trajectories of the Hénon–Heiles system. The fundamental reason for the difference in the behavior of the two systems lay in the fact that the Hé non–Heiles system is a Hamiltonian system while the Rössler system is a dissipative system. The phase space volume occupied by the Hénon–Heiles system was then a constant, while the phase space volume of the Rössler system was not. The fact that the Hénon–Heiles system is integrable for certain energies resulted also in the fact that the trajectory resides on tori for those limited energies. We found that the rate of change of the phase space volume of the Rössler system could be positive or negative depending on the choice of parameters. We showed that the regions of expansion and contraction were determined by the values of c for fixed a and b and that this agreed with our basic predictions.

We were able to find a folding of the portions of the attractor into one another that indicated a unique geometry that was further revealed in the Poincaré sections of the attractor. This, and the apparent self-similarity of the dynamics on different time scales, led us to consider the dimension of the attractor surface. The folding particularly indicated that the surface may have a noninteger dimension. We then introduced the mathematical concept of dimension and, by means of the Cantor set, showed that this can be noninteger. Using the box counting concept we then established that the dimension of the Rössler attractor is fractal. From our work on the dissipative properties of the Rössler system we know that this dimension is dependent on the values of the parameters.

In this chapter our intention was to provide some understanding of dynamical chaos by investigating some of the properties that are often associated with this type of motion. The system we chose made it possible to investigate these properties through numerical studies that were fairly transparent. We, however, made no pretense at anything resembling an exhaustive treatment.

7.8 Exercises

7.1 The equations for the Rössler system are

$$\frac{\mathrm{d}x}{\mathrm{d}t} = -y - z,$$

$$\frac{\mathrm{d}y}{\mathrm{d}t} = x + ay,$$

and

$$\frac{dz}{dt} = bx - cz + xz.$$

Following the argument in the chapter show that the rate of change of the Rössler system volume in phase space may be either positive or negative and that

$$\frac{d}{dt}\Omega = (a - c)\,\Omega + F_{\Omega}(x, y, z),$$

where

$$F_{\Omega}(x, y, z) = \int_{\Omega} x d\Omega.$$

7.2 We consider a simplified model of the water molecule with the oxygen atom at the coordinate x_2 and the two hydrogen atom at coordinates x_1 and x_3 along the same axis. We consider that the vibration of the hydrogen atoms is small enough that the potential energy of the bound hydrogen atoms can be approximated by a quadratic with a spring constant k. The mass of the hydrogen atom is m and the mass of the oxygen is μm. (a) Obtain the Hamiltonian of this model water molecule. (b) Is the Hamiltonian separable? (c) What are the frequencies of vibration of this model water molecule and the corresponding eigenvectors?

7.3 In the text we encountered the algebraic system of equations

$$\begin{aligned} 0 &= -y - z \\ 0 &= x + ay \\ 0 &= bx - cz + xz. \end{aligned}$$

for the condition that the velocity components for the Rössler system vanished, i.e. $dx/dt = dy/dt = dz/dt = 0$. Obtain the two solutions to this set of equations.

Chapter 8
Special Relativity

> Invention is not the product of logical thought, even though the final product is tied to a logical structure.
>
> Albert Einstein

8.1 Introduction

Much of our presentation here will be an abbreviation of the treatment provided in the text *The Classical Theory of Fields* [[45], pp. 273–306]. Because of the importance that must be attached to an understanding of the basis of special relativity, which Einstein called *der Schritt* (the step) [[89], p. 163], we have elected not to simply summarize the results.

Until this point we have accepted the validity of Euclidean geometry and the Newtonian concept of absolute space and time, which were separate and placed in the universe prior to the existence of anything else [[87], p. 6]. And we have allowed ourselves to imagine that we can observe all frames of reference from some separate at rest position in the universe. Einstein's realization that the Newtonian concept of time was seriously flawed led to a new understanding of time, space, and motion that was formalized as a geometry of four dimensions by Hermann Minkowski [[24], pp. 75–91].

In this chapter we will consider the modification of Analytical Mechanics that is required by the special theory of relativity, which was published by Albert Einstein in 1905 [[24], pp. 37–65]. Although the mathematical theory, accompanied by some descriptive notes, can be presented rather briefly, we choose a more careful approach. Physics faced a crisis at the end of the 19th century, which some of the greatest minds tried to resolve by adjusting the laws of mechanics and electrodynamics. Einstein

C.S. Helrich, *Analytical Mechanics*, Undergraduate Lecture Notes in Physics, DOI 10.1007/978-3-319-44491-8_8

identified the problem in our misconception of the meaning of time. Because of the importance of this step we will outline the original development of the special theory.

Our mathematical development will be based on Minkowski's formalism and his realization that we must speak now of space-time rather than space and time. Our vectors must now be four dimensional and we must accept that space-time is not Euclidean. Tensor algebra is natural for our treatment. So we present a brief introduction to tensor algebra providing what is needed for our treatment of Analytical Mechanics and our understanding of the covariance of the Lagrange–Hamilton approach under Lorentz transformation.

We will find that more than an algebraic modification is needed to establish the laws of mechanics in relativistic form. The basic concepts of mass, momentum and energy must be modified. Our first covariant law will be the conservation of momentum. And this simple law alone will provide us an understanding of some fundamental physics, which we can apply, before we attempt to understand forces. We will then base our treatment of forces on electrodynamics, accepting Wolfgang Pauli's argument that the properties of the Lorentz electrodynamic force are representative of a large class of pure forces.

We will close the chapter with a complete discussion of the Euler–Lagrange and Hamiltonian mechanics, which has been the basis of our study from the beginning.

Throughout this chapter we will use the Einstein sum convention for repeated Greek indices [[24], p. 122] (see Sect. 4.3). We will note any exception that occurs.

In writing this chapter we have relied particularly on the treatments of this subject by P.G. Bergmann [3], W. Pauli [92], A. Einstein [24], and H. Weyl [123].

8.2 The New Kinematics

Einstein begins his paper on the special theory of relativity (*On the Electrodynamics of Moving Bodies*) pointing to the asymmetry in the understanding of Faraday's Law and the unsuccessful attempts to discover the motion of the earth relatively to the aether, which was supposedly the medium for the transport of the transverse light waves. The aether had been central in the thought of James Clerk Maxwell. The experimental failure of Albert A. Michelson and Edward W. Morley to detect the motion of the earth through the aether in 1887 [79] was a particularly disturbing result.

The Irish physicist George Francis FitzGerald and the Dutch physicist Hendrik Antoon Lorentz suggested a hypothesis that could reconcile the results of the Michelson–Morley experiment and preserve Newtonian space and time. There was a shortening of the length of material bodies as they moved in the aether. In a 1904 paper Lorentz presented what are now known as the Lorentz transformation equations based on shortening due to electromagnetic interactions in matter [[24], pp. 11–34]. The claim was that the problem lay in the formulation of Newtonian Mechanics, not in Newtonian space and time.

It is a point of some interest in the history of physics that technical expert third class at the Swiss Patent Office in Bern, Albert Einstein, did not consider the Michelson–Morley experiment, beyond his indication that attempts to measure the motion of the earth relatively to the aether had been unsuccessful [see [89], p. 172]. However, beyond the historical fact that Einstein was not attempting an explanation of the Michelson–Morley experiment, the fact that he approached the problem in a different way is of considerable importance for our understanding of the physics. Einstein recognized that any mathematical description of motion has no physical meaning unless we understand time. And he pointed out that all judgements in which time plays a part are judgements of *simultaneous events*. An event occurs at a certain time t_e if the occurrence of that event and the event that the time t_e appears on our clock or timepiece[1] are simultaneous [[24], p. 39].

To make this concept universal Einstein needed to consider the synchronization of timepieces in reference frames moving uniformly with respect to one another. These are called inertial frames. This realization led Einstein to a new kinematics and a subsequent proof that the equations of Maxwell were independent of transformation between inertial frames. The transformation equations from one inertial frame to another, which Einstein obtained, were those derived in 1904 by Lorentz. But the basis was entirely different from that of Lorentz. And Einstein was unfamiliar with any of Lorentz' work after 1895 [[89], p. 121].

Einstein based his development on only two postulates [[24], pp. 37–38]

1. The same laws of electrodynamics and optics will be valid for all frames of reference for which the equations of mechanics hold good.
2. Light is always propagated in empty space with a definite velocity c which is independent of the of the state of motion of the emitting body.

The first postulate, Einstein called the "Principle of Relativity."

8.2.1 *Time*

To expand the concept of the simultaneity of events from our immediate vicinity to an entire inertial reference frame Einstein defined a synchronization process for two timepieces at various points in the inertial frame. In Fig. 8.1 we have drawn a picture of the synchronization process. By Einstein's second postulate we know that the time required for a light pulse to travel from A to B is always the same as the time required for a pulse to travel from B to A.[2] We consider a light pulse emitted from A at time t_{A}. This pulse is then reflected from point B at time t_{B}, and arrives back at point A at the time t'_{A}. The timepieces are synchronized if

[1]The terminology Einstein used was *clock*. We have used *timepiece* because the clock with hands is becoming less common.

[2]If this time of transit were not the same synchronization would not be possible and time would lose its meaning for points that are not in our immediate neighborhood.

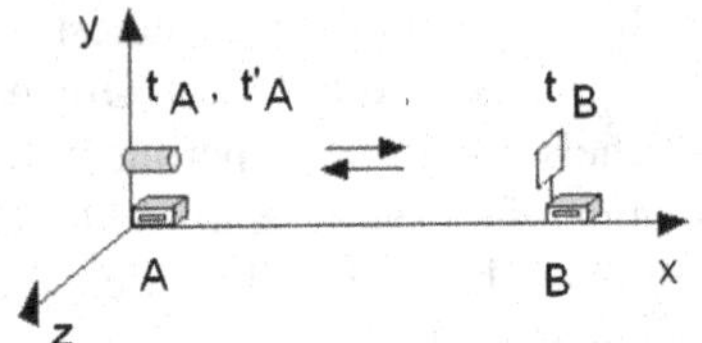

Fig. 8.1 Synchronization of stationary timepieces

$$t_{\mathrm{B}} - t_{\mathrm{A}} = t'_{\mathrm{A}} - t_{\mathrm{B}} \tag{8.1}$$

In this fashion we may synchronize all timepieces in a single inertial frame in which all points A and B are stationary with respect to one another. Because all timepieces in the frame are synchronized, we can speak about simultaneity of events in a particular inertial frame.

To connect the times measured in two inertial frames, Einstein devised a thought experiment.[3] He asked how a time synchronization experiment, conducted by a person in a moving inertial frame, would appear if observed by a person[4] in a stationary inertial frame.

We designate the stationary frame as k and the moving frame as k'. Frame k has coordinates (x, y, z) and the time t and the moving frame k' has coordinates (ξ, η, ζ) and the time τ. We choose the axes x and ξ to be aligned with one another and with the velocity v. We have drawn the inertial frames and the synchronization experiment in Fig. 8.2. The light source is located at the origin of frame k'. A person in frame k measures the distance between the light source and the reflector as

$$x' = x - vt. \tag{8.2}$$

At time τ_0, registered on a timepiece located at the origin of k', a light pulse is sent from the light source down the ξ-axis. This pulse is reflected from a mirror at a point on the ξ-axis. A timepiece at this point registers the time τ_1. The pulse returns to the origin arriving at time τ_2. The synchronization equation (8.1) requires that

$$\tau_1 = \frac{1}{2}\left(\tau_0 + \tau_2\right) \tag{8.3}$$

The person in frame k seeks a functional relationship between the time τ of the moving frame k' in terms of measurements made in frame k. In general this will be

$$\tau = \tau\left(x', y, z, t\right). \tag{8.4}$$

[3]From the German *Gedankenexperiment*. In a thought experiment it must be possible to construct the required apparatus and to perform all the measurements. A thought experiment is not fanciful.

[4]The standard term is "observer" for the German *Beobachter*. The use of person seems less awkward here.

With modern timepieces a single person can gather the data.

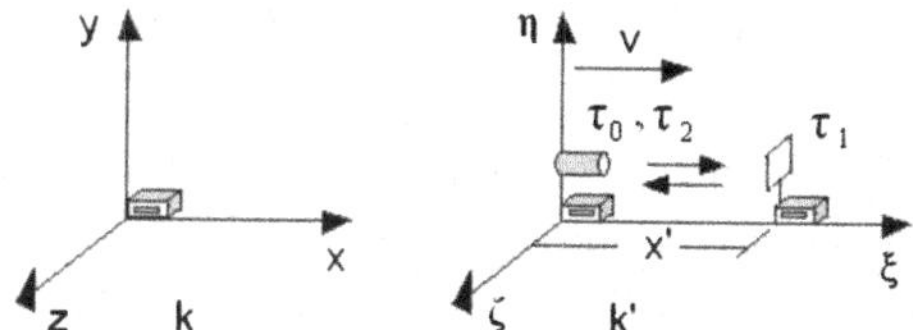

Fig. 8.2 Timepiece synchronization experiment conducted in the moving inertial system k' and observed from the stationary system k

Because of the situation being considered here, the x-coordinate is replaced by a point x', which is at rest in frame k'. Because space and time are homogeneous, this relationship, Einstein claimed, will be linear.

For the experiment $y = z = 0$. The person in frame k records a time t for the beginning of the experiment, and observes that the light pulse moves down the x-axis at a velocity $c - v$ relatively to the apparatus in k'arriving at the reflector in k' at time $t + x'/(c - v)$. This person in k then observes that the returning pulse moves at a velocity $c + v$ relatively to the apparatus in k' arriving at the origin of k' at time $t + x'/(c - v) + x'/(c + v)$. The experimental data recorded by the person in frame k result in three values for the function τ. These are

$$\begin{aligned}\tau_0 &= \tau(0, 0, 0, t)\\ \tau_1 &= \tau\left(x', 0, 0, t + \frac{x'}{c - v}\right) = \tau_0 + x'\frac{\partial\tau}{\partial x'} + \frac{x'}{c - v}\frac{\partial\tau}{\partial t}\\ \tau_2 &= \tau\left(0, 0, 0, t + \frac{x'}{c - v} + \frac{x'}{c + v}\right) = \tau_0 + \left(\frac{x'}{c - v} + \frac{x'}{c + v}\right)\frac{\partial\tau}{\partial t}.\end{aligned}$$

With Einstein, we now choose x' to be infinitesimal. Then (8.3) becomes

$$\boxed{\partial\tau/\partial x' + \left[v/\left(c^2 - v^2\right)\right]\partial\tau/\partial t = 0} \qquad (8.5)$$

This is a linear partial differential equation with constant coefficients. Since τ is a linear function, for a specific v the solution of (8.5) is

$$\tau = a\left(t - \frac{v}{c^2 - v^2}x'\right), \qquad (8.6)$$

where a is a function of the velocity v. Equation (8.6) is the functional relationship (8.4).

Using similar thought experiments, Einstein found the relationships among the spatial coordinates (x, y, z) in frame k and (ξ, η, ζ) in k'. We will not consider the details of these thought experiments here.

8.2.2 *Lorentz Transformation*

The complete transformation equations for the time and spatial coordinates between the inertial frames k and k' are the *Lorentz Transformation* equations as Einstein presented them. We shall now, however, replace (τ, ξ, η, ζ) with $\left(t', x', y', z'\right)$ to obtain a more modern representation.

$$\boxed{\begin{aligned} t' &= \gamma\left(t - \beta x/c\right) \\ x' &= \gamma\left(x - \beta ct\right) \\ y' &= y \\ z' &= z, \end{aligned}} \tag{8.7}$$

where

$$\gamma = \frac{1}{\sqrt{1-\beta^2}}, \tag{8.8}$$

and $\beta = v/c$. We shall usually consider the frame k to be the stationary reference frame and the frame k' as translating along the axis x at a velocity v relatively to k. Of course no frame is actually at rest, which is a concept meaningful only in Newtonian space.

8.3 Minkowski Space

The space of four coordinates, in which time is treated on an equal footing with spatial coordinates, is referred to as *Minkowski Space* because it was first proposed by Hermann Minkowski[5] (1864–1909). Minkowski's paper *Space and Time*, delivered to the 80th Assembly of German Natural Scientists and Physicians, at Cologne, September, 1908, is printed in translated form in [[24], pp. 75–91]. There Minkowski claimed that

> Henceforth space by itself, and time by itself, are doomed to fade away into mere shadows, and only a kind of union of the two will preserve an independent reality.

Minkowski's formalism is a great simplification to special relativity. Initially, however, Einstein was unimpressed. He called this "superfluous learnedness" [[89], p. 152]. But later, Einstein adopted the Minkowski formalism. And in 1916 he acknowledged his debt to Minkowski. The formalism was instrumental in the transition from special to general relativity. The Minkowski formalism will also simplify our study of relativistic mechanics.

[5]Hermann Minkowski was a German mathematician of Lithuanian Jewish descent. He was one of Einstein's professors at the Eidgenössische Polytechnikum in Zürich.

In this section we will place Minkowski's ideas into a mathematical form intended to simplify our work. Our tools will be standard matrix form and the non-Euclidean nature of the space will not hinder our development.

8.3.1 Four Dimensions

We define the coordinates of Minkowski Space using a scheme which preserves x^1, x^2, and x^3 for the spatial coordinates and identifies x^0 as the time coordinate, which is ct. That is we designate the coordinates of a four-dimensional vector in Minkowski Space as

$$\begin{aligned} x^0 &= ct \\ x^1 &= x \\ x^2 &= y \\ x^3 &= z. \end{aligned} \tag{8.9}$$

Specifically we then have the four dimensional position vectors

$$\boldsymbol{x} = \begin{bmatrix} ct \\ x \\ y \\ z \end{bmatrix} = \begin{bmatrix} x^0 \\ x^1 \\ x^2 \\ x^3 \end{bmatrix} \text{ and } \boldsymbol{x}' = \begin{bmatrix} ct' \\ x' \\ y' \\ z' \end{bmatrix} = \begin{bmatrix} x'^0 \\ x'^1 \\ x'^2 \\ x'^3 \end{bmatrix} \tag{8.10}$$

for points in four dimensional inertial frames k and k'. We will designate the elements of these 4-vectors with superscripts, which is a notation introduced by Gregorio Ricci-Curbastro (Ricci) and Tullio Levi-Civita [[24], p. 122]. This is the standard notation presently used in physics.

Points in this four dimensional space are called *world points*. In three dimensional terms a world point joins or associates a spatial point (x, y, z) with a temporal point ct registered on a timepiece. The world point is then an *event* in Einstein's terminology of Sect. 8.2. World points are connected by *world lines*. For example the first part of a time synchronization experiment consists of the events (1) light pulse leaves point A at time t_A and (2) light pulse arrives at point B at time t_B. The world line connects these two events.

In the initial work using Minkowski Space time was introduced as the fourth dimension.

8.3.2 Four Vectors

The vectors $\boldsymbol{x}$ and $\boldsymbol{x}'$ that we have introduced in (8.10) are called *four vectors* or 4-*vectors*.[6] This designation is not simply to indicate that they have four dimensions. It indicates that the components of the 4-vectors $\boldsymbol{x}$ and $\boldsymbol{x}'$, referenced to the two inertial frames k and k', are related to one another by the Lorentz Transformation (8.7). Specifically we obtain the components of the 4-vector $\mathbf{x}'$ from those of the 4-vector $\boldsymbol{x}$ as

$$\boxed{\begin{aligned} x'^0 &= \gamma\left(x^0 - \beta x^1\right) \\ x'^1 &= \gamma\left(x^1 - \beta x^0\right) \\ x'^2 &= x^2 \\ x'^3 &= x^3. \end{aligned}} \tag{8.11}$$

The inverse of the Lorentz Transformation may be found by simply replacing β with $-\beta$ and exchanging the k and k' coordinates, since a person in frame k' sees k receding in the x'^1 direction at a velocity v. The result is

$$\boxed{\begin{aligned} x^0 &= \gamma\left(x'^0 + \beta x'^1\right) \\ x^1 &= \gamma\left(x'^1 + \beta x'^0\right) \\ x^2 &= x'^2 \\ x^3 &= x'^3. \end{aligned}} \tag{8.12}$$

To keep our development from becoming unwieldy, we shall formalize this transformation. People in the inertial reference frames k and k' observe two events separated as points on two world lines, one for each frame. We assume that the two events are very close together in both frames, but not simultaneous in either frame. Then we may designate the differential lengths of the two world lines as the 4-vectors $\mathrm{d}\boldsymbol{x}$ (in k) and $\mathrm{d}\boldsymbol{x}'$ (in k'). Because there is a functional relationship between the components of the two world lines, we may write

$$\mathrm{d}x'^\nu = \frac{\partial x'^\nu}{\partial x^\mu}\mathrm{d}x^\mu. \tag{8.13}$$

And we may calculate the coefficients $\partial x'^\nu/\partial x^\mu$ from (8.11). As we noted in Sect. 8.1, we use the Einstein sum convention. The repeated index μ on the right hand side of Eq. (8.13) indicates a summation over the index $\mu = 0, 1, 2, 3$.

There are 16 such coefficients, which form the elements of a transformation matrix we shall call $\mathbf{A}$. That is

$$(\mathbf{A})^\alpha_\beta = \frac{\partial x'^\alpha}{\partial x^\beta}. \tag{8.14}$$

[6]The designation 4-vector is that used by Jackson. We choose it here as well.

The inverse of this transformation matrix we can obtain from (8.12). The elements are

$$\left(\mathbf{A}^{-1}\right)^{\alpha}_{\beta} = \frac{\partial x^{\alpha}}{\partial x'^{\beta}}. \tag{8.15}$$

In this notation the *Kronecker delta* is

$$\delta^{\alpha}_{\beta} = \frac{\partial x'^{\alpha}}{\partial x^{\lambda}}\frac{\partial x^{\lambda}}{\partial x'^{\beta}} = \frac{\partial x^{\alpha}}{\partial x'^{\lambda}}\frac{\partial x'^{\lambda}}{\partial x^{\beta}}. \tag{8.16}$$

For translation of frame k' along the axis x^1 of frame k the matrices $\mathbf{A}$ and $\mathbf{A}^{-1}$ are

$$\boxed{\mathbf{A} = \begin{bmatrix} \gamma & -\gamma\beta & 0 & 0 \\ -\gamma\beta & \gamma & 0 & 0 \\ 0 & 0 & 1 & 0 \\ 0 & 0 & 0 & 1 \end{bmatrix}} \tag{8.17}$$

and

$$\boxed{\mathbf{A}^{-1} = \begin{bmatrix} \gamma & \gamma\beta & 0 & 0 \\ \gamma\beta & \gamma & 0 & 0 \\ 0 & 0 & 1 & 0 \\ 0 & 0 & 0 & 1 \end{bmatrix}}. \tag{8.18}$$

The choice of the spatial axis along which we orient the relative velocity of the inertial frames is arbitrary. In this chapter we will stay with Einstein's original choice of the x-axis.

8.3.3 *The Minkowski Axiom*

Not all differential 4-vectors are allowed in Minkowski space. To find the limitation we begin by applying the Lorentz Transformation equations (8.11) to the differential lengths of a world line, observed from two inertial frames. By straightforward calculation we find that

$$\begin{aligned} &\pm\left[\left(\mathrm{d}x^{0}\right)^{2} - \left(\mathrm{d}x^{1}\right)^{2} - \left(\mathrm{d}x^{2}\right)^{2} - \left(\mathrm{d}x^{3}\right)^{2}\right] \\ &= \pm\left[\left(\mathrm{d}x'^{0}\right)^{2} - \left(\mathrm{d}x'^{1}\right)^{2} - \left(\mathrm{d}x'^{2}\right)^{2} - \left(\mathrm{d}x'^{3}\right)^{2}\right] \end{aligned} \tag{8.19}$$

(see exercises). The equality in (8.19) holds regardless of the sign we may attach to the square bracket. Minkowski introduced a fundamental axiom, which we shall refer to as the *Minkowski Axiom*, that requires the positive sign to be chosen. Minkowski said

> The substance at any world point may always, with the appropriate determination of space and time, be looked upon as at rest.
>
> [[24], p. 80].

That is in *some* inertial frame k' we will have $\mathrm{d}x' = \mathrm{d}y' = \mathrm{d}z' = 0$ for a substantive, material body. Then, since $c^2\mathrm{d}t'^2 > 0$, we realize that the square of the differential world line in frame k' is $\mathrm{d}\mathbf{s}'^2 = c^2\mathrm{d}t'^2 > 0$. But from (8.19) we know that $\mathrm{d}\mathbf{s}^2 = \mathrm{d}\mathbf{s}'^2$. That is $\mathrm{d}\mathbf{s}^2$ is an *invariant scalar* on Lorentz Transformation between inertial frames, which we must take to be positive for a material body.

There is no frame in which a light beam stands still, as Einstein realized when he was 16 [[89], pp. 130, 131]. For the world line of a light beam, then, we can never have $\mathrm{d}x' = \mathrm{d}y' = \mathrm{d}z' = 0$ in any frame whatsoever. The light beam in vacuum always propagates at c. Therefore, for the world line of a light beam,

$$dx'^2 + dy'^2 + dz'^2 = c^2dt'^2 \tag{8.20}$$

and $\mathrm{d}s'^2 = 0$. We may then write the mathematical form of the Minkowski Axiom as

$$\boxed{ds^2 = \left(\mathrm{d}x^0\right)^2 - \left(\mathrm{d}x^1\right)^2 - \left(\mathrm{d}x^2\right)^2 - \left(\mathrm{d}x^3\right)^2 \geq 0.} \tag{8.21}$$

Using (8.9), the inequality (8.21) requires that

$$c^2 \geq \left(\frac{\mathrm{d}x}{\mathrm{d}t}\right)^2 + \left(\frac{\mathrm{d}y}{\mathrm{d}t}\right)^2 + \left(\frac{\mathrm{d}z}{\mathrm{d}t}\right)^2.$$

for any material body. Therefore, according to the Minkowski Axiom, the velocity c of light is a *limiting velocity* for material bodies.[7] And the limiting case $\mathrm{d}s^2 = 0$ holds only for light.

This limit on velocity can be retrieved from results for the motion of material bodies that we shall develop. And elementary texts normally use those results to establish the limiting value of c for the velocity of a material body. The Minkowski Axiom is, however, the foundational statement of this limitation.

8.3.4 *Metric for Minkowski Space*

The measure of the distance between two points in a space is specified by the *metric* or measure of the space. If we consider that the vector $\mathrm{d}\boldsymbol{s}$ connects two points an infinitesimal distance from one another, we choose the scalar product $\mathrm{d}\boldsymbol{s}\cdot\mathrm{d}\boldsymbol{s}$, which is positive definite ($\mathrm{d}\boldsymbol{s}\cdot\mathrm{d}\boldsymbol{s} \geq 0$), as our measure of the square of the distance between the points ($\mathrm{d}s^2$). Designating the components of $\mathrm{d}\boldsymbol{s}$ as $\mathrm{d}x^\mu$ we write this measure generally as

---

[7] Separate inertial frames must contain (material) measuring instruments, i.e. rods and timepieces.

$$ds^2 = dx^\alpha g_{\alpha\beta} dx^\beta, \tag{8.22}$$

where the matrix with elements $g_{\alpha\beta}$ is called the metric of the space. In the Euclidean space of Newton the terms $g_{\alpha\beta}$ are all unity, i.e. $g_{\alpha\beta} = \delta_{\alpha\beta}$ with $\alpha, \beta = 1, 2, 3$. And the distance between points is specified by a generalization of Pythagoras' theorem.

The surface of a sphere is an example of a space for which the elements of the metric are not unity. If we define the polar and azimuthal angles as ϕ and ϑ the distance on the surface of the sphere is $d\mathbf{s} = R d\phi \hat{e}_\phi + R \sin\phi d\vartheta \hat{e}_\vartheta$ and $ds^2 = R^2 d\phi^2 + R^2 \sin^2\phi d\vartheta^2$. That is, for the spherical surface

$$g_{\text{sphere}} = \begin{bmatrix} R^2 & 0 \\ 0 & R^2 \sin^2\phi \end{bmatrix}.$$

The metric for the spherical surface is, then, not constant. It is a function of the polar angle ϕ.

We can identify the metric for Minkowski space in (8.21). Because this is the only metric we shall deal with here, we designate this simply as g. That is

$$\boxed{\mathbf{g} = \begin{bmatrix} 1 & 0 & 0 & 0 \\ 0 & -1 & 0 & 0 \\ 0 & 0 & -1 & 0 \\ 0 & 0 & 0 & -1 \end{bmatrix}} \tag{8.23}$$

is the metric for the four dimensional Minkowski space.

8.3.5 *The Light Cone*

We cannot picture Minkowski four dimensional space. We can, however, picture the Minkowski space representation of the motion of a material particle and a light pulse in a two dimensional spatial plane. We choose the motion to be in the x^1, x^2 plane and construct the time axis x^0 of our Minkowski space perpendicular to this plane. Then the two-dimensional motion of our material particle is represented in a three-dimensional Minkowski space. And we can visualize this quite well.

In two dimensional Cartesian space the wave front of a light pulse emitted from the origin forms an expanding circle of radius ct. In our limited Minkowski space the wave front of the light pulse emitted from the origin $\mathbf{0} = \left(x^0, x^1, x^2\right) = (0, 0, 0)$ is represented by a circular cone with axis x^0. We call this the *light cone*. The Minkowski Axiom requires that the world line of a material particle passing through the origin must lie within the light cone. If we extend the light cone into the *past* $\left(x^0 < 0\right)$ the earlier world line of the material particle must lie within this extension of the light cone.[8]

[8]The geometrical definition of a cone includes both $x^0 > 0$ and $x^0 < 0$.

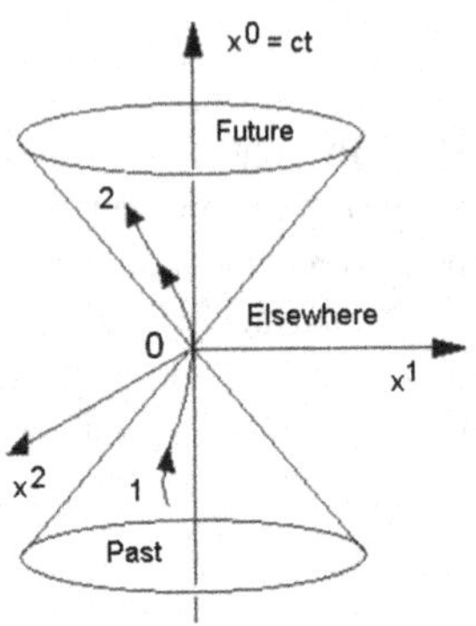

Fig. 8.3 Minkowski space with two spatial dimensions. The third dimension is ct. The world line for a material particle is that from 1 to 2

In Fig. 8.3 we have drawn this limited, three dimensional, Minkowski space, and have drawn a representative world line of a material particle within the light cone passing from point 1 in the past, through the origin, to point 2 in the future.

Intervals on the light cone, for which $\mathrm{d}s^2 = 0$, are accessible only by light. We call these *lightlike intervals*. If $\mathrm{d}s^2 > 0$ we call the interval a *timelike interval*. Timelike intervals satisfy the Minkowski Axiom and lie inside the light cone in Fig. 8.3. World points on a timelike interval are possible *future* world points for the particle. All points within the extension of the light cone along the negative x^0 axis are possible *past* world points for a particle. If $\mathrm{d}s^2 < 0$ we call the interval a *spacelike interval*. Spacelike intervals violate the Minkowski Axiom and are not accessible to material particles. We refer to these world points collectively as *elsewhere* [[51], p. 519].

8.4 Time and Space Measurements

With our basic understanding of Minkowski space and the Lorentz Transformation we are now in a position to discover the differences between specific world lines observed in different inertial frames. In this section we will consider two specific world lines that will provide an understanding of the measurements of time and length in different inertial frames.

8.4.1 Time Dilation

We consider two inertial frames k and k' such as pictured in Fig. 8.2. The times t and t' are measured by stationary timepieces in frames k and k'. These times are appropriate to either of those frames and may be called *local times* for those frames. We can most easily find the relationship between these local times from the invariance of $\mathrm{d}s^2$.

The people in frame k' observe an event occurring at, or in the immediate vicinity of the origin of k' as having a duration $\mathrm{d}t'$ measured by the timepiece at the origin. According to the people in k' the differential world line of the event is

$$\mathrm{d}s' = \begin{bmatrix} c\mathrm{d}t' \\ 0 \\ 0 \\ 0 \end{bmatrix}. \tag{8.24}$$

The people in k observe the same event. They measure the duration of this event to be $\mathrm{d}t$ on their timepiece at the origin of k and determine that, during the event in question, the origin of k' has moved a spatial distance $\hat{e}_x \mathrm{d}x + \hat{e}_y 0 + \hat{e}_z 0$. The event at the origin of k' then has the differential world line

$$\mathrm{d}s = \begin{bmatrix} c\mathrm{d}t \\ \mathrm{d}x \\ 0 \\ 0 \end{bmatrix} \tag{8.25}$$

for the people in k. The invariance of $\mathrm{d}s^2$ requires that

$$c^2 \mathrm{d}t'^2 = c^2 \mathrm{d}t^2 - \mathrm{d}x^2. \tag{8.26}$$

Then

$$dt' = dt\sqrt{1-\beta^2}, \tag{8.27}$$

is the relationship between the differential local times measured in the two inertial frames. This is the relationship obtained by Einstein [[24], p. 49]. Since Eq. (8.27) requires that $\mathrm{d}t' < \mathrm{d}t$, we conclude that the time interval recorded by the timepiece in k' is shorter than that recorded by the timepiece in k. This is referred to as *time dilation*.

The term *local time* was first used by Lorentz [[24], p. 15]. However, as Minkowski points out, Einstein first recognized that the times t and t' are equivalent [[24], p. 82]. Minkowski then defines

$$\boxed{d\tau = dt\sqrt{1-\beta^2}} \tag{8.28}$$

as the *proper time* of the world point along the world line in Fig. 8.3. This is the time indicated by a timepiece at rest with respect to the material particle on the world line. The time interval $\mathrm{d}t$ is the corresponding time measured in an inertial frame considered to be at rest.

8.4.2 Space Contraction

To discover the effect of motion on the dimensions of a body we consider that in frame k' there is a rod of length L_0 lying along the x' axis (refer to Fig. 8.2). We consider that the length of the rod is measured by someone in frame k' and also by someone in frame k. For simplicity we assume that the measurements begin when the origins of k and k' coincide.

To make the measurement the person in frame k' sends a pulse of light from the origin of k' down the rod and records the time $\mathrm{d}t'$ taken for the pulse to reach the end of the rod. In frame k' the rod length is $L_0 = c\mathrm{d}t'$. A person in frame k observes that light pulse as traversing a rod of length $L = (c - v)\mathrm{d}t$ in a time $\mathrm{d}t$, since the person in frame k observes the light pulse to move at a velocity $c - v$ relatively to the rod. The world line in both frames is lightlike with $\mathrm{d}s^2 = \mathrm{d}s'^2 = 0$. The invariance of $\mathrm{d}s^2$ then tells us only what we already know.

We, therefore, turn directly to the Lorentz transform (8.11). In the frame k' the time component of the differential world line is $\mathrm{d}x'^0 = c\mathrm{d}t' = L_0$ and in k the time component is $\mathrm{d}x^0 = c\mathrm{d}t = L/(1-\beta)$ and the first spatial component is $\mathrm{d}x^1 = v\mathrm{d}t + L = L/(1-\beta)$. The Lorentz transform of the time component, which is the first component of (8.11), is then

$$L_0 = \gamma \frac{L}{1-\beta} - \gamma\beta \frac{L}{1-\beta} = \gamma L. \tag{8.29}$$

Therefore the relationship between the length of the rod as seen by people in frames k and k' is

$$\boxed{L = L_0\sqrt{1-\beta^2}} \tag{8.30}$$

The length of the rod, which is stationary in the frame k', appears shorter to the person in frame k than it does to the person in k', who is moving with the rod. The dimensions of the body in the directions perpendicular to the relative velocity are not affected. So a moving sphere will appear as flattened along the direction of motion, as Einstein pointed out [[24], p. 48]. This is referred to as *length contraction*.

Equation (8.30) is the *FitzGerald–Lorentz contraction*, which we discussed in Sect. 8.2. Minkowski correctly considered this hypothesis ungrounded, claiming it had been introduced "as a gift from above [[24], p. 81]." The resolution is in Einstein's idea regarding time.

8.5 Velocities

The velocity of a material body is defined in terms of the displacement of the body from one spatial point to another and the times recorded when the body is at each point. We then expect that the value of the velocity of a moving particle will depend on the inertial frame in which the measurement is made.

Let us consider that we are in the inertial frame k, which we may consider as stationary. Someone in the inertial frame k' moving at the constant velocity $\boldsymbol{v} = v\hat{e}_{\mathrm{x}}$ with respect to us observes a material body moving with a velocity $\boldsymbol{u}'$ with spatial components $\left(u'_{\mathrm{x}}, u'_{\mathrm{y}}, u'_{\mathrm{z}}\right)$. In a short time $\mathrm{d}t'$ that person observes the differential world line $\mathrm{d}\boldsymbol{s}'$ of the body to be

$$\mathrm{d}\boldsymbol{s}' = \begin{bmatrix} \mathrm{d}x^{0\prime} \\ \mathrm{d}x^{1\prime} \\ \mathrm{d}x^{2\prime} \\ \mathrm{d}x^{3\prime} \end{bmatrix} = \begin{bmatrix} c \\ u'_{\mathrm{x}} \\ u'_{\mathrm{y}} \\ u'_{\mathrm{z}} \end{bmatrix} \mathrm{d}t'. \tag{8.31}$$

In the frame k we observe the differential world line $\mathrm{d}\boldsymbol{s}$ of this body to be

$$\mathrm{d}\boldsymbol{s} = \begin{bmatrix} c \\ u_{\mathrm{x}} \\ u_{\mathrm{y}} \\ u_{\mathrm{z}} \end{bmatrix} \mathrm{d}t, \tag{8.32}$$

where $\left(u_{\mathrm{x}}, u_{\mathrm{y}}, u_{\mathrm{z}}\right)$ are the components of the body's velocity as we measure them in frame k. The displacements (8.31) and (8.32) are related by the Lorentz Transformation

$$\mathrm{d}\boldsymbol{s} = \mathbf{A}^{-1} \cdot \mathrm{d}\boldsymbol{s}'. \tag{8.33}$$

Carrying out the matrix multiplication we have

$$\mathrm{d}\boldsymbol{s} = \begin{bmatrix} \gamma & \gamma\beta & 0 & 0 \\ \gamma\beta & \gamma & 0 & 0 \\ 0 & 0 & 1 & 0 \\ 0 & 0 & 0 & 1 \end{bmatrix} \begin{bmatrix} c \\ u'_{\mathrm{x}} \\ u'_{\mathrm{y}} \\ u'_{\mathrm{z}} \end{bmatrix} \mathrm{d}t' = \begin{bmatrix} \gamma\left(1 + \beta\beta'_{\mathrm{x}}\right) c \\ \gamma\left(\beta'_{\mathrm{x}} + \beta\right) c \\ \beta'_{\mathrm{y}} c \\ \beta'_{\mathrm{z}} c \end{bmatrix} \mathrm{d}t', \tag{8.34}$$

where we have introduced $\beta_{\mathrm{j}} = u_{\mathrm{j}}/c$ for $j = 1, 2, 3$. Equating (8.32) and (8.34), we have

$$\begin{bmatrix} c \\ \beta_{\mathrm{x}} c \\ \beta_{\mathrm{y}} c \\ \beta_{\mathrm{z}} c \end{bmatrix} \mathrm{d}t = \begin{bmatrix} \gamma\left(1 + \beta\beta'_{\mathrm{x}}\right) c \\ \gamma\left(\beta'_{\mathrm{x}} + \beta\right) c \\ \beta'_{\mathrm{y}} c \\ \beta'_{\mathrm{z}} c \end{bmatrix} \mathrm{d}t'. \tag{8.35}$$

By equating components in (8.35), using (8.27), and solving for the velocity components in the inertial frame k we have

$$u_{\mathrm{x}} = \frac{u'_{\mathrm{x}} + c\beta}{1 + \beta u'_{\mathrm{x}}/c}, \tag{8.36}$$

$$u_{\mathrm{y}} = \frac{u'_{\mathrm{y}}}{\gamma\left(1+\beta u'_{\mathrm{x}}/c\right)}, \tag{8.37}$$

and

$$u_{\mathrm{z}} = \frac{u'_{\mathrm{z}}}{\gamma\left(1+\beta u'_{\mathrm{x}}/c\right)}. \tag{8.38}$$

The Eqs. (8.36)–(8.38) relate the velocity of a material body as measured by people in two different inertial frames k and k'.

The Minkowski Axiom requires that the velocities of material bodies are always less than the speed of light (see exercises).

8.5.1 Four-Velocity

If we return to (8.31) and (8.32) we can define a 4-vector

$$\frac{\mathrm{d}s}{\mathrm{d}t} = \begin{bmatrix} c \\ u_{\mathrm{x}} \\ u_{\mathrm{y}} \\ u_{\mathrm{z}} \end{bmatrix}, \tag{8.39}$$

which is the representation in Minkowski space of the velocity of a material body along the world line. The 4-vector in (8.39) has meaning for someone in a particular inertial frame, in this case k. But (8.39) is not a reasonable definition for a representation of velocity to be used in relativistic mechanics because the time t is recorded on a timepiece at an arbitrary location relative to the moving body. If we use the proper time (8.28) in place of $\mathrm{d}t$ in (8.32) we have

$$\boldsymbol{U} \equiv d\boldsymbol{x}/\,d\tau = \gamma_{\mathrm{u}} \begin{bmatrix} c \\ u_{\mathrm{x}} \\ u_{\mathrm{y}} \\ u_{\mathrm{z}} \end{bmatrix}, \tag{8.40}$$

where $\gamma_{\mathrm{u}} = \left[1-\left(u_{\mathrm{x}}^2+u_{\mathrm{y}}^2+u_{\mathrm{z}}^2\right)/c^2\right]^{-1/2}$. This $\boldsymbol{U}$ is what we will define as the *4-velocity*. Consistent with the fact that the elements $\mathrm{d}x^{\mu}/\mathrm{d}\tau$ of $\boldsymbol{U}$ have superscripts, in spite of the subscripts on the u_{j}, we designate the elements of $\boldsymbol{U}$ as U^{μ}.

Using the Minkowski Axiom (8.21) and (8.22), the square of the magnitude of the 4-velocity is

$$\begin{aligned} \mathbf{U}^2 = U^{\mu} g_{\mu\nu} U^{\nu} &= \gamma_{\mathrm{u}}^2\left(c^2-u_{\mathrm{x}}^2-u_{\mathrm{y}}^2-u_{\mathrm{z}}^2\right) \\ &= c^2, \end{aligned} \tag{8.41}$$

which is a scalar invariant.

8.6 Mass, Momentum and Energy

Mass, momentum and energy are mechanical, not simply kinematical, concepts. At this point we must, then, ask about the laws of mechanics. Newton's Laws are not covariant with the Lorentz Transformation, which is what we require of the correct laws of mechanics. So we must begin at a more fundamental point.

We should always begin with the simplest situation, which is the motion of a free material body under no force. For this simplest situation we are guided by Newton's first law, which is the claim that the momentum of a free material body must be constant. Extending this to a system of free particles presents no problem, and we have conservation of (4-vector) momentum. Collisions between particles may present difficulties, that we can avoid if we consider the collisions to be instantaneous, which is what we shall do at this point. So we claim *conservation of* 4-*momentum* in collision to be a foundational covariant law of (relativistic) mechanics.

8.6.1 Mass

Einstein considered the velocity dependence of the mass of a material body in the last section of his paper on special relativity. From the requirements of Newton's Second law and the transformation of the electric field he concluded that the mass was dependent upon motion of the body (an electron) parallel or perpendicular to the axis of translation of k'. These he termed longitudinal and transverse masses. This peculiarity he noted was a result of the "definition of force and acceleration" he had chosen [[24], pp. 61–63].

Gilbert N. Lewis and Richard C. Tolman picked up the discussion of relativistic mass in 1909 [[56], p. 161; [68]]. Lewis and Tolman based their discussion on *conservation of momentum* and concluded that the mass must depend on magnitude of the velocity v generally as

$$\boxed{m\left(v\right)=m_{0}\gamma_{\mathrm{v}}=m_{0}/\sqrt{1-\beta_{\mathrm{v}}^{2}}.} \tag{8.42}$$

This is then the dependence of mass on velocity consistent with our most fundamental law of mechanics. This has also been verified by innumerable experiments, including those in undergraduate laboratories.

The mass $m\left(v\right)$ is the *relativistic inertial mass*, or simply relativistic mass, and m_0 is the *rest mass* of the material body. The rest mass is the mass of a material body measured by someone at rest with respect to the body.

8.6.2 Momentum

With (8.42) the *spatial components* of what we may call the 4-momentum are

$$p^{\mu} = m_0 \gamma_{\mathrm{u}} \mathrm{d}\mathbf{x}^{\mu}/\mathrm{d}t = m_0 U^{\mu}, \tag{8.43}$$

where $\mu = 1, 2, 3$. With (8.40) we may then define the 4-*momentum* as

$$\boldsymbol{P} = m_0 \boldsymbol{U}, \tag{8.44}$$

which, in matrix form, is

$$\boldsymbol{P} = \begin{bmatrix} P^0 \\ P^1 \\ P^2 \\ P^3 \end{bmatrix} = m_0 \gamma_{\mathrm{u}} \begin{bmatrix} c \\ u_{\mathrm{x}} \\ u_{\mathrm{y}} \\ u_{\mathrm{z}} \end{bmatrix}. \tag{8.45}$$

Again we note the superscripts on the elements consistent with the definition (8.43) and the traditional use of subscripts on the velocity components. Traditionally we also often designate the spatial components of the momentum P^{μ} ($\nu = 1, 2, 3$) as $p_{\mathrm{j}} = m_0 \gamma_{\mathrm{u}} u_{\mathrm{j}}$ ($j = x, y, z$).

The Lorentz scalar invariant, which is the square of the magnitude of the 4-momentum, is

$$\begin{aligned} \mathbf{P}^2 &= P^{\mu} g_{\mu\nu} P^{\nu} = m_0^2 U^{\mu} g_{\mu\nu} U^{\nu} \\ &= m_0^2 c^2. \end{aligned} \tag{8.46}$$

The rest mass m_0 is then a scalar invariant.

8.6.3 Energy

Einstein introduced the relationship between mass and energy in September of 1905 with the publication entitled *Does the Inertia of a Body Depend on its Energy Content* [[24], pp. 69–71]. This appeared three months after the paper on special relativity, which was published in June. In this three page paper he showed that if the energy of a body changes through the emission of electromagnetic radiation there is a proportional loss in the inertial mass of the body.

In the June paper Einstein found equations for shifts in frequency and energy of a light wave as observed by a person in a moving frame. Then, in September, he considered what would happen if electromagnetic waves were emitted from a body stationary in the inertial frame k. He considered that two light waves, each carrying an energy $L/2$, were emitted from the body in opposite directions. The total energy emitted by the body is then L. Because of the way in which he framed this thought experiment, Einstein was able to calculate a change in the kinetic energy of the material body without resorting to the use of the laws of mechanics. He required only two inertial frames and the covariance of Maxwell's Equations, which he had established in the June paper. In the September paper he showed that the emission of radiation from a material body resulted in a loss of inertial mass of the body $\Delta m = L/c^2$, and he concluded that

> The mass of a body is a measure of its energy content; if the energy changes by an amount L, the mass changes in the same sense by $L/9 \times 10^{20}$, the energy being measured in ergs and the mass in grammes.
>
> [[24], p. 71]

In May of 1906 Einstein wrote

The Law of conservation of mass is a special case of the law of conservation of energy.

[[89], p. 148]

And then in 1907 he claimed that

> In regard to inertia, a mass m is equivalent to an energy content … mc^2. This result is of extraordinary importance since [it implies that] the inertial mass and the energy of a physical system appear as equivalent things.
>
> [[89], p. 148]

In his classic monograph *Theory of Relativity*, written when he was 21, Wolfgang Pauli produced a crisp derivation of a relativistic formula for the kinetic energy of particle with rest mass m_0 acted on by a 4-vector force with spatial component $\boldsymbol{f}$ [[92], pp. 116–117]. He began with the assumption that Newtonian Mechanics are valid in a stationary system, which is the proposal that in the frame k the force law is

$$\frac{\mathrm{d}}{\mathrm{d}t}\left(m_0\gamma_{\mathrm{u}}\boldsymbol{u}\right) = \boldsymbol{F},$$

and the claim that the Lorentz transformation will allow us to deduce the equations of motion unambiguously in another inertial frame k'. For continuity of the discussion we shall only say here that Pauli's development brought him to the result that

$$\frac{\mathrm{d}}{\mathrm{d}t}\left(m_0\gamma_{\mathrm{u}}c^2\right) = \boldsymbol{f}\cdot\boldsymbol{u}, \tag{8.47}$$

where $\boldsymbol{f}$ is the spatial component of the 4-vector force $\boldsymbol{F}$. Because $\boldsymbol{f}\cdot\boldsymbol{u}$ is the rate of change of the kinetic energy, we see that the kinetic energy of the particle is then

$$\mathcal{E}_{\mathrm{kin}} = m_0\gamma_{\mathrm{u}}c^2 + \text{ constant.} \tag{8.48}$$

To identify the constant we expand γ_{u} in powers of β_{u}. Carrying the expansion to second order in β_{u} we have

$$\mathcal{E}_{\text{kin}} \approx m_0 c^2 + \frac{1}{2} m_0 u^2 + \text{constant}. \tag{8.49}$$

We then retrieve the known classical result for the kinetic energy if we choose the constant to be $-m_0 c^2$. Then (8.48) becomes

$$\mathcal{E}_{\text{kin}} = m_0 \gamma_{\text{u}} c^2 - m_0 c^2. \tag{8.50}$$

With The Lewis–Tolman result we can write (8.50) as

$$\mathcal{E}_{\text{kin}} = mc^2 - m_0 c^2. \tag{8.51}$$

Pauli's argument was ostensibly based on the Lorentz electromagnetic force $f_{\text{L}} = Q(\boldsymbol{E} + \boldsymbol{v} \times \boldsymbol{B})$. Here Q is the electric charge[9] on the particle, $\boldsymbol{E}$ and $\boldsymbol{B}$ are the electric field and the magnetic field induction, and $\boldsymbol{v}$ is the particle velocity. This seems limiting. However, Pauli points out that "all kinds of forces transform in the same way [as the Lorentz force]." This follows from the fact that two forces which compensate for one another (are equal and opposite) in the inertial frame k must also compensate for one another in all other frames k' [[92], p. 116]. Pauli's result then holds for all forces for which $\mathrm{d}m_0/\mathrm{d}t = 0$. Such forces are called *pure* forces.

Equation (8.50) had been obtained previously by Einstein and was the mathematical basis for his claim, regarding inertia, a mass m is equivalent to an energy content ... mc^2. If we accept that the energy is mc^2 for a particle in motion and, therefore, $m_0 c^2$ for the same particle at rest, which we term the *rest energy* of the particle, then (8.51) clearly claims that the difference is the kinetic energy. This is the concept we have of kinetic energy from Newtonian mechanics. But with Einstein's formulation we understand mass now as measure of energy content and the difference of mass m and rest mass m_0 as a measure of the particle kinetic energy. Wolfgang Rindler points out that Einstein's identity of *all* [emphasis in original] mass with energy required an act of "aesthetic faith" that was characteristic of Einstein [[99], p74].

There is no question that Einstein brought a new flare to this aesthetic faith, as Abraham Pais points out [89]. The aesthetic faith itself must be and has been a critical ingredient in the approach of the theoretical physicist, as we have seen in our studies here. We can point to that aesthetic faith from Plato to Jacobi and to the mathematician Minkowski. A deep appreciation of this must embody our approach to our present topic.

We shall then accept Einstein's identity and his claim that mass and energy are equivalent. We then write

$$\boxed{\mathcal{E} = m_0 \gamma_{\text{u}} c^2 = mc^2} \tag{8.52}$$

as a universal equation for the total energy $\mathcal{E}$.

[9]Here, as in our discussion of electrodynamics, we use Q rather than the traditional q to designate electrical charge. In this text q is reserved for the generalized coordinates.

We are now able to identify the component P^0 in the 4-momentum (8.45) as

$$P^0 = m_0 \gamma_u c = \frac{\mathcal{E}}{c}. \tag{8.53}$$

Then (8.45) becomes

$$\boldsymbol{P} = \begin{bmatrix} \mathcal{E}/c \\ p_x \\ p_y \\ p_z \end{bmatrix}, \tag{8.54}$$

where we have used the traditional notation $p = m\boldsymbol{u}$ for the spatial components of momentum. Conservation of 4-momentum then requires conservation of total energy $\mathcal{E}$ as the zeroth component of the 4-momentum. And relativistic mass m is conserved, since $\mathcal{E} = mc^2$. Conservation of rest mass is, however, no longer a law of physics.

The condition of scalar invariance of $P^\mu g_{\mu\nu} P^\nu$ (8.46) then becomes

$$\boxed{\mathcal{E}^2 = p^2c^2 + m_0^2c^4,} \tag{8.55}$$

which is a general relationship between momentum and energy in relativistic mechanics.

Because of the understanding of the time component, the 4-vector in (8.54) is sometimes called the *energy-momentum* 4-*vector* [[37], p. 510; [56], p. 164, [3], p. 93].

8.7 Tensors

Einstein called the matrix **g** in (8.23), which we introduced as the metric or measure of Minkowski Space, the *fundamental tensor* in his 1916 paper on the general theory of relativity. There Einstein showed that $g_{\alpha\beta}$ describes the gravitational field. Gravitation is, therefore, exceptional in that it defines the metrical properties of space and time [[24], p. 120].

With this paper on general relativity Einstein brought tensors to the attention of physicists. The concept of tensors as mathematical quantities, however, originated with Carl Friedrich Gauss and was developed by the mathematical community before Einstein encountered them, with the help of his friend Marcel Grossmann [[89], p. 216]. The 4-vectors we have encountered are *tensors*. Specifically $\mathrm{d}x^\alpha$ are the differential elements of a tensor.

Although we shall continue to use the designation 4-vector, our consideration of electromagnetic forces, which are of great importance in relativistic mechanics, requires a familiarity with some basic properties of tensors. We provide that here, following Einstein.

8.7.1 The Fundamental Tensor

If we call

$$\mathrm{d}x_{\alpha} = g_{\alpha\beta}\mathrm{d}x^{\beta} \tag{8.56}$$

we can write (8.22) our scalar product $\mathrm{d}\mathbf{s}^2$ as

$$\mathrm{d}\mathbf{s}^2 = \mathrm{d}x^{\alpha}\mathrm{d}x_{\alpha}. \tag{8.57}$$

We then have identified the general scalar product as a product between different forms of the 4-vectors. We designate these two forms by superscript and subscript (see Sect. 8.3.2) [[24], p. 123]. From the matrix product of $\mathbf{g}$ with $\mathrm{d}\boldsymbol{s}$ we see that the mathematical operation in (8.56) places negative signs on the spatial components of $\mathrm{d}x^{\alpha}$, as required by the form of $\mathrm{d}s^2$ in the Minkowski Axiom (8.21). Inverting the process in (8.56) we are led to

$$\mathrm{d}x^{\beta} = g^{\beta\alpha}\mathrm{d}x_{\alpha}. \tag{8.58}$$

That is

$$\mathrm{d}x^{\beta} = g^{\beta\alpha}g_{\alpha\nu}\mathrm{d}x^{\nu}, \tag{8.59}$$

which requires that

$$g^{\beta\alpha}g_{\alpha\nu} = \delta^{\beta}_{\nu}, \tag{8.60}$$

where δ^{β}_{ν} is the Kronecker δ written with the super and subscripts of the tensor notation. The space of the special theory is what is termed *flat* and is characterized mathematically by the form of g in (8.23).

8.7.2 Scalar Product

We define a scalar product between two general 4-vectors $\boldsymbol{C}$ and $\boldsymbol{D}$ in the same way we have defined the scalar product forming $\mathrm{d}s^2$. That is

$$\boldsymbol{C}\cdot\boldsymbol{D} = C_{\alpha}D^{\alpha}. \tag{8.61}$$

We have already seen that the square of the magnitudes of the 4-velocity and the 4-momentum, obtained from scalar products, are Lorentz invariants. We now ask for the conditions that must be satisfied for the general scalar product to be an invariant under Lorentz Transformation. That is, writing the 4-vector components in the frame k without a prime and those in k' with a prime, we wish to know the transformation properties of the 4-vectors that will result in the equality

$$C_\alpha D^\alpha = C'_\nu D'^\nu. \tag{8.62}$$

We first require that the 4-vector with the superscript D^α transforms in the same way as the coordinates $\mathrm{d}x^\alpha$ (see (8.13)). That is

$$D'^\nu = \frac{\partial x'^\nu}{\partial x^\alpha} D^\alpha = (\mathbf{A})^\nu_\alpha \, D^\alpha, \tag{8.63}$$

using (8.14). The inverse of (8.63) is

$$D^\alpha = \frac{\partial x^\alpha}{\partial x'^\nu} D'^\nu = \left(\mathbf{A}^{-1}\right)^\alpha_\nu \, D'^\nu, \tag{8.64}$$

using (8.15). Then (8.62) becomes

$$C_\alpha D^\alpha = C_\alpha \frac{\partial x^\alpha}{\partial x'^\nu} D'^\nu = C_\alpha \left(\mathbf{A}^{-1}\right)^\alpha_\nu \, D'^\nu. \tag{8.65}$$

If C_α, the 4-vector with the subscript, transforms as

$$C'_\sigma = \frac{\partial x^\alpha}{\partial x'^\sigma} C_\alpha = \left(\mathbf{A}^{-1}\right)^\alpha_\sigma \, C_\alpha, \tag{8.66}$$

using (8.15), with inverse

$$C_\alpha = \frac{\partial x'^\sigma}{\partial x^\alpha} C'_\sigma = (\mathbf{A})^\sigma_\alpha \, C'_\sigma, \tag{8.67}$$

using (8.14), then Eq. (8.65) becomes

$$\begin{aligned} C_\alpha D^\alpha &= (\mathbf{A})^\sigma_\alpha \left(\mathbf{A}^{-1}\right)^\alpha_\nu C'_\sigma D'^\nu \\ &= \delta^\sigma_\nu C'_\sigma D'^\nu = C'_\nu D'^\nu, \end{aligned} \tag{8.68}$$

where we have used (8.60).[10] Therefore scalar invariance of the general scalar product results if one 4-vector transforms as (8.63) and the other as (8.66).

A 4-vector that transforms as (8.63) is called *contravariant* and a 4-vector that transforms as (8.66) is called *covariant*. The components of a contravariant 4-vector are indicated by a superscript and those of a covariant 4-vector by a subscript [[24], p. 123]. This is the convention of Ricci-Curbastro and Levi-Civita as we noted in Sect. 8.3.1. The scalar product is then defined as the sum of the products of the elements of a contravariant and a covariant 4-vector. And the scalar product of two 4-vectors is a Lorentz scalar invariant.

[10] The terms in (8.68) are all simply numbers and commute.

From (8.56) and (8.58) we see that the fundamental tensor (metric) $g_{\alpha\beta}$ converts a contravariant tensor to a covariant tensor and $g^{\alpha\beta}$ converts a covariant tensor to a contravariant tensor. In more picturesque language, the fundamental tensor raises or lowers indices on a tensor. That is

$$C_\alpha = g_{\alpha\beta}C^\beta \text{ and } C^\alpha = g^{\alpha\beta}C_\beta. \tag{8.69}$$

Similarly a tensor $V_{\alpha\beta}$ with two indices can be converted from covariant to contravariant form using two fundamental tensors as

$$V^{\mu\nu} = g^{\mu\alpha}V_{\alpha\beta}g^{\beta\nu} = g^{\mu\alpha}g^{\beta\nu}V_{\alpha\beta} \tag{8.70}$$

and from contravariant to covariant form by

$$V_{\mu\nu} = g_{\mu\alpha}V^{\alpha\beta}g_{\beta\nu} = g_{\mu\alpha}g_{\beta\nu}V^{\alpha\beta}. \tag{8.71}$$

An invariant scalar is a tensor of rank zero. A contravariant or covariant 4-vector is a tensor of rank one. The rank of the tensor is the number of indices required in its definition. Accordingly we have *contravariant* and *covariant tensors* of rank two, which transform as

$$C'^{\sigma\tau} = \frac{\partial x'^\sigma}{\partial x^\mu}C^{\mu\nu}\frac{\partial x'^\tau}{\partial x^\nu} \tag{8.72}$$

and

$$C'_{\sigma\tau} = \frac{\partial x^\mu}{\partial x'^\sigma}C_{\mu\nu}\frac{\partial x^\nu}{\partial x'^\tau} \tag{8.73}$$

respectively.

Tensors may also be of higher rank and we may have mixed tensors, which transform as, for example

$$C'^{\tau}_{\sigma} = \frac{\partial x'^\tau}{\partial x^\nu}C^{\nu}_{\mu}\frac{\partial x^\mu}{\partial x'^\sigma}. \tag{8.74}$$

With the exception of our treatment of electromagnetic forces we will be primarily concerned here with 4-vectors.

We note that we now write our scalar product for a general 4-vector Q as

$$Q \cdot Q = Q^\mu g_{\mu\nu}Q^\nu = Q^\mu Q_\mu$$

or

$$Q \cdot Q = Q_\mu g^{\mu\nu}Q_\nu = Q^\nu Q_\nu.$$

8.7.3 Four-Vector Shorthand

The form of the general 4-vectors $\boldsymbol{V}$ and $\boldsymbol{W}$ is

$$\boldsymbol{V} = \begin{bmatrix} v^0 \\ v^1 \\ v^2 \\ v^3 \end{bmatrix};\ \boldsymbol{W} = \begin{bmatrix} w^0 \\ w^1 \\ w^2 \\ w^3 \end{bmatrix}.$$

The scalar product of these two 4-vectors

$$\begin{aligned} \boldsymbol{V} \cdot \boldsymbol{W} &= V^{\mu} g_{\mu\nu} W^{\nu} \\ &= \left(v^0 w^0\right) - \left(v^1 w^1 + v^2 w^2 + v^3 w^3\right) \end{aligned}$$

invites the general representation of the contravariant and covariant forms of a general 4-vector $\boldsymbol{C}$ as

$$\boldsymbol{C} = C^{\mu} = \left(c^0, \boldsymbol{c}\right) \text{ for the contravariant form,}$$

and, using (8.69),

$$\begin{aligned} C_{\nu} &= g_{\nu\mu} C^{\mu} \\ &= \left(c^0, -\boldsymbol{c}\right) \text{ for the covariant form.} \end{aligned}$$

where

$$\boldsymbol{c} = c^1 \hat{e}^1 + c^2 \hat{e}^2 + c^3 \hat{e}^3$$

is the standard representation of a spatial vector in the Cartesian basis $\left(\hat{e}^1, \hat{e}^2, \hat{e}^3\right)$.

Specifically we may write the 4-velocity as

$$\boldsymbol{U} = \gamma_{\mathrm{u}} \left(c, \boldsymbol{u}\right) \tag{8.75}$$

and the 4-momentum as

$$\begin{aligned} \boldsymbol{P} &= m_0 \boldsymbol{U} \\ &= \left(\mathcal{E}/c, \boldsymbol{p}\right). \end{aligned} \tag{8.76}$$

This notation will make most of our calculations in relativistic Analytical Mechanics much less unwieldy. We must only remember that the scalar product always involves a contravariant and a covariant vector, i.e. involves the fundamental tensor g.

8.7.4 Differential Operators

The partial derivative with respect to the four *contravariant components*, i.e. $\partial/\partial x^\alpha$ transforms as a *covariant vector operator*. We see this if we compare

$$\frac{\partial \Phi}{\partial x'^\alpha} = \frac{\partial x^\beta}{\partial x'^\alpha}\frac{\partial \Phi}{\partial x^\beta} \tag{8.77}$$

with (8.66), i.e. $C'_\sigma = \left(\partial x^\alpha/\partial x'^\sigma\right) C_\alpha$.

We then define the *covariant differential operator*, in our shorthand notation, as

$$\partial_\alpha \equiv \frac{\partial}{\partial x^\alpha} = \left(\frac{\partial}{c\partial t}, \text{grad}\right). \tag{8.78}$$

And for a scalar Φ the 4-vector $\partial_\alpha \Phi$ is *covariant.*

From (8.78) we obtain the *contravariant vector operator*, in our shorthand notation, as

$$\partial^\alpha \equiv \frac{\partial}{\partial x_\alpha} = g^{\alpha\beta}\partial_\beta = \left(\frac{\partial}{c\partial t}, -\text{grad}\right). \tag{8.79}$$

We have written ∂^α as $\partial/\partial x_\alpha$, with subscripts on the coordinates, *only* to distinguish ∂^α from ∂_α, which we have written as $\partial/\partial x^\alpha$, with the standard superscripts. Some authors refer to the operation ∂^α as differentiation with respect to *covariant* components $\left(x^0, -x^1, -x^2, -x^3\right)$ as distinguished from differentiation with respect to the *contravariant* components $\left(x^0, x^1, x^2, x^3\right)$ of the standard 4-vector (see Sect. 8.3.1) [cf. [51], pp. 535–536]. This is helpful provided we remember that the derivatives appearing in both ∂^α and ∂_α are actually with respect to ordinary time and spatial variables and that the distinction is only the negative sign in (8.79).

Because $\partial_\beta \Phi$ is a covariant 4-vector, then $\partial^\alpha \Phi = g^{\alpha\beta}\partial_\beta \Phi$ is a contravariant 4-vector.

These operators (8.78) and (8.79) will be important to us in our consideration of the electromagnetic field (Lorentz) force [cf. [45], pp. 305–311].

8.7.5 Notation

In our shorthand notation there can be confusion about whether the elements displayed in he shorthand form of a 4-vector are the contravariant or covariant elements of the 4-vector. For this reason we have been careful about the notation as we developed the 4-vector formalism. We shall follow the standard notation, which requires that when we display the elements of a 4-vector we display its *contravariant elements* [cf. [99], p. 56; [51], pp. 535–536]]. Specifically our shorthand for the 4-velocity $\boldsymbol{U}$ and the 4-momentum $\boldsymbol{P}$ in (8.75) and (8.76) are *contravariant* 4-vectors.

8.8 Electromagnetism

8.8.1 Current and Potential Vectors

Although we are intuitively inclined to think in terms of electric and magnetic fields, it is mathematically more convenient to base our treatment of electrodynamics on the scalar potential φ and the vector potential $\boldsymbol{A}$. In terms of the potentials φ and $\boldsymbol{A}$ the Maxwell Equations take on the form of two separate wave equations with sources. The charge density ρ is the source term in the equation for φ and current density $\boldsymbol{J}$ is the source term in the equation for $\boldsymbol{A}$. If we are interested in Analytical Mechanics our formulation still, however, requires knowledge of the fields, which act on charged bodies. We obtain the electric field $\boldsymbol{E}$ and magnetic field induction $\boldsymbol{B}$ from $\boldsymbol{E} = -\text{grad}\varphi - \partial \boldsymbol{A}/\partial t$ and $\boldsymbol{B} = \text{curl}\boldsymbol{A}$, that is from application of the differential tensor operators introduced in the preceding section [[45], pp. 255–257].

The potentials φ and $\boldsymbol{A}$ appear in the *4-potential vector*

$$\boxed{\mathsf{A} = (\varphi/c, \boldsymbol{A})\,,} \tag{8.80}$$

and the sources of the potentials, the charge density ρ and the current density $\boldsymbol{J}$, appear in the *current (density) 4-vector*

$$\boxed{\mathsf{J} = (c\rho, \boldsymbol{J})\,.} \tag{8.81}$$

The *Maxwell Equations*, written in terms of scalar and vector potentials, then take on a particularly elegant form

$$\boxed{\Box \mathsf{A} = \mu_0 \mathsf{J},} \tag{8.82}$$

where $\mu_0$is the permeability of free space and

$$\Box = \partial_\mu \partial^\mu = \frac{\partial^2}{\partial x^0 \partial x^0} - \nabla^2 \tag{8.83}$$

is the *d'Alembertian* (differential operator). When they are written in this form, it is particularly easy to show that the Maxwell Equations are invariant under Lorentz Transformation [[45], pp. 302–303].

8.8.2 Field Strength Tensor

In his 1916 paper on *The Foundation of the General Theory of Relativity*, Einstein developed what he called the *six-vector* of the electromagnetic field in empty space [[24], pp. 153–157]. We now call this the *Field Strength tensor* [[45], p. 305; [51], p. 550]. It is an *antisymmetric tensor of rank two*.

The representation of the electric and magnetic fields in terms of an antisymmetric tensor of rank two is natural. Vector calculus allows the existence of *polar* and *axial* vectors. Axial vectors are those which change sign on coordinate reflection, i.e. upon change from a right-handed to a left-handed coordinate system. Vectors formed by vector or cross products are axial vectors, as are those formed by the curl operation. Vectors which do not change direction on reflection are polar vectors. Polar vectors may then be considered to be "normal" vectors.

The electric field, obtained from the gradient and time derivative of the components of the 4-potential, is a polar vector. The magnetic field induction vector, obtained from the curl of the spatial components of the 4-potential is an axial vector. Faraday's and Ampère's Laws relate the time derivatives of $\boldsymbol{B}$ and $\boldsymbol{E}$ to curl$\boldsymbol{E}$ and curl$\boldsymbol{B}$. An axial vector may then be obtained from the curl of a polar vector and vice versa. We can show that equations in which axial vectors appear can be written in a covariant manner using antisymmetric tensors [[3], pp. 56–57].

In covariant form the antisymmetric Field Strength tensor of rank two $\mathcal{F}^{\alpha\beta}$ is defined as

$$\mathcal{F}^{\alpha\beta} = \partial^{\alpha}\mathcal{A}^{\beta} - \partial^{\beta}\mathcal{A}^{\alpha} = -\mathcal{F}^{\beta\alpha}. \tag{8.84}$$

The elements of this tensor are the electric and magnetic fields. From $\boldsymbol{E} = -\text{grad}\varphi - \partial\boldsymbol{A}/\partial t$ we anticipate that the elements of the electric field $\boldsymbol{E}$ will involve $\partial^0\mathcal{A}^\beta$ and $\partial^\beta\mathcal{A}^0$ where $\beta = 1, 2, 3$. And from $\boldsymbol{B} = \text{curl}\boldsymbol{A}$ we anticipate that the elements of the magnetic induction $\boldsymbol{B}$ will involve neither ∂^0 nor $\mathcal{A}^0$. Using (8.79) and (8.80), we find, for example, that $\mathcal{F}^{10} = E_x/c$ and $\mathcal{F}^{12} = -B_z$.

In matrix form the *contravariant Field Strength* tensor is

$$\boxed{\mathcal{F}^{\alpha\beta} = \begin{bmatrix} 0 & -E_x/c & -E_y/c & -E_z/c \\ E_x/c & 0 & -B_z & B_y \\ E_y/c & B_z & 0 & -B_x \\ E_z/c & -B_y & B_x & 0 \end{bmatrix}} \tag{8.85}$$

and the covariant form is (see exercises)

$$\mathcal{F}_{\alpha\beta} = \begin{bmatrix} 0 & E_x/c & E_y/c & E_z/c \\ -E_x/c & 0 & -B_z & B_y \\ -E_y/c & B_z & 0 & -B_x \\ -E_z/c & -B_y & B_x & 0 \end{bmatrix}. \tag{8.86}$$

The contravariant Field Strength tensor (8.85) transforms from an inertial frame k, which we consider to be at rest, to an inertial frame k' as (8.72), which in matrix form is

$$\mathcal{F}' = \mathbf{A}\mathcal{F}\mathbf{A}. \tag{8.87}$$

Carrying out the matrix multiplication indicated in (8.87) and using the Lorentz transformation matrix (8.17),

$$\mathcal{F}'^{\alpha\beta} = \begin{bmatrix} 0 & -E_x/c & -(\gamma/c)(E_y - c\beta B_z) & -(\gamma/c)(E_z + c\beta B_y) \\ \frac{1}{c}E_x & 0 & -\gamma(B_z - (\beta/c)E_y) & \gamma(B_y + (\beta/c)E_z) \\ (\gamma/c)(E_y - c\beta B_z) & \gamma(B_z - (\beta/c)E_y) & 0 & -B_x \\ (\gamma/c)(E_z + c\beta B_y) & -\gamma(B_y + (\beta/c)E_z) & B_x & 0 \end{bmatrix}. \tag{8.88}$$

This result gives us a picture of how the electric and magnetic fields measured by people in the frame k and k' differ from one another.

Although the same laws of electrodynamics are valid for all inertial frames of reference, the electric and magnetic fields a person measures depend on the relative state of motion of the person. This we see in (8.88). For example, if in the stationary inertial frame k we have only an electrical field $\boldsymbol{E} = E_z\hat{e}_z$ then in an inertial frame k' moving with a velocity $\boldsymbol{v} = v\hat{e}_x$ a person will detect an electric field $\boldsymbol{E}' = \gamma E_z\hat{e}_z$ and a magnetic field induction $\boldsymbol{B}' = (\beta/c)\,\gamma E_z\hat{e}_y$. We may identify these fields by comparing the form of $\mathcal{F}'$ that we have in final matrix in equation (8.88) with $\mathcal{F}$ in (8.85).

Similarly if we have only a magnetic field with induction $\boldsymbol{B} = B_y\,\hat{e}_y$ in the stationary frame k then in the moving frame k' a person will detect a magnetic field with induction $\boldsymbol{B}' = \gamma B_y\hat{e}_y$ and an electric field $\boldsymbol{E}' = c\beta B_y\hat{e}_z$.

8.8.3 Electromagnetic Force

The electromagnetic (Lorentz) force on a particle with charge Q is $Q\,(\boldsymbol{E} + \boldsymbol{v} \times \boldsymbol{B})$. We can extract this force from the Field Strength tensor if we take the scalar product of $\mathcal{F}$ with the 4-velocity $\boldsymbol{U}$. That is, with the contravariant and covariant forms of $\mathcal{F}$ in (8.85) and (8.86) and the 4-velocity in (8.40) we have

$$\mathcal{F}^{\alpha\beta}U_\beta = \gamma_u \begin{bmatrix} 0 & -E_x/c & -E_y/c & -E_z/c \\ E_x/c & 0 & -B_z & B_y \\ E_y/c & B_z & 0 & -B_x \\ E_z/c & -B_y & B_x & 0 \end{bmatrix} \begin{bmatrix} c \\ -u_x \\ -u_y \\ -u_z \end{bmatrix} = \gamma_u \begin{bmatrix} E_x u_x/c + E_y u_y/c + E_z u_z/c \\ E_x - B_y u_z + B_z u_y \\ E_y + B_x u_z - B_z u_x \\ E_z - B_x u_y + B_y u_x \end{bmatrix}. \tag{8.89}$$

and

$$\mathcal{F}_{\alpha\beta}U^\beta = \gamma_u \begin{bmatrix} 0 & E_x/c & E_y/c & E_z/c \\ -E_x/c & 0 & -B_z & B_y \\ -E_y/c & B_z & 0 & -B_x \\ -E_z/c & -B_y & B_x & 0 \end{bmatrix} \begin{bmatrix} c \\ u_x \\ u_y \\ u_z \end{bmatrix} = \gamma_u \begin{bmatrix} E_x u_x/c + E_y u_y/c + E_z u_z/c \\ B_y u_z - E_x - B_z u_y \\ B_z u_x - B_x u_z - E_y \\ B_x u_y - E_z - B_y u_x \end{bmatrix} \tag{8.90}$$

We can identify the work done on a charged particle in the time component of each of these 4-vectors. And in the spatial components of (8.89) we can identify the positive and in (8.90) the negative electromagnetic force fields. Therefore the contravariant and covariant 4-vector electromagnetic (Lorentz) force on a particle with charge Q are

$$F^{\alpha} = Q\gamma_{u}^{-1}\mathcal{F}^{\alpha\beta}U_{\beta} \text{ and } F_{\alpha} = Q\gamma_{u}^{-1}\mathcal{F}_{\alpha\beta}U^{\beta}.$$

Using our shorthand notation and defining the Lorentz force on a body carrying a charge Q as

$$\boldsymbol{f}_{\mathrm{L}} = Q\left(\boldsymbol{E} + \boldsymbol{v} \times \boldsymbol{B}\right),$$

we may write the contravariant and covariant forms of the 4-vector electromagnetic force as

$$F^{\alpha} = \gamma_{u}\left(\frac{1}{c}\boldsymbol{f}_{\mathrm{L}} \cdot \boldsymbol{u}, \boldsymbol{f}_{\mathrm{L}}\right), \tag{8.91}$$

and

$$F_{\alpha} = \gamma_{u}\left(\frac{1}{c}\boldsymbol{f}_{\mathrm{L}} \cdot \boldsymbol{u}, -\boldsymbol{f}_{\mathrm{L}}\right). \tag{8.92}$$

The components of the (relativistic) 4-vector Lorentz force then contain a factor γ_{u}. We may, therefore, think of γ_{u} as providing an indication of the deviation between Newtonian mechanics and what we shall find as a correct relativistic formulation. If $\gamma_{u} \approx 1$ we may use ordinary Newtonian mechanics. However, in laboratory studies of collisions between elementary particles γ_{u} may be 10^{4} and γ_{u} is as high as 10^{11} for cosmic ray protons [[99], p. 70]. So we are not dealing with situations in which Newtonian mechanics is slightly in error. Newtonian mechanics is simply wrong in high energy regimes. We must, therefore, begin anew.

8.9 Relativistic Mechanics

8.9.1 The Laws of Mechanics

As we saw at the end of the preceding section, Newtonian mechanics cannot be easily adjusted to accommodate situations in which velocities are close to that of light. In a certain sense, our position at the beginning of the 20th century was similar to Newton's in the 17th century. We had some guideposts, but no general formulation.

With Minkowski Space we must deal with 4-vectors rather than simply spatial vectors. We may be guided by what we have learned from our study of Newtonian mechanics. But our new 4-vector laws must be invariant under Lorentz Transformation, i.e. Lorentz invariant.

8.9.2 Conservation of Momentum

In Sect. 8.6 we (tentatively) proposed conservation of momentum as a foundational law of relativistic mechanics. At least the (Lewis and Tolman) concept of relativistic mass, which has been verified experimentally, is based on conservation of momentum in a collision. We shall then begin with a 4-vector formulation of conservation of momentum, which includes collisions. We avoid discussions of interactive forces between collision partners with the implicit assumption that collisions are instantaneous. The identity of the particles may, however, change in a collision.

Specifically, if $\boldsymbol{P}_{\mathrm{j}}$ is the 4-momentum of the jth colliding particle, we make the Ansatz that

$$\sum_{\text{initial}} \boldsymbol{P}_{\mathrm{j}} = \sum_{\text{final}} \boldsymbol{P}_{\mathrm{j}}. \tag{8.93}$$

That is the 4-vector sum of the momenta before the collision is equal to the 4-vector sum of the momenta after the collision. This is the mathematical statement of our first law of relativistic mechanics.

Each component of the 4-vector law (8.93) must be independently valid. For the time component

$$\sum_{\text{initial}} \mathcal{E}_{\mathrm{j}} = \sum_{\text{final}} \mathcal{E}_{\mathrm{j}}, \tag{8.94}$$

since c is an invariant. Our law then includes conservation of total energy. And, for each of the spatial components ($\mu = 1, 2, 3$),

$$\sum_{\text{initial}} m_{0\mathrm{j}} U_{\mathrm{j}}^{\mu} = \sum_{\text{final}} m_{0\mathrm{j}} U_{\mathrm{j}}^{\mu}. \tag{8.95}$$

Introducing the relativistic masses

$$m_{\mathrm{j}} = m_{0\mathrm{j}} \gamma_{\mathrm{j}},$$

where $\gamma_{\mathrm{j}} = \gamma(u)$ for $\boldsymbol{u} = \boldsymbol{u}_{\mathrm{j}}$, we may write (8.95) as

$$\sum_{\text{initial}} m_{\mathrm{j}} \boldsymbol{u}_{\mathrm{j}} = \sum_{\text{final}} m_{\mathrm{j}} \boldsymbol{u}_{\mathrm{j}}, \tag{8.96}$$

which, with $p_{\mathrm{j}} = m_{\mathrm{j}} \boldsymbol{u}_{\mathrm{j}}$, is the Newtonian expression for conservation of momentum. We then recover conservation of energy and momentum in what appear to be standard Newtonian forms from our proposed law of conservation of 4-momentum. We claim that the Newtonian form is only apparent because the energy $\mathcal{E} = m_0 \gamma_{\mathrm{u}} c^2 = mc^2$ is the relativistic energy and the mass $m = m_0 \gamma_{\mathrm{u}}$ is the relativistic mass.

We know that the scalar product of two 4-vectors is a Lorentz scalar invariant (see (8.62)). And we have already shown that for a single material body (particle) $\mathbf{P}^2 = m_0^2 c^2$ (see (8.46)). For two separate particles we can always define the frame k to be the frame in which one of the two particles is at rest and consider that the second particle is moving with a velocity $\boldsymbol{u}$ in k. The 4-momenta of the particles are then $\boldsymbol{P}_1 = m_{01}(c, \mathbf{0})$ and $\boldsymbol{P}_2 = m_{02} \gamma_{\mathrm{v}}(c, \boldsymbol{u})$ and the scalar product is

$$\begin{aligned} \boldsymbol{P}_1 \cdot \boldsymbol{P}_2 &= m_{01} m_{02} \gamma_{\mathrm{u}} c^2 = m_{01} m_2 c^2 = m_{02} m_1 c^2 \\ &= m_{01} \mathcal{E}_2 = m_{02} \mathcal{E}_1 . \end{aligned} \tag{8.97}$$

We can then obtain the form of

$$\left(\sum_{\text{initial}} \boldsymbol{P}_{\mathrm{j}} \right)^2 = \left(\sum_{\text{final}} \boldsymbol{P}_{\mathrm{j}} \right)^2 \tag{8.98}$$

in terms of rest masses and total energies.

Conservation of (total) energy (8.94) may be written as

$$\sum_{\text{initial}} m_{\mathrm{j}} = \sum_{\text{final}} m_{\mathrm{j}} \tag{8.99}$$

since c^2 is a scalar invariant. Relativistic mass is, therefore, conserved in a collision. Mass and energy are equivalent, as Einstein pointed out.

Example 8.9.1 (*Elastic Collision*) An example of the application of the law of conservation of 4-momentum is the elastic collision, which is one in which the rest masses of the colliding particles do not vary. This is, perhaps, the simplest example of a collision between two masses moving at velocities requiring the application of relativistic mechanics. We consider the elastic collision between particles with masses m_{01} and m_{02} and velocities u_1 and u_2. The initial 4-momenta are

$$\begin{aligned} \boldsymbol{P}_1 &= m_{01} \gamma_1 \left(c, \boldsymbol{u}_1 \right) \\ \boldsymbol{P}_2 &= m_{02} \gamma_2 \left(c, \boldsymbol{u}_2 \right) . \end{aligned}$$

We have designated γ_1 for the velocity $\boldsymbol{u}_1$ and γ_2 for the velocity $\boldsymbol{u}_2$. The final 4-momenta are

$$\begin{aligned} \boldsymbol{P}_1' &= m_{01} \gamma_1' \left(c, \boldsymbol{u}_1' \right) \\ \boldsymbol{P}_2' &= m_{02} \gamma_2' \left(c, \boldsymbol{u}_2' \right) . \end{aligned}$$

And our law of mechanics is

$$\boldsymbol{P}_1 + \boldsymbol{P}_2 = \boldsymbol{P}_1' + \boldsymbol{P}_2' .$$

Squaring both sides we have

$$P_1^2 + P_2^2 + 2\boldsymbol{P}_1 \cdot \boldsymbol{P}_2 = P_1'^2 + P_2'^2 + 2\boldsymbol{P}_1' \cdot \boldsymbol{P}_2' .$$

Because they are scalar invariants in an elastic collision

$$P_1^2 = P_1'^2$$
$$P_2^2 = P_2'^2.$$

Then

$$\boldsymbol{P}_1 \cdot \boldsymbol{P}_2 = \boldsymbol{P}_1' \cdot \boldsymbol{P}_2'$$

for any elastic collision between two material bodies.

8.10 Center of Momentum Frame

We simplify our calculations by defining a Center of Momentum frame (CM)[11] in which the spatial component of the momentum of the system of masses (particles) interacting with one another vanishes.

We consider a system of particles in a particular inertial frame k, which we may consider to be at rest. The total 4-momentum $\bar{\mathbf{P}}$ of the system is defined by the 4-vector sum

$$\bar{\mathbf{P}} = \sum \boldsymbol{P}_{\mathrm{j}} \tag{8.100}$$

over the 4-momenta of all of the particles. The components of the 4- momentum $\bar{\mathbf{P}}$ provide then the definitions of total relativistic mass

$$\bar{\mathbf{m}} = \sum_{\mathrm{j}} m_{\mathrm{j}} = \sum_{\mathrm{j}} m_{0\mathrm{j}} \gamma_{\mathrm{j}} \tag{8.101}$$

and total spatial momentum

$$\bar{\mathbf{p}} = \sum_{\mathrm{j}} \boldsymbol{p}_{\mathrm{j}} \tag{8.102}$$

for the system of particles. Then

$$\bar{\mathbf{P}} = \sum_{\mathrm{j}} \left(m_{\mathrm{j}} c, \boldsymbol{p}_{\mathrm{j}}\right) = (\bar{\mathbf{m}} c, \bar{\mathbf{p}}) \tag{8.103}$$

$$= \left(\bar{\mathcal{E}}/c, \bar{\mathbf{p}}\right), \tag{8.104}$$

where $\bar{\mathcal{E}} = \bar{\mathbf{m}} c^2$ is the total energy of the system of particles.

The CM frame k_{CM} is that in which the spatial component $\bar{\mathbf{p}}$ of the total 4-momentum $\bar{\mathbf{P}}$ in (8.102) vanishes. The velocity of this frame $\boldsymbol{u}_{\mathrm{CM}}$, measured in

[11] We note that CM designates center of mass in nonrelativistic (Newtonian) mechanics. The designation here as center of momentum is standard.

the system k, is the spatial component of the 4-velocity of the system of particles measured in the system k

$$\boldsymbol{u}_{\mathrm{CM}} = \frac{\bar{\mathbf{p}}}{\bar{\mathbf{m}}}. \tag{8.105}$$

In k_{CM} the spatial component of the sum of the 4-momenta of the particles is zero. With (8.105) we may write (8.103) as

$$\bar{\mathbf{P}} = \bar{\mathbf{m}}\,(c, \boldsymbol{u}_{\mathrm{CM}})\,. \tag{8.106}$$

The 4-velocity $\boldsymbol{U}_{\mathrm{CM}}$ of the frame k_{CM} relatively to the frame k is

$$\boldsymbol{U}_{\mathrm{CM}} = \gamma\,(u_{\mathrm{CM}})\,(c, \boldsymbol{u}_{\mathrm{CM}})\,, \tag{8.107}$$

where $\boldsymbol{u}_{\mathrm{CM}}$ is defined in (8.105) as a measurable spatial vector in the frame k and, therefore, $\gamma\,(u_{\mathrm{CM}})$ is also well defined in k. We may then express the 4-momentum $\bar{\mathbf{P}}$ in terms of well-defined spatial quantities as

$$\bar{\mathbf{P}} = \bar{\mathbf{m}}\gamma^{-1}\,(u_{\mathrm{CM}})\,\boldsymbol{U}_{\mathrm{CM}}. \tag{8.108}$$

The square of the 4-momentum $\bar{\mathbf{P}}$ is then

$$\bar{\mathbf{P}}^2 = \left[\bar{\mathbf{m}}\gamma^{-1}\,(u_{\mathrm{CM}})\right]^2 c^2, \tag{8.109}$$

which is an invariant. Therefore $\bar{\mathbf{m}}\gamma^{-1}\,(u_{\mathrm{CM}})$ is an invariant.

Analogously to the single particle rest mass we may define the rest mass of the system of particles as

$$\bar{\mathbf{m}}_0 = \bar{\mathbf{m}}\gamma^{-1}\,(u_{\mathrm{CM}}) \tag{8.110}$$

Then (8.109) becomes

$$\bar{\mathbf{P}}^2 = \bar{\mathbf{m}}_0^2 c^2 \tag{8.111}$$

and $\bar{\mathbf{P}} = \bar{\mathbf{m}}_0 \boldsymbol{U}_{\mathrm{CM}}$. We then have CM quantities $\bar{\mathbf{m}}$, $\bar{\mathbf{m}}_0$, and $\boldsymbol{U}_{\mathrm{CM}}$, which are system analogues of the single particle quantities m, m_0, and $\boldsymbol{U}$, and well-defined measurable quantities in the frame k.

If we identify the kinetic energy $\bar{\mathbf{T}}$ of the system of particles as the difference between the total and the rest energy of the system then

$$\begin{aligned}\bar{\mathbf{T}} &= \bar{\mathcal{E}} - \bar{\mathbf{m}}_0 c^2 = (\bar{\mathbf{m}} - \bar{\mathbf{m}}_0)\, c^2 \\ &= [\gamma\,(u_{\mathrm{CM}}) - 1]\,\bar{\mathbf{m}}_0 c^2.\end{aligned}$$

And we have the general result that the mass of the system of moving particles always exceeds the sum of the rest masses of the particles. We have included no interactive forces in this development. So we cannot push this result toward any conclusions regarding atomic or nuclear systems. We are still dealing simply with the 4-momentum of a system of separate, i.e. unbound particles. We realize, however, that if particles collide with sufficient kinetic energy it is possible to produce additional particles. That is a portion of the kinetic energy of the colliding particles may be converted into the mass of an additional particle produced in the collision.

In the 21st century this process is common. It is the process by which elementary particles are identified. Nevertheless we may ask what our theory has to say about the minimal kinetic energy required to produce a specific particle from a collision. We choose a collision between two protons, with rest masses 938.27 MeV/c^2, that results in a pion π^0 with a rest mass 135 MeV/c^2 [[117], pp. 435, 529]. Because the proton is stable this interaction avoids any questions of particle transformation.

Example 8.10.1 (*Threshold Energy*) The reaction to produce a pion π^0 from a collision between two protons is

$$p + p \to p + p + \pi^0.$$

We shall formulate this example in a general fashion considering the two initial particles to have rest masses M_0 and the particle produced in the collision to have rest mass m_0. We consider that one of the initial particles is at rest and the other is moving toward it with velocity $\boldsymbol{v} = -v\hat{e}_x$. The at rest particle may be considered the target and the moving particle the projectile, or bullet. After the collision we have all the particles moving apart. We reference these to the CM frame k_{CM}.

The 4-momenta of the initial states are

$$\boldsymbol{P}_1 = M_0 \left(c, \boldsymbol{0}\right)$$

for the target and

$$\boldsymbol{P}_2 = M_0 \gamma_v \left(c, -v\hat{e}_x\right)$$

for the projectile. After the collision the 4-momentum of the particles in the CM frame k_{CM} is $\bar{\mathbf{P}}$. This 4-momentum is a constant during the collision. We may, therefore, calculate the value of $\bar{\mathbf{P}}$ from the conditions before the collision referenced to the frame k. That is

$$\bar{\mathbf{P}} = \boldsymbol{P}_1 + \boldsymbol{P}_2.$$

Squaring this we have

$$\bar{\mathbf{P}}^2 = P_1^2 + P_2^2 + 2\boldsymbol{P}_1 \cdot \boldsymbol{P}_2.$$

With

$$\boldsymbol{P}_1 \cdot \boldsymbol{P}_2 = M_0^2 \gamma_v \left(c, \boldsymbol{0}\right) \cdot \left(c, -v\hat{e}_x\right) = M_0^2 \gamma_v c^2$$

and $P_1^2 = P_2^2 = M_0^2 c^2$ we have

$$\bar{\mathbf{P}}^2 = 2M_0^2 c^2 + 2M_0^2 \gamma_v c^2.$$

From (8.111) we have $\bar{\mathbf{P}}^2 = \bar{\mathbf{m}}_0^2 c^2$. Then

$$2M_0^2 c^2 + 2M_0^2 \gamma_{\mathrm{v}} c^2 = \bar{\mathbf{m}}_0^2 c^2.$$

We do not know the velocities of the particles in k_{CM} after the collision. Therefore we do not know $\bar{\mathbf{m}}_0$. We do, however, have a minimum value for $\bar{\mathbf{m}}_0$, which is the sum of the rest masses $2M_0 + m_0$. Using this minimum value for $\bar{\mathbf{m}}_0$ will provide the minimum kinetic energy of the projectile required.

We then consider

$$2M_0^2 c^2 + 2M_0^2 \gamma_{\mathrm{v}} c^2 = (2M_0 + m_0)^2 c^2.$$

Solving for γ_{v},

$$\gamma_{\mathrm{v}} = 1 + \frac{m_0^2}{2M_0^2} + \frac{2m_0}{M_0}.$$

The kinetic energy of the projectile is then

$$\begin{aligned} \mathcal{E}_{\mathrm{kin}} &= \mathcal{E} - M_0 c^2 = (\gamma_{\mathrm{v}} - 1)\, M_0 c^2 \\ &= \left(\frac{m_0}{2M_0} + 2 \right) m_0 c^2. \end{aligned}$$

This is the energy that goes into creating m_0 in this limiting case. The actual energy in the rest mass is $m_0 c^2$. The efficiency of the reaction is the ratio of the energy that actually goes into m_0 divided by the incoming kinetic energy of the projectile, or

$$\begin{aligned} \text{efficiency} &= \frac{m_0 c^2}{(m_0/2M_0 + 2)\, m_0 c^2} \\ &= \frac{2}{4 + (m_0/M_0)}. \end{aligned}$$

For the proton and the pion, $m_0/M_0 = 0.143\,88$ and

$$\text{efficiency} = \frac{2}{4 + 0.143\,88} \approx 0.48$$

and the minimum kinetic energy required is

$$\mathcal{E}_{\mathrm{kin}} = \left(\frac{0.143\,88}{2} + 2 \right) 135 = 279.71\,\mathrm{MeV}.$$

8.11 Waves

Plane waves are periodic (sinusoidal) disturbances which propagate in a single direction. In mathematical terms the disturbance extends uniformly to infinity in the two spatial coordinates perpendicular to the direction of propagation. The plane wave must then be recognized as a mathematical idealization, which cannot be localized in a finite spatial region, but which can be used as a basis for constructing general wave disturbances [[45], pp. 249–254]. Here we will develop a Minkowski Space description of plane waves as frequency 4-vectors [cf. [99], pp. 60–65]. Then we will use the frequency 4-vector to represent the photon, which is the nonlocalizable quantum of the electromagnetic field, and the basis for the de Broglie matter wave.

8.11.1 Frequency 4-Vector

In time $\mathrm{d}t$ a plane wave moving in the direction $\hat{n}$ travels a distance $\hat{n}\cdot\,\mathrm{d}\boldsymbol{r}$ where $\mathrm{d}\boldsymbol{r}$ is a spatial element $(\mathrm{d}x^1,\mathrm{d}x^2,\mathrm{d}x^3)$. Then

$$\hat{n}\cdot\mathrm{d}\boldsymbol{r} = w\mathrm{d}t, \tag{8.112}$$

in which w is the speed of the wave. Introducing the speed of light in vacuum c we may write (8.112) as

$$\nu\mathrm{d}x^0 - \frac{\nu c}{w}\hat{n}\cdot\mathrm{d}\boldsymbol{r} = 0, \tag{8.113}$$

where ν is the frequency of the wave and $c/w = 1$ for light.

We may contain this description in the *frequency 4-vector* for a plane wave $\boldsymbol{N}$, which we define as

$$N^{\mu} = \nu\left(1, \frac{c}{w}\hat{n}\right). \tag{8.114}$$

And the covariant frequency 4-vector is $N_{\mu} = \nu\left(1, -\frac{c}{w}\hat{n}\right)$. Then

$$\begin{aligned} \boldsymbol{N}\cdot\mathrm{d}\boldsymbol{x} &= N^{\alpha}g_{\alpha\beta}\mathrm{d}x^{\beta} = \nu\left(1, \frac{c}{w}\hat{n}\right)\cdot(c\mathrm{d}t, \mathrm{d}\boldsymbol{r}) \\ &= \nu c\mathrm{d}t - \frac{\nu c}{w}\hat{n}\cdot\mathrm{d}\boldsymbol{r}. \end{aligned} \tag{8.115}$$

And using (8.113) we see that the plane wave is defined in terms of $\boldsymbol{N}$ by

$$N_{\mu}\mathrm{d}x^{\mu} = 0. \tag{8.116}$$

Equation (8.116) also shows that the frequency 4-vector $\boldsymbol{N}$ is orthogonal[12] to the differential 4-vector $\mathrm{d}\boldsymbol{x}$.

---

[12]In Newtonian, 3 dimensional space this would mean perpendicular.

Dividing (8.116) by c we have

$$\frac{1}{c}N_{\mu}\mathrm{d}x^{\mu}=\frac{1}{\tau}\mathrm{d}t-\frac{1}{\lambda}\hat{n}\cdot\mathrm{d}\boldsymbol{r}=0,$$

where $\tau=1/\nu$ is the period of the wave and $\lambda=w/\nu$ is the wavelength. We now choose X to be some function whose differential is

$$\mathrm{d}X=\frac{\partial X}{\partial x^{\mu}}\mathrm{d}x^{\mu}=\frac{1}{c}N_{\mu}\mathrm{d}x^{\mu}. \tag{8.117}$$

Since N_{μ} is independent of x^{μ}, (8.117) can be easily integrated to obtain

$$X\left(ct,\boldsymbol{r}\right)=\frac{1}{c}N_{\mu}x^{\mu}=\left(\nu t-\frac{\hat{n}\cdot\boldsymbol{r}}{\lambda}\right). \tag{8.118}$$

For constant t we see that X decreases by an integer as $\boldsymbol{r}$ advances by one wavelength λ in the direction of $\hat{n}$. X is then a *counting index* for the wave crests. The index is such that the earlier waves produced by the source have lower indices than those produced later, which is a logical convention. At a point in space the waves passing us increase in index as time increases.

For a light wave $w=c$ and the frequency 4-vector becomes

$$N^{\mu}=\nu\left(1,\hat{n}\right). \tag{8.119}$$

And for a light wave,

$$N_{\mu}N^{\mu}=0. \tag{8.120}$$

We emphasize that our treatment here is of plane waves (phase waves) of specific frequency ν and wavelength λ.

8.11.2 The Photon

The idea that black body radiation can be considered as composed of quanta first appeared in Einstein's paper in March of 1905 *On a Heuristic Viewpoint Concerning the Production and Transformation of Light*. [23] Einstein considered the same physical situation Max Planck had considered. The radiation was contained in a cavity (Hohlraum) enclosed by perfectly reflecting walls containing also a system of oscillators. The Hohlraum was full of a gas (air) and radiation. The oscillators were in equilibrium with the air molecules. He concluded that [[58], p. 26]

> Monochromatic radiation of low density (within the limits of the Wien law) behaves with respect to thermal phenomena as if it were composed of independent energy quanta of magnitude $h\nu$.

In the March, 1905, paper Einstein did not actually write the energy quantum as $h\nu$, but as $(R\beta/N)\,\nu$ in which R and N are the gas constant and Avogadro's Number and $\beta = (h/k_{\mathrm{B}})$. These are the same numerically, but not in spirit. Einstein did not use Planck's energy relationship in this paper. He cited Planck's formula for the spectrum, but used the Wien spectrum in his development, which was thermodynamic. He derived the entropy of the radiation field and compared his result to Ludwig Boltzmann's result for the entropy of a collection of atoms [23]. His claim above is based on a this comparison. In 1905, then, Planck's quanta are states of the oscillator, while Einstein's are of the radiation field.

Einstein's identification of the photon momentum as $p = h\nu/c$ appeared first in his work on the absorption and emission (spontaneous and induced) in 1916 [[89], pp. 405–407]. In 1939 Eugene Wigner identified the photon as a state of the electromagnetic field [127].

It is very tempting to move quickly from Einstein's March paper to a picture of the photon as a massless particle, turn to our general relationship for material particles (8.55), and find the consequences if we set $m_0 = 0$. If we were to do so we would find that

$$\mathcal{E} = pc, \tag{8.121}$$

from which the momentum of the photon would follow as $p = h\nu/c$. Although this is a correct result, as we now realize, Einstein, in spite of his bold approach in the March paper, did not take this step. The reasons are very logical, as we have indicated, and should be appreciated.

From the Planck–Einstein energy relationship $\mathcal{E} = h\nu$ for the photon, and from the momentum of the photon $p = h\nu/c$, we can write a 4-momentum for the photon $\boldsymbol{P}_\gamma$ as (see (8.54))

$$\boldsymbol{P}_\gamma = \frac{h\nu}{c}\left(1, \hat{n}\right), \tag{8.122}$$

where $\hat{n}$ is the direction of propagation of the wave. Comparing (8.122) with (8.114) we see that for a light wave

$$c\boldsymbol{P}_\gamma = h\boldsymbol{N} \tag{8.123}$$

for the 4-vectors $\boldsymbol{P}_\gamma$ and $\boldsymbol{N}$. Because the photon has zero rest mass,

$$\mathbf{P}_\gamma^2 = m_0 c^2 = 0. \tag{8.124}$$

Although the photon has zero rest mass we can introduce the concept of relativistic photon mass from $\mathcal{E} = mc^2$. That is

$$m = \frac{h\nu}{c^2}. \tag{8.125}$$

Example 8.11.1 (*Compton Effect (1922)*) Arthur Compton studied x-ray reflection from a solid. He observed that the x-rays coming from the sample varied in frequency depending on scattering angle. He analyzed the experiment considering that Einstein's photon and the principles of relativistic mechanics were correct. Because the energy of the x-ray photon was a thousand times that of the bound electron he considered that the electron was a free particle at the time of collision. So initially we have an electron at rest at the origin of k and a photon coming toward it. The 4-momentum of the electron is

$$\boldsymbol{P}_{\mathrm{e}} = m_0 \left(c, \boldsymbol{0} \right)$$

and of the photon is

$$\boldsymbol{P}_{\gamma} = \frac{h\nu}{c} \left(1, \hat{n} \right).$$

The electron receives energy and momentum from the x-ray photon and leaves the collision site at an angle ϕ to the x-axis. The photon leaves with a frequency ν' and at an angle ϑ to the x-axis. After the collision, then, the 4-momenta are

$$\boldsymbol{P}'_{\mathrm{e}} = m_0 \gamma_{\mathrm{v}} \left(c, \boldsymbol{v} \right)$$

and of the photon'

$$\boldsymbol{P}'_{\gamma} = \frac{h\nu'}{c} \left(1, \hat{n}' \right),$$

where $\boldsymbol{v}$ is the velocity of the scattered electron, ν' is the frequency of the scattered x-ray, and $\hat{n}'$ is at an angle ϑ with respect to $\hat{n}$. Our law of mechanics is then

$$\boldsymbol{P}_{\mathrm{e}} + \boldsymbol{P}_{\gamma} = \boldsymbol{P}'_{\mathrm{e}} + \boldsymbol{P}'_{\gamma}.$$

We know nothing about the scattered electron except that the value of the square of its 4-momentum is $m_0^2 c^2$. We then isolate the 4-momentum of the scattered electron and square the resulting expression. That is

$$\left(\boldsymbol{P}_{\mathrm{e}} + \boldsymbol{P}_{\gamma} - \boldsymbol{P}'_{\gamma} \right)^2 = \left(\boldsymbol{P}'_{\mathrm{e}} \right)^2,$$

or

$$P_{\mathrm{e}}^2 + P_{\gamma}^2 + P_{\gamma}^{\prime 2} + 2\boldsymbol{P}_{\mathrm{e}} \cdot \left(\boldsymbol{P}_{\gamma} - \boldsymbol{P}'_{\gamma} \right) - 2\boldsymbol{P}_{\gamma} \cdot \boldsymbol{P}'_{\gamma} = P_{\mathrm{e}}^{\prime 2} = m_0^2 c^2. \tag{8.126}$$

Now, noting that the scalar product requires $g_{\alpha\beta}$,

$$\begin{aligned} \boldsymbol{P}_{\mathrm{e}} \cdot \left(\boldsymbol{P}_{\gamma} - \boldsymbol{P}'_{\gamma} \right) &= P_{\mathrm{e}}^{\alpha} g_{\alpha\beta} \left(P_{\gamma}^{\beta} - P_{\gamma}^{\prime\beta} \right) \\ &= m_0 \frac{h}{c} \left(c, \boldsymbol{0} \right) \cdot \left[\nu \left(1, \hat{n} \right) - \nu' \left(1, \hat{n}' \right) \right] \\ &= m_0 h \left(\nu - \nu' \right), \end{aligned}$$

and

$$\begin{aligned}\boldsymbol{P}_\gamma \cdot \boldsymbol{P}'_\gamma &= P_\gamma^\alpha g_{\alpha\beta} P_\gamma^\beta \\ &= \left(\frac{h}{c}\right)^2 \nu\nu' \left(1, \hat{n}\right) \cdot \left(1, \hat{n}'\right) \\ &= \left(\frac{h}{c}\right)^2 \nu\nu' \left(1 - \cos\vartheta\right),\end{aligned}$$

since $\hat{n} \cdot \hat{n}' = \cos\vartheta$. For the photons before and after collision

$$P_\gamma^2 = P_\gamma'^2 = 0,$$

and for the electron

$$P_{\mathrm{e}}^2 = m_0^2 c^2.$$

Then (8.126) becomes

$$\frac{h}{c^2}\nu\nu' \left(1 - \cos\vartheta\right) = m_0 \left(\nu - \nu'\right).$$

The theoretical analysis agreed with the experimental results making this a classic result because of the use of the photon concept and relativity.

8.11.3 De Broglie Particle Waves

The idea of Louis Victor Pierre Raymond, prince de Broglie on particle waves was an important step in the development of the quantum theory. Pais presents an outline of the relationship between the ideas of de Broglie and those of Einstein at the time of de Broglie's doctoral defence [[89], pp. 435–438].

Louis was the younger brother of Louis César Victor Maurice, the 6th duc de Broglie [[6], vol. 4, p. 262]. In spite of the family's disdain for science, Maurice de Broglie was able to establish a respected reputation in experimental physics, while simultaneously engaging in a more respectable naval career. Louis de Broglie first took a degree in ancient history at the University of Paris (1910) and then moved toward science [[49], p. 447].

After 1913 Maurice was involved in the study of x-rays. The British father and son team, William and Lawrence Bragg, had come to recognize that neither the wave nor particle picture of x-rays was adequate. Maurice shared the opinion of the Braggs. And his brother wrote that Maurice considered x-rays to be a combination of wave and particle. However, Louis also noted that Maurice was not a theoretician and had no clear ideas on this matter.

Louis de Broglie served in the French army during the First World War. And in 1919, after the war, he joined the physics laboratory headed by Maurice [[89], p. 435]. For a while the two brothers worked together on experiments studying the behavior of electrons when x-rays were scattered from solids. Louis wrote that he and his brother had long discussions about the interpretation of the experiments, and Louis was "led to profound meditations on the need of always associating the aspects of waves with that of particles" [[13], p. 276]. The ideas we consider here appeared in Louis de Broglie's dissertation in 1924 [[49], p. 447].

Our intention here is only to outline the development of Louis de Broglie's ideas in the language of the 4-momenta of particles and waves. De Broglie's treatment of matter waves was relativistic. And at our present stage we can carry out the basic relativistic development. De Broglie equated the product of the particle 4-momentum with c, which makes the terms in the particle 4-momentum all energy terms, to the product of the (phase wave) frequency 4-vector with Planck's constant h, which, from the Planck–Einstein formula, makes the terms in the frequency 4-vector all energy terms. *De Broglie's Equation* [[99], p. 82] is then

$$c\boldsymbol{P} = h\boldsymbol{N} \tag{8.127}$$

or

$$(\mathcal{E}, pc) = \left(h\nu, h\nu\frac{c}{w}\hat{n}\right). \tag{8.128}$$

De Broglie's Equation (8.128) is identical in form to (8.123) for the photon. In (8.128), however, the 4-momentum is that for a material particle, while in (8.123) the 4-momentum is that for a photon. The appearance of the phase wave frequency 4-vector in the de Broglie Equation means that the results of the derivation will be for the particle phase wave associated with a particle having a definite momentum.

From the time (zeroth) element of (8.128) we have

$$\mathcal{E} = mc^2 = h\nu \tag{8.129}$$

for material particles, which serves to identify a frequency ν to be associated with the matter wave. And from the spatial elements of (8.128) we have

$$\boldsymbol{p} = h\frac{\nu}{w}\hat{n} = \frac{h}{\lambda}\hat{n}. \tag{8.130}$$

With (8.129) and $p = mu$, (8.130) becomes

$$\frac{\mathcal{E}}{c^2} = \frac{h\nu}{wu}. \tag{8.131}$$

And with (8.129), Eq. (8.131) becomes

$$w = \frac{c^2}{u}, \tag{8.132}$$

which relates the phase velocity of the particle wave w to the particle velocity u.

From the Minkowski Axiom we know that for material particles $u < c$. Therefore the phase velocity of a de Broglie wave associated with a material particle is $> c$. Phase waves, however, are not the carriers of information. So this is not a violation of the limit on the transmission of information. If de Broglie waves are associated with a particle, the disturbance representing the moving particle must be constructed from these phase waves and the velocity of the disturbance will be the group velocity, not the phase velocity [cf. [45], pp. 330, 373, 385, 388].

8.12 Relativistic Forces

In addition to conservation of momentum we must also formulate a law of mechanics that can treat the motion of particles under the action of forces, such as those in the plasma filling the cosmos or in high energy accelerators. The forces acting on these particles are electromagnetic (field) forces, for which we developed the relativistic theory in Sect. 8.8.3. So we must at least have an understanding of the relativistic motion of charged particles under the influence of electromagnetic forces.

The force laws we propose for mechanics must be Lorentz invariant. And we know from the results of Einstein's investigation in the paper on special relativity that the standard form of Newton's Second Law will not fulfill this requirement. We know as well that Newton's Third Law cannot be expected to hold for field forces because simultaneity is no longer a universal concept. Any laws we may propose must, however, reduce to Newtonian laws for spatial components in our (at rest) inertial frame.

Relativistic forces may be divided into those which do not change the rest mass of the body on which they act and those which will change the rest mass of a body by their action. We will call those forces which do not result in a change in rest mass *pure* forces. And those which result in a change in the rest mass we will call *impure* forces. Not surprisingly, perhaps, we will see that nuclear forces result in a change in the rest mass of nucleons. Electromagnetic forces, however, do not result in a change of rest mass.

We may argue that devoting primary consideration to the electromagnetic force is justified on the basis of practicality. Pauli, however, had a more logical argument for considering only electromagnetic forces in detail. He pointed out that numerous forces transform in the same way as do electromagnetic forces, since we may always compensate for another force with an electromagnetic force. The compensation will then be independent of inertial frame [[92], p. 116]. We will then claim that our treatment of the electromagnetic force encompasses any and all pure forces.

We will find that a formulation in terms of Hamilton's Principal Function is a Lorentz invariant approach. And this will be the final approach we take.

8.12.1 Pure Forces

A general 4-vector formulation of Newton's Second Law is

$$\begin{aligned} \boldsymbol{F} &= \frac{\mathrm{d}}{\mathrm{d}\tau}\boldsymbol{P} \\ &= \frac{\mathrm{d}}{\mathrm{d}\tau}(mc, \boldsymbol{p}), \end{aligned} \tag{8.133}$$

where $\boldsymbol{F}$ is what we shall term a 4-vector force and τ is the proper time for the world line of the body (see (8.28)). We know only that the spatial component of $\boldsymbol{F}$, which we shall identify as $\boldsymbol{f}$, satisfies Newton's Second Law in our at rest frame so that

$$\boldsymbol{f} = \frac{\mathrm{d}}{\mathrm{d}t}\boldsymbol{p}, \tag{8.134}$$

where $\boldsymbol{p}$ is the spatial component of the 4-momentum and t is the time for the reference frame k, which we may choose to be at rest.

We showed in Sect. 8.7 that the scalar product of two 4-vectors is a Lorentz invariant. If we take the scalar product of $\boldsymbol{F}$ with the 4-velocity $\boldsymbol{U}$ we have the invariant $\boldsymbol{U}\cdot\boldsymbol{F} = U_\alpha F^\alpha$, which with (8.133) is

$$\boldsymbol{U}\cdot\boldsymbol{F} = \boldsymbol{U}\cdot\left(c\frac{\mathrm{d}}{\mathrm{d}\tau}m, \frac{\mathrm{d}}{\mathrm{d}\tau}\boldsymbol{p}\right). \tag{8.135}$$

Then, with (8.28), (8.75) and (8.134), Eq. (8.135) becomes

$$\boldsymbol{U}\cdot\boldsymbol{F} = \gamma_{\mathrm{u}}^2\left(c^2\frac{\mathrm{d}}{\mathrm{d}t}m - \boldsymbol{u}\cdot\boldsymbol{f}\right). \tag{8.136}$$

In the frame in which the mass is at rest at a particular instant $\boldsymbol{u} = \boldsymbol{0}$, $\gamma_{\mathrm{u}} = 1$, and $m = m_0$. We choose this to be our rest frame k and study the subsequent differential step of the body's motion along its world line. In this frame (8.136) is

$$\boldsymbol{U}\cdot\boldsymbol{F} = c^2\frac{\mathrm{d}m_0}{\mathrm{d}\tau}. \tag{8.137}$$

Because of the Lorentz invariance of the scalar product of 4-vectors, (8.136) and (8.137) are equal. Then, with $\mathrm{d}\tau = \mathrm{d}t/\gamma_{\mathrm{u}}$

$$c^2\left(\frac{\mathrm{d}m}{\mathrm{d}t} - \frac{1}{\gamma_{\mathrm{u}}}\frac{\mathrm{d}m_0}{\mathrm{d}t}\right) = \boldsymbol{u}\cdot\boldsymbol{f}. \tag{8.138}$$

Equation (8.138) is a general expression relating the difference between the rates of change of relativistic mass (total energy) and rest mass (rest energy) to the rate at which work is done on the mass by the spatial component of the 4-vector force $\boldsymbol{F}$. From (8.137) we see that whether or not the rest mass is conserved is determined by whether or not $\boldsymbol{U} \cdot \boldsymbol{F}$ vanishes. A *pure force*, which conserves the rest mass, i.e. for which $\mathrm{d}m_0/\mathrm{d}t = 0$, is one for which $\boldsymbol{U} \cdot \boldsymbol{F} = 0$. An *impure force* is one for which $\mathrm{d}m_0/\mathrm{d}t \neq 0$ and $\boldsymbol{U} \cdot \boldsymbol{F} \neq 0$ [[99], p. 92].

From (8.138), with $\mathcal{E} = mc^2$, we see that for a pure force

$$\frac{\mathrm{d}\mathcal{E}}{\mathrm{d}t} = \boldsymbol{u} \cdot \boldsymbol{f}, \tag{8.139}$$

That is a pure force results in a change in the total energy (relativistic mass) of the body.

8.12.2 Impure Forces

We may gain some insight into the character of an impure force if we assume that it results from a general potential $\Psi(ct, \boldsymbol{r})$ in such a manner that the (covariant) force F_μ is (see (8.78))

$$F_\mu = \partial_\mu \Psi(ct, \boldsymbol{r}) = \frac{\partial}{\partial x^\mu} \Psi(ct, \boldsymbol{r}). \tag{8.140}$$

Then $\boldsymbol{U} \cdot \boldsymbol{F}$ is

$$\begin{aligned} U_\alpha F^\alpha &= g_{\alpha\nu} \frac{\mathrm{d}x^\nu}{\mathrm{d}\tau} g^{\alpha\mu} \frac{\partial}{\partial x^\mu} \Psi(ct, \boldsymbol{r}) \\ &= \delta^\mu_\nu \frac{\mathrm{d}x^\nu}{\mathrm{d}\tau} \frac{\partial}{\partial x^\mu} \Psi(ct, \boldsymbol{r}) \\ &= \frac{\mathrm{d}\Psi(ct, \boldsymbol{r})}{\mathrm{d}\tau} \end{aligned} \tag{8.141}$$

From (8.137) we then have the differential equation

$$\frac{\mathrm{d}\Psi(ct, \boldsymbol{r})}{\mathrm{d}\tau} = c^2 \frac{\mathrm{d}m_0}{\mathrm{d}\tau}, \tag{8.142}$$

which we may immediately integrate to obtain

$$m_0 c^2 = \Psi(ct, \boldsymbol{r}) + \text{constant}.$$

The space-time variation of the rest mass is then determined by the space-time variation contained in the scalar potential [cf. [99], p. 92].

An example of this sort of variation in rest mass is provided by the meson theory of the strong nuclear force, first proposed by Hideki Yukawa in 1934. The nuclear force is mediated by the exchange of π-mesons between nucleons. The π-mesons have charges $\pm e$ or 0. The exchange of π-mesons between nucleons (protons and neutrons) may then result in the exchange of charge and, hence, rest mass, since the masses of the proton and neutron differ. The short-range Yukawa potential has the same form as the (Debye) screened Coulomb potential [[78], p. 297]

$$V(r) = V_0 \frac{\exp(-\alpha r)}{\alpha r}. \tag{8.143}$$

The importance of finite volume effects and the inelastic threshold are discussed, e.g., by Bernard, et al. [2].

The first discovery of a change in the rest mass of a body resulted from the beautiful radiochemistry experiments of Otto Hahn and Friedrich Strassmann in the winter of 1938–1939 at the Kaiser Wilhelm Institute in Berlin. Hahn and Strassmann discovered that after ${}^{238}_{92}\mathrm{U}$ was bombarded by neutrons Barium (Ba) could be found in the products. Hahn's colleague Lise Meitner and her nephew Otto Frisch analyzed these experiments (sitting on a log in a snowy woods at Kungälv, Sweden). Meitner and Frisch were both then refugees from Nazi Germany. Meitner imagined that the ${}^{238}_{92}\mathrm{U}$ nucleus had split. She knew the difference in masses of ${}^{238}_{92}\mathrm{U}$ and ${}^{137}_{56}\mathrm{Ba}$ + ${}^{83}_{46}\mathrm{Kr}$. And she knew Einstein's mass-energy relationship $\mathcal{E} = \Delta mc^2$, where Δm, is the (rest) mass lost. The result was (about) 200 MeV. This matched the potential energy of ${}^{137}_{56}\mathrm{Ba}$ and ${}^{83}_{46}\mathrm{Kr}$ nuclei located at twice the nuclear radius apart, which would be the potential energy just after splitting. In keeping with the fact that the Lorentz force could not be responsible for the loss of rest mass, we can claim that the Hahn–Strassmann experiments and the Meitner–Frisch analysis, indicated that a non-electromagnetic force, such as the Yukawa force, must be present in the nucleus[13] [[47], pp. 10–11; [100], pp. 257–264].

8.12.3 *Lorentz Force*

Using (8.75) and the contravariant form of the Lorentz force (8.91), we see that

$$\boldsymbol{U} \cdot \boldsymbol{F} = U_\alpha F^\alpha = \gamma_{\mathrm{u}}^2 (c, -\boldsymbol{u}) \left(\frac{1}{c} \boldsymbol{f}_{\mathrm{L}} \cdot \boldsymbol{u}, \boldsymbol{f}_{\mathrm{L}} \right) = 0. \tag{8.144}$$

The Lorentz force is then a pure force and can have no effect on the rest mass of a body.

With Pauli we may, however, take the Lorentz force

$$Q\mathcal{F}^{\alpha\beta} U_\beta = Q \left(\partial^\alpha \mathcal{A}^\beta - \partial^\beta \mathcal{A}^\alpha \right) U_\beta \tag{8.145}$$

[13] Of course this was already clear from the fact that a non-electromagnetic force must hold the nucleus together.

as a model for forces acting on charged bodies outside of the nucleus. Along the world line of the charged body the force law, in Lorentz invariant form, will then be

$$\begin{aligned} m_0 g_{\mu\alpha} d^2 x^\alpha / d\tau^2 &= \mathcal{Q} g_{\mu\alpha} \left(\partial^\alpha \mathcal{A}^\beta - \partial^\beta \mathcal{A}^\alpha \right) g_{\beta\lambda} g^{\lambda\nu} U_\nu \\ &= \mathcal{Q} \left(\partial_\mu \mathcal{A}_\lambda - \partial_\lambda \mathcal{A}_\mu \right) U^\lambda . \end{aligned} \tag{8.146}$$

And we will use (8.146) as a guide in our search for a general covariant form of the laws of mechanics valid in special relativity.

8.13 Relativistic Analytical Mechanics

8.13.1 Hamilton's Principle

We begin our search for generally covariant laws of mechanics with Hamilton's Principle expressed by the variation of Hamilton's Principal Function

$$\mathbf{S} = \int_{t_1}^{t_2} \mathrm{d}t \, \Lambda^{(\mathrm{t})} \left(x^\mu, \dot{x}^\mu \right), \tag{8.147}$$

which we write in bold font for the relativistic case. Here we have introduced the notation $\Lambda^{(\mathrm{t})}$ for the *relativistic Lagrangian* based on the local time t of the frame k. We have also introduced the contravariant vectors x^μ in our Lagrangian and chosen to consider conservative systems for which the Lagrangian is not explicitly a function of the time t.

We know that Hamilton's Principle produces the correct equations of motion in an inertial frame when we use the local time as the basis for our formulation of the Principal Function **S**. The proper time τ for the body is, however, the time we actually want to consider in our formulation. So we convert (8.147) to an integral over τ, which is related to the time measured by a stationary timepiece t in the reference frame k as $\mathrm{d}t/\mathrm{d}\tau = \gamma_{\mathrm{u}}$, where u is the velocity of the body (see (8.28)). We will keep the dot notation for the derivative with respect to t and use a prime to denote derivatives with respect to τ, i.e. $\mathrm{d}x^\mu/\mathrm{d}\tau = x^{\mu\prime}$. The change of variables then results in

$$\mathbf{S} = \int_{\tau_1}^{\tau_2} \mathrm{d}\tau \left[\gamma_{\mathrm{u}} \Lambda^{(\mathrm{t})} \left(x^\mu, x^{\mu\prime}/\gamma_{\mathrm{u}} \right) \right]. \tag{8.148}$$

If we define the Lagrangian $\Lambda^{(\tau)}$ as

$$\Lambda^{(\tau)} = \gamma_{\mathrm{u}} \Lambda^{(\mathrm{t})} \left(x^\mu, x^{\mu\prime}/\gamma_{\mathrm{u}} \right), \tag{8.149}$$

Hamilton's Principal Function then retains its original form

$$\mathbf{S} = \int_{\tau_1}^{\tau_2} \mathrm{d}\tau \Lambda^{(\tau)} \left(x^{\mu}, x^{\mu\prime}/\gamma_{\mathrm{u}} \right), \tag{8.150}$$

and Hamilton's Principle $\delta \mathbf{S} = 0$ can be required based on the proper time τ.

8.13.2 Euler–Lagrange Equations

From (8.147) and (8.150) the Euler–Lagrange Equations for the local and the proper time are

$$\frac{\partial \Lambda^{(\mathrm{t})}}{\partial x^{\mu}} - \frac{\mathrm{d}}{\mathrm{d}t}\left(\frac{\partial \Lambda^{(\mathrm{t})}}{\partial \dot{x}^{\mu}} \right) = 0, \tag{8.151}$$

and

$$\frac{\partial \Lambda^{(\tau)}}{\partial x^{\mu}} - \frac{\mathrm{d}}{\mathrm{d}\tau}\left(\frac{\partial \Lambda^{(\tau)}}{\partial x^{\mu\prime}} \right) = 0, \tag{8.152}$$

We know that the Euler–Lagrange Equations based on the local time t in (8.151) are valid. We must now show that the Euler–Lagrange Equations based on the proper time τ in (8.152) are equivalent to (8.151) before we can accept the change from local to proper time. To do this we transform the dependence of all terms from τ to t.

For the derivative of $\Lambda^{(\tau)}$ with respect to x^{μ}, with $\mu = 1, 2, 3$, we have

$$\frac{\partial \Lambda^{(\tau)}}{\partial x^{\mu}} = \frac{\partial}{\partial x^{\mu}} \left[\gamma_{\mathrm{u}} \Lambda^{(\mathrm{t})} \right] = \gamma_{\mathrm{u}} \frac{\partial \Lambda^{(\mathrm{t})}}{\partial x^{\mu}}. \tag{8.153}$$

For the derivative of $\Lambda^{(\tau)}$ with respect to $x^{\mu\prime}$ we note that $\partial \dot{x}^{\nu}/\partial x^{\mu\prime} = (1/\gamma_{\mathrm{u}})\, \delta^{\nu}_{\mu}$. Then

$$\begin{aligned} \frac{\partial \Lambda^{(\tau)}}{\partial x^{\mu\prime}} &= \gamma_{\mathrm{u}} \frac{\partial \dot{x}^{\nu}}{\partial x^{\mu\prime}} \frac{\partial}{\partial \dot{x}^{\nu}} \Lambda^{(\mathrm{t})} \\ &= \frac{\partial \Lambda^{(\mathrm{t})}}{\partial \dot{x}^{\mu}}. \end{aligned} \tag{8.154}$$

That is the partial derivatives $\partial \Lambda^{(\tau)}/\partial x^{\mu\prime}$ are the canonical momenta for the frame k.

Then, from $\mathrm{d}x^{\mu}/\mathrm{d}\tau = x^{\mu\prime}$, (8.153), (8.154) and (8.152) we have

$$\frac{\partial \Lambda^{(\tau)}}{\partial x^{\mu}} - \frac{\mathrm{d}}{\mathrm{d}\tau}\left(\frac{\partial \Lambda^{(\tau)}}{\partial x^{\mu\prime}} \right) = \gamma_{\mathrm{u}} \left[\frac{\partial \Lambda^{(\mathrm{t})}}{\partial x^{\mu\prime}} - \frac{\mathrm{d}}{\mathrm{d}t}\left(\frac{\partial \Lambda^{(\mathrm{t})}}{\partial \dot{x}^{\mu}} \right) \right] = 0.$$

The Euler–Lagrange Equations written for the proper time τ are then completely equivalent to those written for the local time t.

8.13.3 Force-Free Lagrangian

To develop the relativistic Lagrangian we consider first the force-free situation. We know that the relativistic spatial momenta for a free material body are the spatial components of $\boldsymbol{P} = (mc, \boldsymbol{p})$. That is

$$p^{\mu} = \frac{\partial \Lambda^{(\mathrm{t})}}{\partial \dot{x}^{\mu}} = \frac{m_0 \dot{x}^{\mu}}{\sqrt{1 - \dot{x}^{\nu}\dot{x}^{\nu}/c^2}}, \tag{8.155}$$

with ν and $\mu = 1, 2, 3$. We may integrate (8.155) to obtain

$$\Lambda^{(\mathrm{t})} = m_0 c^2 \left(K - \sqrt{1 - \dot{x}^{\nu}\dot{x}^{\nu}/c^2} \right), \tag{8.156}$$

where K is an integration constant.[14] If we accept for the moment that (8.156) is the correct relativistic Lagrangian for the free material body, and introduce

$$\sqrt{1 - \dot{x}^{\nu}\dot{x}^{\nu}/c^2} = \frac{\mathrm{d}\tau}{\mathrm{d}t}, \tag{8.157}$$

then Hamilton's Principle Function becomes

$$\begin{aligned} \mathbf{S} &= \int_{t_1}^{t_2} \mathrm{d}t \left[m_0 c^2 \left(K - \frac{\mathrm{d}\tau}{\mathrm{d}t} \right) \right] \\ &= m_0 c^2 \left[K \left(t_2 - t_1 \right) - \tau \left(P_2 - P_1 \right) \right], \end{aligned} \tag{8.158}$$

where $P_{1,2}$ indicate two points on the world line of the body. If K were zero (8.158) would be Lorentz invariant because the proper time between two points on a body's world line is independent of inertial frame. But the term $m_0 c^2 K \left(t_2 - t_1 \right)$ presents a problem. As a difference between local times it is not Lorentz invariant. However, since in Hamilton's Principle we hold the end points fixed, this term is merely an additive constant with no effect on the mechanics. We, therefore, set $K = 0$ with no consequences and propose the Lorentz invariant Principal Function as

$$\mathbf{S} = - \int_{t_1}^{t_2} \mathrm{d}t \left(m_0 c^2 \sqrt{1 - \dot{x}^{\nu}\dot{x}^{\nu}/c^2} \right). \tag{8.159}$$

[14]We can verify this result by partial differentiation.

The relativistic Lagrangian is then

$$\Lambda^{(t)} = -m_0 c^2 \sqrt{1 - \dot{x}^\nu \dot{x}^\nu / c^2} \tag{8.160}$$

To convert the relativistic Lagrangian $\Lambda^{(t)}$ to $\Lambda^{(\tau)}$ we use (8.149) and we note that in the conversion from $\dot{x}^\nu \dot{x}^\nu$ with $\nu = 1, 2, 3$ to the 4-velocity (8.75), we have

$$\begin{aligned} 1 - \frac{u^2}{c^2} &= 1 - \dot{x}^\nu \dot{x}^\nu / c^2 \\ &= \frac{1}{c^2 \gamma_{\mathrm{u}}^2} g_{\mu\nu} U^\mu U^\nu. \end{aligned}$$

Then, using the prime notation for derivatives with respect to τ, (8.160) results in the proper time based, force-free relativistic Lagrangian

$$\Lambda^{(\tau)}\left(x^\mu, x^{\mu\prime}\right) = -m_0 c \sqrt{g_{\mu\nu} x^{\mu\prime} x^{\nu\prime}}. \tag{8.161}$$

8.13.4 Lagrangian for Lorentz Force

We choose the potential energy for the Lorentz force to be $-Q\mathcal{A}_\nu \mathrm{d}x^\nu / \mathrm{d}\tau = -Q\mathcal{A}U^\nu$ because of the form of the force in (8.145). We then write our relativistic Lagrangian as

$$\Lambda^{(\tau)}\left(x^\mu, x^{\mu\prime}\right) = -m_0 c \sqrt{g_{\mu\nu} x^{\mu\prime} x^{\nu\prime}} - Q\mathcal{A}^\mu g_{\mu\nu} x^{\nu\prime}. \tag{8.162}$$

for a charged material body moving under an electromagnetic (Lorentz) force. We have already shown that the first term in the Lagrangian (8.162) is a Lorentz invariant. The second term is the scalar product of two 4-vectors, which is also a Lorentz scalar invariant. Therefore the integral of the Lagrangian (8.162) over the proper time τ is Lorentz invariant. We may then propose the relativistic form of Hamilton's Principal Function to be

$$\mathbf{S} = \int_{\tau_1}^{\tau_2} \left(-m_0 c \sqrt{g_{\mu\nu} x^{\mu\prime} x^{\nu\prime}} - Q\mathcal{A}^\mu g_{\mu\nu} x^{\nu\prime} \right) \mathrm{d}\tau. \tag{8.163}$$

From (8.163) the canonical 4-momenta are

$$\begin{aligned} P_{\mathrm{C},\alpha}^{(\tau)} \equiv \frac{\partial \Lambda^{(\tau)}}{\partial x^{\alpha\prime}} &= -\frac{1}{2} m_0 c \frac{1}{\sqrt{g_{\mu\nu} x^{\mu\prime} x^{\nu\prime}}} \left(\frac{\partial}{\partial x^{\alpha\prime}} g_{\alpha\beta} x^{\beta\prime} x^{\alpha\prime} \right) - Q\mathcal{A}^\mu g_{\mu\alpha} \\ &= -P_\alpha - Q\mathcal{A}_\alpha. \end{aligned} \tag{8.164}$$

With (8.54) and (8.80) these canonical 4-momenta can be written as

$$\boldsymbol{P}_{\mathrm{C}}^{(\tau)}=\left(-mc-Q\frac{\varphi}{c},\, \boldsymbol{p}+Q\boldsymbol{A}\right), \tag{8.165}$$

recalling that we display the contravariant elements (see Sect. 8.7.3). Differentiating (8.164) with respect to τ we have

$$\frac{\mathrm{d}}{\mathrm{d}\tau}P_{\mathrm{C},\alpha}^{(\tau)}=-m_0x_\alpha''-Q\frac{\partial \mathcal{A}_\alpha}{\partial x^\beta}x^{\beta\prime}. \tag{8.166}$$

The partial derivative of the Lagrangian with respect to the components of the 4-vector x^α results in

$$\frac{\partial \Lambda^{(\tau)}}{\partial x^\alpha}=-Q\frac{\partial \mathcal{A}_\beta}{\partial x^\alpha}x^{\beta\prime} \tag{8.167}$$

Combining (8.166) and (8.167) the Euler–Lagrange Equations are

$$m_0x_\alpha''=Q\left(\frac{\partial \mathcal{A}_\beta}{\partial x^\alpha}-\frac{\partial \mathcal{A}_\alpha}{\partial x^\beta}\right)x^{\beta\prime} \tag{8.168}$$

or

$$m_0x_\alpha''=Q\left(\partial_\alpha\mathcal{A}_\beta-\partial_\beta\mathcal{A}_\alpha\right)x^{\beta\prime}, \tag{8.169}$$

which is the covariant form of our force law (8.146). We have then shown that the Lagrangian (8.162) is an appropriate relativistic Lagrangian for the electromagnetic (Lorentz) force when the time variable is the proper time τ.

8.13.5 Hamiltonian

Although τ is the most natural time, prior to our introduction of special relativity we formulated Analytical Mechanics in the local time t. We must then convert $\Lambda^{(\tau)}$ to this local time to obtain the relativistic Hamiltonian for the particle moving under the electromagnetic (Lorentz) force.

To obtain a Lagrangian based on the time t we return to Hamilton's Principal Function (8.147), where the Lagrangian $\Lambda^{(\mathrm{t})}$ is now

$$\begin{aligned}\Lambda^{(\mathrm{t})}&=\frac{1}{\gamma_{\mathrm{u}}}\Lambda^{(\tau)}\left(x^\mu,\gamma_{\mathrm{u}}\dot{x}^\mu\right)\\&=-m_0c\sqrt{g_{\mu\nu}\dot{x}^\mu\dot{x}^\nu}-Q\mathcal{A}^\mu g_{\mu\nu}\dot{x}^\nu,\end{aligned} \tag{8.170}$$

and now the indices take on the values 0, 1, 2, 3 and the coordinates are elements of 4-vectors. Hamilton's Principal Function is now

$$\mathbf{S} = \int_{t_1}^{t_2} \mathrm{d}t \left(-m_0 c \sqrt{g_{\mu\nu} \dot{x}^\mu \dot{x}^\nu} - Q\mathcal{A}^\mu g_{\mu\nu} \dot{x}^\nu \right) \tag{8.171}$$

For $\Lambda^{(\mathrm{t})}$ in (8.170) the covariant canonical momenta are $P_{\mathrm{C},\alpha}^{(\mathrm{t})} \equiv \partial \Lambda^{(\mathrm{t})} / \partial \dot{x}^\alpha$, which are equal to $P_{\mathrm{C},\alpha}^{(\tau)}$ (see (8.154)). That is, displaying the contravariant form,

$$\boldsymbol{P}_{\mathrm{C}}^{(\mathrm{t})} = \left(-mc - Q\frac{\varphi}{c}, \boldsymbol{p} + Q\boldsymbol{A} \right). \tag{8.172}$$

The spatial components of (8.172) $p_{\mathrm{C},\alpha}^{(\mathrm{t})}$ for $\alpha = 1, 2, 3$ are then (see Sect. 8.7.4, and (8.78))

$$p_{\mathrm{C},\alpha}^{(\mathrm{t})} = \frac{\partial \Lambda^{(\mathrm{t})}}{\partial u^\alpha} = \frac{m_0 u^\alpha}{\sqrt{1 - u^2/c^2}} + QA_\alpha. \tag{8.173}$$

Solving (8.173) for u^α $(= \dot{x}^\alpha)$, we have

$$u^\alpha = \frac{1}{m_0} \left(p_{\mathrm{C},\alpha}^{(\mathrm{t})} - QA_\alpha \right) \left[1 + \frac{\left(p_{\mathrm{C},\beta}^{(\mathrm{t})} - QA_\beta \right)^2}{m_0^2 c^2} \right]^{-1/2} \tag{8.174}$$

Then

$$\sqrt{1 - \frac{u^2}{c^2}} = \left[1 + \frac{\left(p_{\mathrm{C},\beta}^{(\mathrm{t})} - QA_\beta \right)^2}{m_0^2 c^2} \right]^{-1/2},$$

and

$$QA_\alpha u^\alpha = \frac{Q}{m_0} A_\alpha \left(p_{\mathrm{C},\alpha}^{(\mathrm{t})} - QA_\alpha \right) \left[1 + \frac{\left(p_{\mathrm{C},\beta}^{(\mathrm{t})} - QA_\beta \right)^2}{m_0^2 c^2} \right]^{-1/2}.$$

The relativistic Lagrangian $\Lambda^{(\mathrm{t})}$ in (8.170) is then

$$\begin{aligned} \Lambda^{(\mathrm{t})} &= -m_0 c^2 \sqrt{1 - u^2/c^2} - Q\varphi + QA_\alpha u^\alpha \\ &= \left[1 + \frac{\left(p_{\mathrm{C},\beta}^{(\mathrm{t})} - QA_\beta \right)^2}{m_0^2 c^2} \right]^{-1/2} \left[-m_0 c^2 + \frac{Q}{m_0} A_\alpha \left(p_{\mathrm{C},\alpha}^{(\mathrm{t})} - QA_\alpha \right) \right] - Q\varphi. \end{aligned} \tag{8.175}$$

From (8.173) and (8.174) the relativistic Hamiltonian, written in terms of the inertial frame time t is then

$$\begin{aligned}\mathcal{H}^{(t)} &= p^{(t)}_{C,\alpha}u^{\alpha} - \Lambda^{(t)} \\ &= m_0c^2\left[1+\frac{\left(p^{(t)}_{C,\alpha} - QA_{\alpha}\right)^2}{m_0^2c^2}\right]^{1/2} + Q\varphi \end{aligned} \tag{8.176}$$

This Eq. (8.176), we note, is the starting point of Dirac's relativistic electron theory [[18], p. 118, p. 255]. We may easily rewrite (8.176) as

$$\left(\frac{\mathcal{H}^{(t)} - Q\varphi}{c}\right)^2 - \left(p^{(t)}_{C,\alpha} - QA_{\alpha}\right)^2 = m_0^2c^2. \tag{8.177}$$

In Sect. 5.5 we obtained the Hamilton–Jacobi Equation for Hamilton's Principal Function S. With $\mathcal{H}^{(t)} = -\partial S/\partial t$ and $p^{(t)}_{C,\alpha} = \partial S/\partial x^{\alpha}$ for $\alpha = 1, 2, 3$ we have the relativistic Hamilton–Jacobi Equation as

$$\left(\frac{\partial S/\partial t + Q\varphi}{c}\right)^2 - \left(\frac{\partial S}{\partial x^{\alpha}} - QA_{\alpha}\right)^2 = m_0^2c^2. \tag{8.178}$$

Finally, in the low energy (non-relativistic) limit the Hamiltonian (8.176) is

$$\mathcal{H}^{(t)} = m_0c^2 + \frac{1}{2m_0}\left(p^{(t)}_{C,\alpha} - QA_{\alpha}\right)^2 + Q\varphi. \tag{8.179}$$

We may, of course, neglect the invariant rest energy m_0c^2. This is the Hamiltonian we used in our earlier (nonrelativistic) studies of the motion of charged particles.

8.14 Summary

This chapter we have presented a rather complete, but brief, outline of special relativity and the application of the basic tensor algebra of Minkowski Space to Analytical Mechanics. We have been careful about Einstein's basic idea concerning time and followed this with Minkowski's concept of the merging of space and time into a non-Euclidean geometry. This led us in a natural manner into a study of the basic tensor algebra required to handle the electrodynamics and Analytical Mechanics of relativity.

Our treatment of mechanics required deliberate care because of the requirements of Lorentz invariance imposed on all physical laws. So we began with conservation of 4-momentum and collisions. Conservation of momentum allowed us to study numerous situations including photons and de Broglie's ideas on matter waves. We studied forces separately, before attempting our final formulation of Analytical Mechanics, primarily because of the fact that relativistic forces separate into pure and impure forces. Our example of an impure force was the nuclear force and our example of a pure force was the Lorentz force. We dealt only with the Lorentz force in our final treatment, citing the claim of Pauli that the Lorentz force is representative of many forces.

In our final steps toward the relativistic Lagrangian and Hamiltonian we also undertook with care, beginning with force-free motion. The final analysis brought us back to the starting point of our study. The treatments of Euler, Lagrange, Hamilton, and Jacobi constitute the basis of any serious treatment of Analytical Mechanics.

8.15 Exercises

8.1. Using the Lorentz Transformation show that

$$\begin{aligned} &\pm\left[\left(\mathrm{d}x^0\right)^2-\left(\mathrm{d}x^1\right)^2-\left(\mathrm{d}x^2\right)^2-\left(\mathrm{d}x^3\right)^2\right] \\ &=\pm\left[\left(\mathrm{d}x'^0\right)^2-\left(\mathrm{d}x'^1\right)^2-\left(\mathrm{d}x'^2\right)^2-\left(\mathrm{d}x'^3\right)^2\right] \end{aligned}$$

8.2. Use the Lorentz Transformation matrix $\mathbf{A}$ to obtain time dilation. Consider that frame k' moves at a velocity v in the direction of the x-axis of k. Then $\mathrm{d}y=\mathrm{d}z=0$ and $\mathrm{d}x=v\mathrm{d}t$, which is the distance that the origin of k' moves in the time $\mathrm{d}t$. The differential world line in k is then

$$\mathrm{d}\mathbf{s}=\begin{bmatrix} c\mathrm{d}t \\ v\mathrm{d}t \\ 0 \\ 0 \end{bmatrix}.$$

This differential world line is transformed into the differential world line $\mathrm{d}\mathbf{s}'$ in k' by

$$\mathrm{d}\mathbf{s}'=\mathbf{A}\cdot\mathrm{d}\mathbf{s}.$$

Find the differential world line $\mathrm{d}\mathbf{s}'$ and from the result show that $\mathrm{d}t'=\sqrt{1-\beta^2}\mathrm{d}t$.

8.3. The Minkowski Axiom requires that the velocities of material bodies are always less than the speed of light. Using $u_{\mathrm{x}}=\left(u'_{\mathrm{x}}+c\beta\right)/\left(1+\beta u'_{\mathrm{x}}/c\right)$ show that if the inertial frame k' has a velocity $v<c$ and if the particle moving in k' also has a velocity $u_{\mathrm{x}}<c$, as it must, the velocity of the particle as measured in k is also $<c$ regardless of how close v and u_{x} are to c.

[Hint: Choose $\beta'_{\mathrm{x}}=1-\kappa$ and $\beta=1-\lambda$]

8.4. Consider a collision between a high energy proton and a photon. The proton is moving at a velocity $\boldsymbol{u}$ along the positive x-axis and the photon of frequency ν is initially moving toward it in the direction of the negative x-axis. After the collision a photon of frequency ν' leaves the site of collision moving along the positive x-axis. Designate the 4-momentum of the proton before and after collision as $\boldsymbol{P}_{\mathrm{p}}$ and $\boldsymbol{P}'_{\mathrm{p}}$ and that of the photons before and after collision as $\boldsymbol{P}_{\gamma}$ and $\boldsymbol{P}'_{\gamma}$. What is the energy of the final photon?

8.5. Using the concept of relativistic mass of the photon, find the relativistic mass of two photons moving in opposite directions with frequencies ν_1 and ν_2.

8.6. A photon with sufficient energy can produce an electron and a positron. The positron is the antiparticle of an electron. It was first predicted by Paul Dirac and first identified in a cloud chamber track by Carl Anderson. For the photon to produce an electron and a positron there must only be another particle present for momentum conservation. We consider a particle of mass m_{01} at rest at the origin of a frame k. A photon with frequency ν approaches this particle. At the point of "collision" the photon disappears and an electron e and a positron e^{+} appear. The electron and the positron both have rest mass $m_{0\mathrm{e}}$. What is the minimum energy of the photon required for pair production?

[After the collision refer all particles to a CM frame k_{CM}.]

8.7. Using the fundamental tensor, show that the covariant form of the Field Strength tensor is

$$\mathcal{F}_{\alpha\beta} = \partial_\alpha \mathcal{A}_\beta - \partial_\beta \mathcal{A}_\alpha,$$

which is, in matrix form is

$$\mathcal{F}_{\alpha\beta} = \begin{bmatrix} 0 & E_{\mathrm{x}}/c & E_{\mathrm{y}}/c & E_{\mathrm{z}}/c \\ -E_{\mathrm{x}}/c & 0 & -B_{\mathrm{z}} & B_{\mathrm{y}} \\ -E_{\mathrm{y}}/c & B_{\mathrm{z}} & 0 & -B_{\mathrm{x}} \\ -E_{\mathrm{z}}/c & -B_{\mathrm{y}} & B_{\mathrm{x}} & 0 \end{bmatrix}$$

8.8. Using the differential operators show that

$$\partial_\sigma \partial^\tau \mathcal{A}^\sigma = \partial^\tau \left(\frac{1}{c^2} \frac{\partial}{\partial t} \varphi + \mathrm{div} \boldsymbol{A} \right).$$

8.9. Show that the elements of the 4-vector

$$\partial_\alpha \mathcal{F}_{\alpha\beta} = 0$$

are Gauss's Law

$$\mathrm{div} \boldsymbol{E} = 0$$

and Ampère's Law

$$\mathrm{curl}\boldsymbol{B} = -\frac{1}{c^2}\frac{\partial}{\partial t}\boldsymbol{E}.$$

of Maxwell's equations for empty space.

8.10. Show that the even permutations of

$$\partial_\rho \mathcal{F}_{\sigma\tau} = 0$$

or

$$\partial^\rho \mathcal{F}^{\sigma\tau} = 0$$

yield Oersted's result

$$\mathrm{div}B = 0$$

and Faraday's Law

$$\mathrm{curl}\boldsymbol{E} = -\frac{\partial}{\partial t}\boldsymbol{B}$$

of Maxwell's Equations in empty space.

8.11. Although we do have the spatial force law in the form of Newton's Second Law

$$f = \frac{\mathrm{d}}{\mathrm{d}t}\boldsymbol{p},$$

if we attempt to formulate this in terms of an acceleration, for the sake of our intuition, we encounter difficulties. For example we may attempt consider only pure forces in the 4-vector form of Newton's Second Law

$$\begin{aligned}\boldsymbol{F} &= \frac{\mathrm{d}}{\mathrm{d}\tau}\boldsymbol{P} \\ &= \frac{\mathrm{d}}{\mathrm{d}\tau}(mc, \boldsymbol{p}),\end{aligned}$$

and use the relation

$$\boldsymbol{P} = m_0\boldsymbol{U},$$

we can write

$$\boldsymbol{F} = m_0\frac{\mathrm{d}}{\mathrm{d}\tau}\boldsymbol{U}.$$

If we then attempt to define a 4-acceleration as

$$\boldsymbol{A} = \frac{\mathrm{d}}{\mathrm{d}\tau}\boldsymbol{U}$$

and write $\boldsymbol{F} = m_0\boldsymbol{A}$ for the pure force we find difficulty. Find the difficulty.

8.12. In our discussion of energy we followed Wolfgang Pauli to obtain the Einstein mass-energy relation. This required a proposal for the equation of motion of a material body, in addition to conservation of momentum. We chose this to be the relativistic form of Newton's Second Law written, using the spatial components of momenta, as

$$\frac{\mathrm{d}}{\mathrm{d}t}(m_0\gamma_\mathrm{u}\boldsymbol{u}) = \boldsymbol{F}.$$

And we later verified that this is the correct covariant form of the force law.

Begin by accepting this force law and obtain Pauli's result that

$$\begin{aligned}\mathcal{E}_\mathrm{kin} &= m_0\gamma_\mathrm{u}c^2 + \text{ constant}\\ &\approx m_0c^2 + \frac{1}{2}m_0u^2 + \text{ constant.}\end{aligned}$$

Then note that we retrieve the known classical result for the kinetic energy if we choose the constant to be $-m_0c^2$. That is

$$\mathcal{E}_\mathrm{kin} = m_0\gamma_\mathrm{u}c^2 - m\,_0c^2.$$

We may then identify identify

$$\mathcal{E} = m_0\gamma_\mathrm{u}c^2$$

as the total energy and

$$\mathcal{E}_\mathrm{kin} = \mathcal{E} - m_0c^2.$$

You will first need to evaluate $\mathrm{d}(m_0\gamma_\mathrm{u}\boldsymbol{u})/\mathrm{d}t$. Then you will need to use your result to obtain an expression for the work done on a material body by the force $\boldsymbol{F}$. This will lead you to the result $\mathrm{d}(m_0\gamma_\mathrm{u}c^2)/\mathrm{d}t = \boldsymbol{F}\cdot\boldsymbol{u}$, which you integrate to obtain the result above for the kinetic energy. The next steps require Einstein's aesthetic insight. They should now follow.

8.13. By direct differentiation show that

$$P\mathrm{can}_\alpha^{(\mathrm{t})} \equiv \frac{\partial \Lambda^{(\mathrm{t})}}{\partial \dot{x}^\alpha}$$

and

$$P\mathrm{can}_\alpha^{(\tau)} \equiv \frac{\partial \Lambda}{\partial x^{\alpha\prime}}$$

are identical, thus verifying

$$\frac{\partial \Lambda^{(\tau)}}{\partial x^{\mu\prime}} = \frac{\partial \Lambda^{(\mathrm{t})}}{\partial \dot{x}^\mu}.$$

8.14. Show that the Canonical Equations from the relativistic Hamiltonian for the Lorentz force

$$\mathcal{H}^{(t)} = m_0 c^2 \left[1 + \frac{\left(\left(p_{\text{can}}^{(t)}\right)_\alpha - QA_\alpha \right)^2}{m_0^2 c^2} \right]^{1/2} + Q\varphi$$

are

$$\frac{\mathrm{d}}{\mathrm{d}t} \left(p_{\text{can}}^{(t)}\right)_\mu = Qu^\beta \left(\partial A_\beta / \partial x^\mu \right) - Q \frac{\partial \varphi}{\partial x^\mu}$$

and

$$\frac{\mathrm{d}}{\mathrm{d}t} x^\mu = u^\beta = \frac{1}{m_0} \left[1 + \frac{\left(\left(p_{\text{can}}^{(t)}\right)_\alpha - QA_\alpha \right)^2}{m_0^2 c^2} \right]^{-1/2} \left(\left(p_{\text{can}}^{(t)}\right)_\mu - QA_\mu \right)$$

where $\left(p_{\text{can}}^{(t)}\right)_\mu$ are the spatial components of the 4-momentum

$$\boldsymbol{P}\text{can}^{(t)} = \left(-mc - Q\frac{\varphi}{c}, \boldsymbol{p} + Q\boldsymbol{A} \right).$$

Appendix A
Differential of S

If we consider a general variation of Hamilton's principal function

$$S = \int_{t_1}^{t_2} \mathrm{d}tL\,(q, \dot{q}, t)$$

we can obtain a differential equation for S rather than only differential equations for the coordinates on which S depends. We shall choose the initial and final times to be t_1 and t_2, rather than 0 and t. And we shall designate the general variation as Δ_{T} will include variations in the coordinates q_{i} at t_1 and t_2 as well as variations in the end points t_1 and t_2. The differential of Hamilton's principle function, which is the difference between the value of S before and after the variation will be the $\mathrm{d}S = \Delta_{\mathrm{T}}S$. That is

$$\Delta_{\mathrm{T}}S = \int_{t_1'}^{t_2'} \mathrm{d}tL'\left(q, \dot{q}, t\right) - \int_{t_1}^{t_2} \mathrm{d}tL\left(q, \dot{q}, t\right), \tag{A.1}$$

where t_1' and t_2' indicate variations in the end point times and $L'\left(q, \dot{q}, t\right)$ is the variation in $L\left(q, \dot{q}, t\right)$ resulting from variations in q_{i} and $\dot{q}_{\mathrm{i}}$.

We shall designate the variations in q_{i} and $\dot{q}_{\mathrm{i}}$ as δq_{i} and $\delta \dot{q}_{\mathrm{i}}$. In the general variation we are considering there are variations in these coordinates and velocities at the end points of the integrals in (A.1). If we designate the actual values of the coordinates and the velocities at the time end points as q_{i}' and $\dot{q}_{\mathrm{i}}'$ then

$$q_{\mathrm{i}}' = q_{\mathrm{i}} + \delta q_{\mathrm{i}} \tag{A.2}$$
$$\dot{q}_{\mathrm{i}}' = \dot{q}_{\mathrm{i}} + \delta \dot{q}_{\mathrm{i}}. \tag{A.3}$$

These variations are arbitrary, except that they must be related by time differentiation.

Similarly we shall designate the times at the end points of the integration, t_1' and t_2', which differ from t_1 and t_2 by the infinitesimal amounts Δt_1 and Δt_2

C.S. Helrich, *Analytical Mechanics*, Undergraduate Lecture Notes in Physics, DOI 10.1007/978-3-319-44491-8

$$
\begin{aligned}
t_1' &= t_1 + \Delta t_1 \\
t_2' &= t_2 + \Delta t_2 .
\end{aligned}
\tag{A.4}
$$

The dependence of δq_i and $\delta \dot{q}_i$ on Δt_1 and Δt_2 will be of second order in infinitesimal quantities, which we neglect. However, $\dot{q}_i'$ is the time derivative of q_i'. That is, from (A.2) we have

$$
\dot{q}_i' = \dot{q}_i + \frac{\mathrm{d}}{\mathrm{d}t}\delta q_i . \tag{A.5}
$$

Comparing (A.5) with (A.3) we see that

$$
\frac{\mathrm{d}}{\mathrm{d}t}\delta q_i = \delta \dot{q}_i = \delta \frac{\mathrm{d}}{\mathrm{d}t} q_i , \tag{A.6}
$$

and the operators $\mathrm{d}/\mathrm{d}t$ and δ commute.

To first order in infinitesimals the variation in the Lagrangian is

$$
\begin{aligned}
L'(q, \dot{q}, t) &= L(q + \delta q, \dot{q} + \delta \dot{q}, t) \\
&= L(q, \dot{q}, t) \\
&\quad + \sum_i \left\{ \frac{\partial L}{\partial q_i}\delta q_i + \frac{\partial L}{\partial \dot{q}_i}\delta \dot{q}_i \right\} .
\end{aligned}
\tag{A.7}
$$

In (A.7) time is not a variable but a dummy in the integration. The variations δq_i and $\delta \dot{q}_i$ are arbitrary variations in the coordinates and velocities at each point along the path of the integration and are not, therefore, functions of time. They are, however, related as indicated above.

The first integral on the right hand side of (A.1) is then

$$
\begin{aligned}
\int_{t_1'}^{t_2'} \mathrm{d}t L'(q, \dot{q}, t) &= \left\{ \int_{t_1'}^{t_1} \mathrm{d}t + \int_{t_1}^{t_2} \mathrm{d}t + \int_{t_2}^{t_2'} \mathrm{d}t \right\} L'(q, \dot{q}, t) \\
&= \int_{t_1}^{t_2} \mathrm{d}t L'(q, \dot{q}, t) \\
&\quad - \int_{t_1}^{t_1+\Delta t_1} \mathrm{d}t L'(q, \dot{q}, t) + \int_{t_2}^{t_2+\Delta t_2} \mathrm{d}t L'(q, \dot{q}, t) ,
\end{aligned}
\tag{A.8}
$$

with L' given by (A.7). Then

$$
\begin{aligned}
\Delta_{\mathrm{T}} S &= \int_{t_1}^{t_2} \mathrm{d}t \left[L'(q, \dot{q}, t) - L(q, \dot{q}, t) \right] \\
&\quad - \int_{t_1}^{t_1+\Delta t_1} \mathrm{d}t L'(q, \dot{q}, t) + \int_{t_2}^{t_2+\Delta t_2} \mathrm{d}t L'(q, \dot{q}, t) .
\end{aligned}
\tag{A.9}
$$

We write this as

$$\Delta_{\mathrm{T}} S=\delta_{\mathrm{T}} \int_{t_1}^{t_2} \mathrm{d}t L(q, \dot{q}, t) - \int_{t_1}^{t_1+\Delta t_1} \mathrm{d}t L'(q, \dot{q}, t)+\int_{t_2}^{t_2+\Delta t_2} \mathrm{d}t L'(q, \dot{q}, t) . \quad (\mathrm{A}.10)$$

Here δ_{T} designates a variation that excludes variations in the end point times, but includes variations in the q and $\dot{q}$ at the endpoints. The variation δ_{T} then differs from the previous δ variation of Hamilton's Principle, but will contain the variation δ.

Since the terms Δt_1 and Δt_2 are infinitesimals, the integrands in the last two integrals in (A.10) are constants over the respective integrations, provided the Lagrangian is a continuous function of its coordinates. Then

$$\begin{aligned}\int_{t_2}^{t_2+\Delta t_2} \mathrm{d}t L'(q, \dot{q}, t)-\int_{t_1}^{t_1+\Delta t_1} \mathrm{d}t L'(q, \dot{q}, t) &= L'(t_2) \Delta t_2-L'(t_1) \Delta t_1 \\ &= \left.L'(t) \Delta t\right]_{t_1}^{t_2} .\end{aligned} \quad (\mathrm{A}.11)$$

Using (A.7) and dropping all terms that are greater than first order in infinitesimals (A.11) becomes

$$\int_{t_2}^{t_2+\Delta t_2} \mathrm{d}t L'(q, \dot{q}, t)-\int_{t_1}^{t_1+\Delta t_1} \mathrm{d}t L'(q, \dot{q}, t)=\left.L(t) \Delta t\right]_{t_1}^{t_2}, \quad (\mathrm{A}.12)$$

The total variation (A.10) is then

$$\Delta_{\mathrm{T}} S=\delta_{\mathrm{T}} \int_{t_1}^{t_2} \mathrm{d}t L(q, \dot{q}, t)+\left.L(t) \Delta t\right]_{t_1}^{t_2} . \quad (\mathrm{A}.13)$$

We must now consider the δ_{T} variation in detail. Carrying terms to first order in infinitesimals, we have

$$\begin{aligned}\delta_{\mathrm{T}} \int_{t_1}^{t_2} \mathrm{d}t L(q, \dot{q}, t) &= \int_{t_1}^{t_2} \mathrm{d}t L'(q, \dot{q}, t)-\int_{t_1}^{t_2} \mathrm{d}t L(q, \dot{q}, t) \\ &= \int_{t_1}^{t_2} \mathrm{d}t \left\{L(q, \dot{q}, t)\right. \\ &\quad \left.+\sum_{\mathrm{i}}\left[\frac{\partial L}{\partial q_{\mathrm{i}}} \delta q_{\mathrm{i}}+\frac{\partial L}{\partial \dot{q}_{\mathrm{i}}} \delta \dot{q}_{\mathrm{i}}\right]-L(q, \dot{q}, t)\right\} \\ &= \int_{t_1}^{t_2} \mathrm{d}t \sum_{\mathrm{i}}\left[\frac{\partial L}{\partial q_{\mathrm{i}}} \delta q_{\mathrm{i}}+\frac{\partial L}{\partial \dot{q}_{\mathrm{i}}} \delta \dot{q}_{\mathrm{i}}\right],\end{aligned} \quad (\mathrm{A}.14)$$

Integrating the second term by parts we have

$$\delta_{\mathrm{T}} \int_{t_1}^{t_2} \mathrm{d}t L(q, \dot{q}, t) = \sum_{\mathrm{i}} \frac{\partial L}{\partial \dot{q}_{\mathrm{i}}} \delta q_{\mathrm{i}} \Bigg]_{t_1}^{t_2} + \int_{t_1}^{t_2} \mathrm{d}t \sum_{\mathrm{i}} \left[\frac{\partial L}{\partial q_{\mathrm{i}}} - \frac{d}{dt} \left(\frac{\partial L}{\partial \dot{q}_{\mathrm{i}}} \right) \right] \delta q_{\mathrm{i}}. \tag{A.15}$$

The second integral on the right hand side of (A.15) is the variation with fixed end points δS of Hamilton's principle function. This must be zero if Newton's Laws are to hold. We are then left with

$$\delta_{\mathrm{T}} \int_{t_1}^{t_2} \mathrm{d}t L(q, \dot{q}, t) = \sum_{\mathrm{i}} \frac{\partial L}{\partial \dot{q}_{\mathrm{i}}} \delta q_{\mathrm{i}} \Bigg]_{t_1}^{t_2}. \tag{A.16}$$

With (A.16), (A.13) becomes

$$\Delta_{\mathrm{T}} S = \sum_{\mathrm{i}} \frac{\partial L}{\partial \dot{q}_{\mathrm{i}}} \delta q_{\mathrm{i}} + L(t) \Delta t \Bigg]_{t_1}^{t_2} \tag{A.17}$$

In this result the separate variations δq_{i} and Δt appear on the right hand side. This is not a convenient final form for our total variation. A more convenient form would be one in which the time variation Δt appears in both terms on the right hand side of (A.17). The generalized coordinate q_{i} is a function of the time t. But, as we have discussed above, the variation δq_{i} is not a function of the time. We may obtain a general form for the variation in the coordinate q_{i} that involves the time variation Δt by writing the variation of q_{i} as a linear approximation involving the variations δq_{i} and the variation in the time Δt. This is

$$\Delta q_{\mathrm{i}} = \delta q_{\mathrm{i}} + \dot{q}_{\mathrm{i}} \Delta t. \tag{A.18}$$

With (A.18) and $p_{\mathrm{i}} = \partial L / \partial \dot{q}_{\mathrm{i}}$, the combination in brackets on the right hand side of (A.17) is

$$\begin{aligned} \sum_{\mathrm{i}} \frac{\partial L}{\partial \dot{q}_{\mathrm{i}}} \delta q_{\mathrm{i}} + L(t) \Delta t &= \sum_{\mathrm{i}} (p_{\mathrm{i}} \Delta q_{\mathrm{i}}) + \left[L(t) - \sum_{\mathrm{i}} p_{\mathrm{i}} \dot{q}_{\mathrm{i}} \right] \Delta t. \\ &= \sum_{\mathrm{i}} (p_{\mathrm{i}} \Delta q_{\mathrm{i}}) - \mathcal{H} \Delta t, \end{aligned} \tag{A.19}$$

using the definition of the Hamiltonian. The result of our general variation $\Delta_{\mathrm{T}} S$ is then

$$\Delta_{\mathrm{T}} S = \left[\sum_{\mathrm{i}} (p_{\mathrm{i}} \Delta q_{\mathrm{i}}) - \mathcal{H} \Delta t \right]_{t_1}^{t_2}. \tag{A.20}$$

This is the differential of Hamilton's principal function which involves variations in the end points. Those variations appear explicitly in Δq_{i} and Δt in (A.20). And $\mathrm{d}S = \Delta_{\mathrm{T}}S$ is the differential we sought.

Appendix B
Hamilton–Jacobi Equation

With $\mathrm{d}S = \Delta_{\mathrm{T}}S$ (A.20) is a Pfaffian and the terms Δq_{i} and Δt become differentials.

$$\begin{aligned} \mathrm{d}q_{\mathrm{i1}} &= \Delta q_{\mathrm{i}}\,(t_1) \\ \mathrm{d}q_{\mathrm{i2}} &= \Delta q_{\mathrm{i}}\,(t_2) \end{aligned} \tag{B.1}$$

and

$$\begin{aligned} \mathrm{d}t_1 &= \Delta t\,(t_1) \\ \mathrm{d}t_2 &= \Delta t\,(t_2)\,. \end{aligned} \tag{B.2}$$

Equation (A.20) is then

$$\mathrm{d}S = \sum_{\mathrm{i=1}}^{\mathrm{n}} p_{\mathrm{i2}}\,\mathrm{d}q_{\mathrm{i2}} - \sum_{\mathrm{i=1}}^{\mathrm{n}} p_{\mathrm{i1}}\mathrm{d}q_{\mathrm{i1}} - \mathcal{H}_2\mathrm{d}t_2 + \mathcal{H}_1\mathrm{d}t_1\,. \tag{B.3}$$

The Pfaffian for $\mathrm{d}S$ in (B.3) is

$$\mathrm{d}S = \sum_{\mathrm{i=1}}^{\mathrm{n}} \left[\left(\frac{\partial S}{\partial q_{\mathrm{i1}}}\right) \mathrm{d}q_{\mathrm{i1}} + \left(\frac{\partial S}{\partial q_{\mathrm{i2}}}\right) \mathrm{d}q_{\mathrm{i2}} \right] + \frac{\partial S}{\partial t_1}\mathrm{d}t_1 + \frac{\partial S}{\partial t_2}\mathrm{d}t_2, \tag{B.4}$$

since S is a function of the variables $(q_{\mathrm{i1}}, q_{\mathrm{i2}}, t_1, t_2)$. By comparing terms in (B.3) with those in (B.4) we identify

$$\frac{\partial S}{\partial q_{\mathrm{i2}}} = p_{\mathrm{i2}};\ \frac{\partial S}{\partial q_{\mathrm{i1}}} = -p_{\mathrm{i1}};\ \frac{\partial S}{\partial t_2} = -\mathcal{H}_2;\ \frac{\partial S}{\partial t_1} = \mathcal{H}_1\,. \tag{B.5}$$

We are considering results at a time t_2 resulting from configurations of a system at some initial time t_1. Then q_{i1} and p_{i1} are constants defined by the initial system configuration. If we call

C.S. Helrich, *Analytical Mechanics*, Undergraduate Lecture Notes in Physics, DOI 10.1007/978-3-319-44491-8

$$q_{i1} = \alpha_i \tag{B.6}$$

then the second Equation in (B.5) is

$$p_{i1} = -\partial S/\partial \alpha_i. \tag{B.7}$$

The final time t_2 is arbitrary. So we drop the subscript 2 on the final configuration and time quantities. We may then write Hamilton's Principal function as

$$S = S(\alpha, q, t),$$

where, for shorthand, we have used

$$\begin{aligned} \alpha &= \{\alpha_i\}_{i=1}^{n} \\ q &= \{q_i\}_{i=1}^{n} \\ t &= \{t_1, t_2\}. \end{aligned}$$

If $S(\alpha, q, t)$ is known, then (B.7) is a set of algebraic equations for the q_is, since the momenta p_{i1} and the coordinates $q_{i1} = \alpha_i$ are known from the initial configuration. The corresponding momenta p_i, may be found from the first of the Eq. (B.5),

$$\frac{\partial S}{\partial q_i} = p_i\,. \tag{B.8}$$

Therefore, if we know $S(\alpha, q, t)$ in functional form our dynamical problem is reduced to an algebraic problem.

We have used the first two equations in the set (B.5) and are left with only the equation

$$\frac{\partial S}{\partial t} = -\mathcal{H}. \tag{B.9}$$

written at the initial and final times. These are the two equations Hamilton cited. And in the text we indicated that Jacobi showed that only the equation at the final time was necessary. If we use (B.8) in (B.9) we have

$$\frac{\partial S}{\partial t} + \mathcal{H}\left(q, \frac{\partial S}{\partial q}, t\right) = 0, \tag{B.10}$$

which is the Hamilton–Jacobi Equation.

Appendix C
With Variables p, q, q̇

If we consider a general variation of Hamilton's principal function

$$S = \int_{t_1}^{t_2} \mathrm{d}tL\left(q, \dot{q}, t\right)$$
$$= \int_{t_1}^{t_2} \mathrm{d}t \left[\sum_{\mathrm{i}} p_{\mathrm{i}}\dot{q}_{\mathrm{i}} - \mathcal{H}\left(q, p, t\right)\right]$$

we can obtain a differential equation for S rather than only differential equations for the coordinates on which S depends. We shall choose the initial and final times to be t_1 and t_2, rather than 0 and t. And we shall designate the general variation Δ_{T} will include variations in the coordinates q_{i} at t_1 and t_2 as well as variations in the end points t_1 and t_2. The differential of Hamilton's principle function, which is the difference between the value of S before and after the variation will be the $\mathrm{d}S = \Delta_{\mathrm{T}}S$. That is

$$\Delta_{\mathrm{T}}S = \int_{t_1'}^{t_2'} \mathrm{d}t \left[\sum_{\mathrm{i}} p_{\mathrm{i}}'\dot{q}_{\mathrm{i}}' - \mathcal{H}'\left(q, p, t\right)\right] - \int_{t_1}^{t_2} \mathrm{d}t \left[\sum_{\mathrm{i}} p_{\mathrm{i}}\dot{q}_{\mathrm{i}} - \mathcal{H}\left(q, p, t\right)\right], \quad \text{(C.1)}$$

where t_1' and t_2' indicate variations in the end point times and $L'\left(q, \dot{q}, t\right)$ is the variation in $L\left(q, \dot{q}, t\right)$ resulting from variations in q_{i} and $\dot{q}_{\mathrm{i}}$.

We shall designate the variations in p_{i}, q_{i} and $\dot{q}_{\mathrm{i}}$ as δp_{i}, δq_{i} and $\delta \dot{q}_{\mathrm{i}}$. If we designate the actual values of the coordinates and the velocities at the end points as q_μ' and $\dot{q}_\mu'$ then

$$p_{\mathrm{i}}' = p_{\mathrm{i}} + \delta p_{\mathrm{i}} \quad \text{(C.2)}$$
$$q_{\mathrm{i}}' = q_{\mathrm{i}} + \delta q_{\mathrm{i}} \quad \text{(C.3)}$$
$$\dot{q}_{\mathrm{i}}' = \dot{q}_{\mathrm{i}} + \delta \dot{q}_{\mathrm{i}}. \quad \text{(C.4)}$$

C.S. Helrich, *Analytical Mechanics*, Undergraduate Lecture Notes in Physics, DOI 10.1007/978-3-319-44491-8

These variations are arbitrary, except that they must be related by time differentiation.

Similarly we shall designate the times at the end points of the integration, t_1' and t_2', which differ from t_1 and t_2 by the infinitesimal amounts Δt_1 and Δt_2

$$t_1' = t_1 + \Delta t_1 \tag{C.5}$$

$$t_2' = t_2 + \Delta t_2. \tag{C.6}$$

The dependence of δq_{i} and $\delta \dot{q}_{\mathrm{i}}$ on Δt_1 and Δt_2 will be of second order in infinitesimal quantities, which we neglect. However, $\dot{q}_{\mathrm{i}}'$ is the time derivative of q_{i}'. That is, from (C.1) we have

$$\dot{q}_{\mathrm{i}}' = \dot{q}_{\mathrm{i}} + \frac{\mathrm{d}}{\mathrm{d}t}\delta q_{\mathrm{i}}. \tag{C.7}$$

Comparing (A.5) with (A.3) we see that

$$\frac{\mathrm{d}}{\mathrm{d}t}\delta q_{\mathrm{i}} = \delta \dot{q}_{\mathrm{i}} = \delta \frac{\mathrm{d}}{\mathrm{d}t} q_{\mathrm{i}}, \tag{C.8}$$

and the operators $\mathrm{d}/\mathrm{d}t$ and δ commute.

To first order in infinitesimals the variation in the Lagrangian is

$$\begin{aligned} &\sum_{\mathrm{i}} p_{\mathrm{i}}' \dot{q}_{\mathrm{i}}' - \mathcal{H}'(q,p,t) \\ &= \sum_{\mathrm{i}} p_{\mathrm{i}} \dot{q}_{\mathrm{i}} - \mathcal{H}(q,p,t) \\ &-\sum_{\mathrm{i}} \left\{ \frac{\partial \mathcal{H}}{\partial q_{\mathrm{i}}}\delta q_{\mathrm{i}} - p_{\mathrm{i}}\delta \dot{q}_{\mathrm{i}} + \left[\frac{\partial \mathcal{H}}{\partial p_{\mathrm{i}}} - \dot{q}_{\mathrm{i}}\right]\delta p_{\mathrm{i}} \right\} \end{aligned} \tag{C.9}$$

Using the canonical equations the term in the bracket $[\cdots]$ vanishes and $(\partial \mathcal{H}/\partial q_{\mathrm{i}})$ $\delta q_{\mathrm{i}} - p_{\mathrm{i}}\delta \dot{q}_{\mathrm{i}} = -\mathrm{d}(p_{\mathrm{i}}\delta q_{\mathrm{i}})/\mathrm{d}t$. Then

$$\sum_{\mathrm{i}} p_{\mathrm{i}}' \dot{q}_{\mathrm{i}}' - \mathcal{H}'(q,p,t) = \sum_{\mathrm{i}} p_{\mathrm{i}} \dot{q}_{\mathrm{i}} - \mathcal{H}(q,p,t) + \frac{\mathrm{d}}{\mathrm{d}t}\left(\sum_{\mathrm{i}} p_{\mathrm{i}}\delta q_{\mathrm{i}}\right) \tag{C.10}$$

The first integral on the right hand side of (C.1) is then

$$\begin{aligned} &\int_{t_1'}^{t_2'} \mathrm{d}t \left[\sum_{\mathrm{i}} p_{\mathrm{i}}' \dot{q}_{\mathrm{i}}' - \mathcal{H}'(q,p,t)\right] \\ &= \int_{t_1}^{t_2} \mathrm{d}t \left[\sum_{\mathrm{i}} p_{\mathrm{i}} \dot{q}_{\mathrm{i}} - \mathcal{H}(q,p,t) + \frac{\mathrm{d}}{\mathrm{d}t}\left(\sum_{\mathrm{i}} p_{\mathrm{i}}\delta q_{\mathrm{i}}\right)\right] \\ &- \int_{t_1}^{t_1+\Delta t_1} \mathrm{d}t \left[\sum_{\mathrm{i}} p_{\mathrm{i}}' \dot{q}_{\mathrm{i}}' - \mathcal{H}'(q,p,t)\right] + \int_{t_2}^{t_2+\Delta t_2} \mathrm{d}t \left[\sum_{\mathrm{i}} p_{\mathrm{i}}' \dot{q}_{\mathrm{i}}' - \mathcal{H}'(q,p,t)\right] \end{aligned}$$

Then

$$\Delta_T S = \int_{t_1}^{t_2} dt \sum_i \frac{d}{dt} (p_i \delta q_i)$$
$$- \int_{t_1}^{t_1+\Delta t_1} dt \left[\sum_i p_i' \dot{q}_i' - \mathcal{H}' (q, p, t) \right] + \int_{t_2}^{t_2+\Delta t_2} dt \left[\sum_i p_i' \dot{q}_i' - \mathcal{H}' (q, p, t) \right]$$

We write this as

$$\Delta_T S = \delta_T \int_{t_1}^{t_2} dt \left[\sum_i p_i \dot{q}_i - \mathcal{H} (q, p, t) \right]$$
$$- \int_{t_1}^{t_1+\Delta t_1} dt \left[\sum_i p_i' \dot{q}_i' - \mathcal{H}' (q, p, t) \right] + \int_{t_2}^{t_2+\Delta t_2} dt \left[\sum_i p_i' \dot{q}_i' - \mathcal{H}' (q, p, t) \right] \tag{C.11}$$

with

$$\delta_T \int_{t_1}^{t_2} dt \left[\sum_i p_i \dot{q}_i - \mathcal{H} (q, p, t) \right] = \int_{t_1}^{t_2} dt \frac{d}{dt} \left(\sum_i p_i \delta q_i \right) \tag{C.12}$$

Here δ_T designates a variation that excludes variations in the end point times, but includes variations in p, q and $\dot{q}$ at the endpoints. The variation δ_T then differs from the previous δ variation of Hamilton's Principle, but will contain the variation δ.

Since the terms Δt_1 and Δt_2 are infinitesimals, the integrands in the last two integrals in (C.11) are constants over the respective integrations, provided the Lagrangian is a continuous function of its coordinates. Then

$$\int_{t_2}^{t_2+\Delta t_2} dt \left[\sum_i p_i' \dot{q}_i' - \mathcal{H}' (q, p, t) \right] - \int_{t_1}^{t_1+\Delta t_1} dt \left[\sum_i p_i' \dot{q}_i' - \mathcal{H}' (q, p, t) \right]$$
$$= \left[\sum_i p_i' \dot{q}_i' - \mathcal{H}' (q, p, t) \right]_{t=t_2} \Delta t_2 - \left[\sum_i p_i' \dot{q}_i' - \mathcal{H}' (q, p, t) \right]_{t=t_1} \Delta t_1$$
$$= L' (t) \, \Delta t \Big]_{t_1}^{t_2} \tag{C.13}$$

where

$$L' (t) = \sum_i p_i' \dot{q}_i' - \mathcal{H}' (q, p, t) \tag{C.14}$$

Using (C.14) and dropping all terms that are greater than first order in infinitesimals (C.13) becomes

$$\int_{t_2}^{t_2+\Delta t_2} \mathrm{d}t \left[\sum_{\mathrm{i}} p_{\mathrm{i}}' \dot{q}_{\mathrm{i}}' - \mathcal{H}'(q,p,t) \right] - \int_{t_1}^{t_1+\Delta t_1} \mathrm{d}t \left[\sum_{\mathrm{i}} p_{\mathrm{i}}' \dot{q}_{\mathrm{i}}' - \mathcal{H}'(q,p,t) \right]$$

$$= \left[\sum_{\mathrm{i}} p_{\mathrm{i}} \dot{q}_{\mathrm{i}} - \mathcal{H}(q,p,t) \right] \Delta t \Big\}_{\mathrm{t}_1}^{\mathrm{t}_2} \tag{C.15}$$

The total variation (C.11) is then

$$\begin{aligned} \Delta_{\mathrm{T}} S &= \int_{t_1}^{t_2} \mathrm{d}t \frac{\mathrm{d}}{\mathrm{d}t} \left(\sum_{\mathrm{i}} p_{\mathrm{i}} \delta q_{\mathrm{i}} \right) + \left[\sum_{\mathrm{i}} p_{\mathrm{i}} \dot{q}_{\mathrm{i}} - \mathcal{H}(q,p,t) \right] \Delta t \Big\}_{\mathrm{t}_1}^{\mathrm{t}_2} \\ &= \left[\sum_{\mathrm{i}} p_{\mathrm{i}} \delta q_{\mathrm{i}} \right] + \left[\sum_{\mathrm{i}} p_{\mathrm{i}} \dot{q}_{\mathrm{i}} - \mathcal{H}(q,p,t) \right] \Delta t \Big\}_{\mathrm{t}_1}^{\mathrm{t}_2} \end{aligned}$$

or

$$\Delta_{\mathrm{T}} S = \sum_{\mathrm{i}} p_{\mathrm{i}} \delta q_{\mathrm{i}} + \left(\sum_{\mathrm{i}} p_{\mathrm{i}} \dot{q}_{\mathrm{i}} - \mathcal{H} \right) \Delta t \Big]_{t_1}^{t_2}$$

In this result the separate variations δq_{i} and Δt appear on the right hand side. If the time end points are fixed, i.e. if $\Delta t_{1,2} = 0$, then $\Delta_{\mathrm{T}} S \to \delta_{\mathrm{T}} S$ and

$$\delta_{\mathrm{T}} S = \sum_{\mathrm{i}} p_{\mathrm{i}} \delta q_{\mathrm{i}} \Big]_{t_1}^{t_2} .$$

Appendix D
Zero-Component Lemma

If a specific component of a 4−vector vanishes in all inertial frames then the 4−vector must be the zero 4−vector. This is known as the *zero-component lemma*. This is a remarkable property of 4−vectors that will be of practical use to us.

To prove this lemma we assume the contrary. That is we take as our hypothesis the fact that one of the components of a 4−vector vanishes in all inertial frames. We then assume that the 4−vector is not a zero 4− vector and prove that this assumption leads to a contradiction.

Suppose that one of the spatial components of the 4−vector vanishes in all inertial frames, but at least one of the other spatial components is not zero in some frame. Then we can rotate the coordinates making either other spatial component nonzero in violation of the hypothesis.

Suppose then that one of the spatial components of the 4−vector vanishes in all inertial frames, but that the time component is nonzero in at least one frame. A Lorentz Transformation of a 4−vector for which all the spatial components are zero, but the time component is nonzero will produce a 4−vector with only a single nonzero spatial component, as we see in the matrix form of the Lorentz Transformation here

$$\begin{bmatrix} \gamma & -\gamma\beta & 0 & 0 \\ -\gamma\beta & \gamma & 0 & 0 \\ 0 & 0 & 1 & 0 \\ 0 & 0 & 0 & 1 \end{bmatrix} \begin{bmatrix} a \\ 0 \\ 0 \\ 0 \end{bmatrix} = \begin{bmatrix} a\gamma \\ -a\beta\gamma \\ 0 \\ 0 \end{bmatrix}. \tag{D.1}$$

We can then, by coordinate rotation, make this any spatial component. Thus a nonzero time component will also contradict the hypothesis that a particular spatial component vanishes in all inertial frames.

Suppose that the time component of a 4−vector is zero in all inertial frames. If there is some inertial frame in which one of the spatial components is nonzero then a Lorentz Transformation of the 4−vector may be found that results in a 4−vector with a nonzero time components we see here

C.S. Helrich, *Analytical Mechanics*, Undergraduate Lecture Notes in Physics, DOI 10.1007/978-3-319-44491-8

$$\begin{bmatrix} \gamma & -\gamma\beta & 0 & 0 \\ -\gamma\beta & \gamma & 0 & 0 \\ 0 & 0 & 1 & 0 \\ 0 & 0 & 0 & 1 \end{bmatrix} \begin{bmatrix} 0 \\ a \\ 0 \\ 0 \end{bmatrix} = \begin{bmatrix} -a\beta\gamma \\ a\gamma \\ 0 \\ 0 \end{bmatrix}. \tag{D.2}$$

This result then violates the hypothesis that the time component is zero in all inertial frames. If the time component is not zero we can perform a Lorentz transformation to make any of the spatial components nonzero in violation of the hypothesis.

We must, therefore, conclude that if any one of the components of a 4−vector vanishes in all inertial frames then the 4−vector is the zero 4−vector. (cf. [99], p. 66).

Appendix E
Maxwell Equations from Field Strength Tensor

Maxwell's Equations can be extracted from the Field Strength tensor. If we operate on the Field Strength tensor $\mathcal{F}^{\sigma\tau}$ with the covariant differential operator ∂_σ in (8.78) the result is

$$\partial_\sigma \mathcal{F}^{\sigma\tau} = \partial_\sigma \partial^\sigma \mathcal{A}^\tau - \partial_\sigma \partial^\tau \mathcal{A}^\sigma. \tag{E.1}$$

From (8.83) we know that

$$\partial_\sigma \partial^\sigma \mathcal{A}^\tau = \Box \mathcal{A}^\tau, \tag{E.2}$$

and $\Box \boldsymbol{A} = \boldsymbol{0}$ is the form of Maxwell's Equations in empty space. We can also show (see exercises) that

$$\partial_\sigma \partial^\tau \mathcal{A}^\sigma = \partial^\tau \left(\frac{1}{c^2} \frac{\partial}{\partial t} \varphi + \mathrm{div} \boldsymbol{A} \right), \tag{E.3}$$

and

$$\frac{1}{c^2} \frac{\partial}{\partial t} \varphi + \mathrm{div} \boldsymbol{A} = 0 \tag{E.4}$$

is the Lorentz Gauge, which must hold for Maxwell's Equations to take the form (8.82) ([45], p. 255). Then if we require

$$\partial_\rho \delta^\rho_\sigma \mathcal{F}^{\sigma\tau} = \partial_\sigma \partial^\sigma \mathcal{A}^\tau - \partial_\sigma \partial^\tau \mathcal{A}^\sigma = 0, \tag{E.5}$$

we have the empty space form of Maxwell's Equations, i.e. with $\boldsymbol{J} = \boldsymbol{0}$, written in terms of the potentials φ and $\boldsymbol{A}$ and requiring the Lorentz Gauge. This elegant result we may consider characteristic of the sort of formulation we are led to expect at this level in our study of theoretical physics.

From

$$\partial_\alpha \mathcal{F}_{\alpha\beta} = 0, \tag{E.6}$$

using the elements of $\mathcal{F}_{\alpha\beta}$ directly we obtain Gauss's Law

C.S. Helrich, *Analytical Mechanics*, Undergraduate Lecture Notes in Physics, DOI 10.1007/978-3-319-44491-8

$$\mathrm{div}\boldsymbol{E} = 0 \tag{E.7}$$

and Ampère's Law (see exercises)

$$\mathrm{curl}\boldsymbol{B} = -\frac{1}{c^2}\frac{\partial}{\partial t}\boldsymbol{E}. \tag{E.8}$$

From the fact that the order of partial differentiation is immaterial Einstein pointed out that (see exercises).

$$\partial_\rho \mathcal{F}_{\sigma\tau} + \partial_\tau \mathcal{F}_{\rho\sigma} + \partial_\sigma \mathcal{F}_{\tau\rho} = 0 \tag{E.9}$$

$$\partial^\rho \mathcal{F}^{\sigma\tau} + \partial^\tau \mathcal{F}^{\rho\sigma} + \partial^\sigma \mathcal{F}^{\tau\rho} = 0. \tag{E.10}$$

We note that the indices on the terms in (E.9) and (E.10) are the even permutations of $\rho\sigma\tau$. For $\rho\sigma\tau = 123$, (E.9) or (E.10) yields Oersted's result $\mathrm{div}\boldsymbol{B} = 0$ and for $\rho\sigma\tau = 023, 013$, and 012 (E.9) or (E.10) produces the three components of Faraday's Law $\partial\boldsymbol{B}/\partial t + \mathrm{curl}\boldsymbol{E} = \boldsymbol{0}$ (see exercise). Einstein indicates this explicitly ([24], p. 154).

Appendix F
Differential Operators

Rectangular Coordinates

$$\text{grad}\Phi = \hat{e}_\text{x}\frac{\partial \Phi}{\partial x} + \hat{e}_\text{y}\frac{\partial \Phi}{\partial y} + \hat{e}_\text{z}\frac{\partial \Phi}{\partial z} \tag{F.1}$$

$$\text{div}\boldsymbol{F} = \frac{\partial F_\text{x}}{\partial x} + \frac{\partial F_\text{y}}{\partial y} + \frac{\partial F_\text{z}}{\partial z} \tag{F.2}$$

$$\begin{aligned}\text{curl}\boldsymbol{F} = {}& \hat{e}_\text{x}\left(\frac{\partial F_\text{z}}{\partial y} - \frac{\partial F_\text{y}}{\partial z}\right) + \hat{e}_\text{y}\left(\frac{\partial F_\text{x}}{\partial z} - \frac{\partial F_\text{z}}{\partial x}\right) \\ & + \hat{e}_\text{z}\left(\frac{\partial F_\text{y}}{\partial x} - \frac{\partial F_\text{x}}{\partial y}\right)\end{aligned} \tag{F.3}$$

$$\nabla^2\Phi = \frac{\partial^2 \Phi}{\partial x^2} + \frac{\partial^2 \Phi}{\partial x^2} + \frac{\partial^2 \Phi}{\partial x^2}. \tag{F.4}$$

Cylindrical Coordinates

$$\text{grad}\Phi = \hat{e}_\text{r}\frac{\partial \Phi}{\partial r} + \hat{e}_\vartheta\frac{1}{r}\frac{\partial \Phi}{\partial \vartheta} + \hat{e}_\text{z}\frac{\partial \Phi}{\partial z} \tag{F.5}$$

$$\text{div}\boldsymbol{F} = \frac{1}{r}\frac{\partial}{\partial r}(rF_\text{r}) + \frac{1}{r}\frac{\partial F_\vartheta}{\partial \vartheta} + \frac{\partial F_\text{z}}{\partial z} \tag{F.6}$$

$$\begin{aligned}\text{curl}\boldsymbol{F} = {}& \hat{e}_\text{r}\left[\frac{1}{r}\frac{\partial F_\text{z}}{\partial \vartheta} - \frac{\partial F_\vartheta}{\partial z}\right] + \hat{e}_\vartheta\left[\frac{\partial F_\text{r}}{\partial z} - \frac{\partial F_\text{z}}{\partial r}\right] \\ & + \hat{e}_\text{z}\frac{1}{r}\left[\frac{\partial}{\partial r}(rF_\vartheta) - \frac{\partial F_\text{r}}{\partial \vartheta}\right]\end{aligned} \tag{F.7}$$

C.S. Helrich, *Analytical Mechanics*, Undergraduate Lecture Notes in Physics, DOI 10.1007/978-3-319-44491-8

$$\nabla^2 \Phi = \frac{1}{r}\frac{\partial}{\partial r}\left(r\frac{\partial \Phi}{\partial r}\right) + \frac{1}{r^2}\frac{\partial^2 \Phi}{\partial \vartheta^2} + \frac{\partial^2 \Phi}{\partial z^2}. \tag{F.8}$$

Spherical Coordinates

$$\text{grad}\Phi = \hat{e}_{\text{r}}\frac{\partial \Phi}{\partial r} + \hat{e}_{\vartheta}\frac{1}{r\sin\phi}\frac{\partial \Phi}{\partial \vartheta} + \hat{e}_{\phi}\frac{1}{r}\frac{\partial \Phi}{\partial \phi} \tag{F.9}$$

$$\text{div}\boldsymbol{F} = \frac{1}{r^2}\frac{\partial}{\partial r}\left(r^2 F_{\text{r}}\right) + \frac{1}{r\sin\phi}\frac{\partial F_{\vartheta}}{\partial \vartheta} + \frac{1}{r\sin\phi}\frac{\partial}{\partial \phi}\left(F_{\phi}\sin\phi\right) \tag{F.10}$$

$$\begin{aligned}\text{curl}\boldsymbol{F} = {} & \hat{e}_{\text{r}}\frac{1}{r\sin\phi}\left[\frac{\partial}{\partial \phi}\left(F_{\vartheta}\sin\phi\right) - \frac{\partial F_{\phi}}{\partial \vartheta}\right] \\ & + \hat{e}_{\vartheta}\frac{1}{r}\left[\frac{\partial}{\partial r}\left(rF_{\phi}\right) - \frac{\partial F_{\text{r}}}{\partial \phi}\right] \\ & + \hat{e}_{\phi}\frac{1}{r}\left[\frac{1}{\sin\phi}\frac{\partial F_{\text{r}}}{\partial \vartheta} - \frac{\partial}{\partial r}\left(rF_{\vartheta}\right)\right]\end{aligned} \tag{F.11}$$

$$\begin{aligned}\nabla^2 \Phi = {} & \frac{1}{r^2}\frac{\partial}{\partial r}\left(r^2\frac{\partial \Phi}{\partial r}\right) + \frac{1}{r^2\sin^2\phi}\frac{\partial^2 \Phi}{\partial \vartheta^2} \\ & + \frac{1}{r^2\sin\phi}\frac{\partial}{\partial \phi}\left(\sin\phi\frac{\partial \Phi}{\partial \phi}\right).\end{aligned} \tag{F.12}$$

Appendix G
Answers to Selected Exercises

2.1 $\mathrm{d}\boldsymbol{r}/\mathrm{d}t = \dot{r}\hat{e}_{\mathrm{r}} + r\dot{\vartheta}\hat{e}_{\vartheta} + \dot{z}\hat{e}_{\mathrm{z}}$.

2.2 $\mathrm{d}\boldsymbol{r}/\mathrm{d}t = \dot{\rho}\hat{e}_{\rho} + \rho\dot{\vartheta}\sin\phi\hat{e}_{\vartheta} + \rho\dot{\phi}\hat{e}_{\phi}$.

2.3 The point on the line is $(-3/5, 1/5)$. The vectors along the line and to this point are $\left(\frac{2}{3}, 2\right)$ and $(-3/5, 1/5)$. The scalar product of these is zero.

2.4 $m_{\mathrm{i}}\mathrm{d}^2\boldsymbol{r}_{\mathrm{i}}/\mathrm{d}t^2 \cdot \delta\boldsymbol{r}_{\mathrm{i}} = \mathrm{d}\left(\frac{1}{2}m_{\mathrm{i}}v_{\mathrm{i}}^2\right)$; $\boldsymbol{F}_{\mathrm{i}} \cdot \mathrm{d}\boldsymbol{r}_{\mathrm{i}} = -\operatorname{grad}V \cdot \mathrm{d}\boldsymbol{r}_{\mathrm{i}} = -\mathrm{d}V$.

2.5 The points on the two curves are $(x_1, y_1) = \left(-\frac{1}{2}, -\frac{9}{2}\right)$ and $(x_2, y_2) = \left(-\frac{23}{10}, -\frac{18}{5}\right)$. The vector from the parabola to the straight line has the components $(x_2 - x_1, y_2 - y_1) = \left(-\frac{18}{5}, \frac{9}{10}\right)$. The vector direction along the straight line is $\left(\frac{1}{2}, 1\right)$. The scalar product of the vectors is then $\left(-\frac{18}{5}, \frac{9}{10}\right) \cdot \left(\frac{1}{2}, 1\right) = 0$. The shortest line between the curves is then perpendicular to the straight line.

2.6 Find extremum of $h(P) = -k_{\mathrm{B}}\sum_{\mathrm{r}} P_{\mathrm{r}}\ln P_{\mathrm{r}} + \alpha\left(\sum_r P_r - 1\right) + \beta \left(\sum_{\mathrm{r}} P_{\mathrm{r}}\mathcal{E}_{\mathrm{r}} - \mathcal{E}\right)$.

2.7 $J[y] = \int_0^1 \mathrm{d}x\,(y') = \int_0^1 \mathrm{d}y = 1$, which has no extremum.

2.8 $J[y] = \int_0^1 \mathrm{d}\left(\frac{1}{2}y^2\right) = \frac{1}{2}$, which has no extremum.

2.9 The Euler–Lagrange Equation is then solved by the function $y(x) = 0$. There is then a minimum for $y = 0$ and this minimum is 0.

2.10 $y'' - y = -\sin(x)$.

2.11 $\delta S[x, y] = \int_{t_1}^{t_2}\mathrm{d}t\left[\left(\partial\Psi/\partial\xi - \mathrm{d}\partial\Psi/\partial\dot{\xi}/\mathrm{d}t\right)h_{\mathrm{x}} + (\partial\Psi/\partial\eta - \mathrm{d}\partial\Psi/\partial\dot{\eta}/\mathrm{d}t)\, h_{\mathrm{y}}\right] = 0$ when each bracket (...) vanishes.

2.12 This is a demonstration.

2.13 This is a demonstration.

2.14 $y = K\cosh\left[(x + C)/K\right]$, where K and C are (integration) constants.

2.15 $y = K\sin^2\phi = \frac{K}{2}(1 - \cos 2\phi)$, which is a cycloid.

2.16 The time derivative of the Hamiltonian is $\mathrm{d}\mathcal{H}/\mathrm{d}t = \frac{p_{\mathrm{x}}}{m}\mathrm{d}p_{\mathrm{x}}/\mathrm{d}t = 0$, since $\mathrm{d}p_{\mathrm{x}}/\mathrm{d}t = 0$ from the Euler–Lagrange equation.

C.S. Helrich, *Analytical Mechanics*, Undergraduate Lecture Notes in Physics, DOI 10.1007/978-3-319-44491-8

2.17 $L = m\dot{y}^2/2 - mgy$, $p_y = m\dot{y}$, $-mg - \mathrm{d}(m\dot{y})/\mathrm{d}t = 0$, $\mathrm{d}\mathcal{H}/\mathrm{d}t = \dot{y}\left(\mathrm{d}p_y/\mathrm{d}t + mg\right) = 0$, $y(t) = y_0 + \dot{y}_0 t - mgt^2/2$.

2.18 $\mathcal{H} = \left(p_x^2 + p_y^2\right)/2m + mgy$, Euler–Lagrange Equation $-mg\alpha - \mathrm{d}\left(m\left(1+\alpha^2\right)\dot{x}\right)/\mathrm{d}t = 0$, $x = x_0 + \dot{x}_0 t - (1/2)\left[g\alpha/\left(1+\alpha^2\right)\right]t^2$, reaction forces with the incline are $f_x = -\lambda\alpha$ and $f_y = +\lambda$ with $\lambda = mg/\left(1+\alpha^2\right)$.

2.19 $-m\ddot{x} + \lambda\left(12x^2 - 10x + 1\right) = 0$ and $-m\ddot{y} + \lambda = 0$.

2.20 The eigenvalues are $\omega^2 = \omega_0^2, 3\omega_0^2$.And the corresponding normalized eigenvectors are

$$\frac{1}{\sqrt{2}}\begin{bmatrix}1\\1\end{bmatrix} \text{ for the eigenvalue } \omega_0^2$$
$$\frac{1}{\sqrt{2}}\begin{bmatrix}-1\\1\end{bmatrix} \text{ for the eigenvalue } 3\omega_0^2.$$

2.21 Equilibrium orbit is with the polar angle $\sin\phi_0 + \left(m^2 g R^3/\ell^2\right)\left(1-\sin^2\phi_0\right)^2 = 0$. The motion around the equilibrium point is then sinusoidal with a frequency

$$\omega = \sqrt{\left(\frac{g}{R}\sin\phi_0 + \left(\frac{\ell}{mR^2}\right)^2 \frac{2\cos^2\phi_0 - 3}{\cos^4\phi_0}\right)}.$$

2.22 Euler–Lagrange Equations are $-m\ell\dot{x}\dot{\vartheta}\sin\vartheta - mg\ell\sin\vartheta - m\left(\ell^2\ddot{\vartheta} + \ell\ddot{x}\cos\vartheta - \ell\dot{x}\dot{\vartheta}\sin\vartheta\right) = 0$ and $-kx - \left((M+m)\ddot{x} + m\ell\ddot{\vartheta}\cos\vartheta - m\ell\dot{\vartheta}^2 \sin\vartheta\right) = 0$. The frequencies for small oscillations are

$$\omega^2 = \begin{cases} \omega_p^2 + \frac{1}{2}\omega_s^2 + \frac{1}{2}\sqrt{4\omega_p^4 + \omega_s^4} \\ \omega_p^2 + \frac{1}{2}\omega_s^2 - \frac{1}{2}\sqrt{4\omega_p^4 + \omega_s^4} \end{cases}.$$

2.23 $\ddot{\phi} = (g/R)\sin\phi$.

2.24 $\cos\phi_0 = g/\left(R\Omega^2\right)$. The frequency is

$$\Omega\sqrt{1 + \frac{g}{R\Omega^2} - 2\left(\frac{g}{R\Omega^2}\right)^2}$$

2.25 $L = m\left\{\frac{1}{6}\left[\ell^2 - 3\ell d + 3d^2\right]\dot{\vartheta}^2 + \frac{1}{2}g\left[\ell - 2d\right]\cos\vartheta\right\}$. The Euler–Lagrange Equation is

$$\ddot{\vartheta} = -\frac{3}{2}\frac{g\left[\ell - 2d\right]}{\left[\ell^2 - 3\ell d + 3d^2\right]}\sin\vartheta.$$

2.26 $t = -\int \frac{\mathrm{d}\vartheta}{\sqrt{3\frac{g}{\ell}(1-\sin\vartheta)}} + K_2$ where K_2 is an integration constant.

2.27 $-\frac{1}{2}mg\ell\cos\vartheta - m\left(\frac{1}{3}\ell^2\ddot{\vartheta} - \frac{1}{2}\ell\ddot{x}\sin\vartheta\right) = 0$ and $m\dot{x} - m\ell\dot{\vartheta}\sin\vartheta =$constant.

2.28 The normal modes are

$$\omega = \omega_0\sqrt{2+\sqrt{2}}$$
$$\omega = \omega_0\sqrt{2-\sqrt{2}}.$$

2.29 Answers provided in text.

2.30 Our equations are the Euler–Lagrange Equations

$$mr\dot{\vartheta}^2 - m\ddot{r} + \lambda_1 = 0$$
$$-mg - m\ddot{z} - \lambda_1\tan\alpha = 0,$$
$$\frac{\mathrm{d}}{\mathrm{d}t}\left(mr^2\dot{\vartheta}\right) + \lambda_2 = 0$$

and the constraint equations

$$r - (\tan\alpha)\,z = 0,$$
$$\vartheta - \Omega\,(t - t_0) = 0.$$

2.31 There is no equilibrium orbit.

2.32 $\vartheta = \cos^{-1}(2/3)$

2.33 Equilibrium point is $r_0 = \left[\ell^2/\left(m^2 g\right)\right]^{1/3}$. The orbit is open.

2.34 Answers provided in text.

3.1 The variation produces

$$0 = \sum_{\mathrm{j}}\int_{\mathrm{t}_1}^{\mathrm{t}_2}\mathrm{d}t\left\{\left[\dot{q}_{\mathrm{j}} - \frac{\partial\mathcal{H}}{\partial p_{\mathrm{j}}}\right]\delta p_{\mathrm{j}} + \left[-\dot{p}_{\mathrm{j}} - \frac{\partial\mathcal{H}}{\partial q_{\mathrm{j}}}\right]\delta q_{\mathrm{j}}\right\}.$$

3.2 At the extrema of u (and ρ) we have $0 = -A\cos\vartheta + B\sin\vartheta$. If we choose this to define $\vartheta = 0$ then $B \equiv 0$.

3.3 For the eigenvalues $i\omega = \pm i\sqrt{k/m}$ the eigenvectors are

$$\begin{bmatrix}\tilde{p}_1\\ \tilde{p}_2\\ \tilde{x}_1\\ \tilde{x}_2\end{bmatrix} = \begin{bmatrix}-im^2\sqrt{\frac{k}{m^3}}\\ -im^2\sqrt{\frac{k}{m^3}}\\ 1\\ 1\end{bmatrix} \text{ and } \begin{bmatrix}im^2\sqrt{\frac{k}{m^3}}\\ im^2\sqrt{\frac{k}{m^3}}\\ 1\\ 1\end{bmatrix}$$

respectively. And for the eigenvalues $i\omega = \pm i\sqrt{3k/m}$ the eigenvectors are

$$\begin{bmatrix} \tilde{p}_1 \\ \tilde{p}_2 \\ \tilde{x}_1 \\ \tilde{x}_2 \end{bmatrix} = \begin{bmatrix} i\sqrt{3}m^2\sqrt{\frac{k}{m^3}} \\ -i\sqrt{3}m^2\sqrt{\frac{k}{m^3}} \\ -1 \\ 1 \end{bmatrix} \text{ and } \begin{bmatrix} -i\sqrt{3}m^2\sqrt{\frac{k}{m^3}} \\ i\sqrt{3}m^2\sqrt{\frac{k}{m^3}} \\ -1 \\ 1 \end{bmatrix}.$$

3.4 Answers for small angles are provided in text. Cannot use undetermined multipliers because the constraint is a function of time.

3.5 The Hamiltonian is conserved $\mathcal{H} = \frac{1}{2}mR^2\dot{\phi}^2 - \frac{1}{2}mR^2\omega^2\sin^2\phi + mgR\cos\phi$, but this is not the total energy. The system is not conservative. There is an equilibrium point provided $R\omega^2 > \sqrt{3}g$.

3.6 The effective potential is $V_{\text{eff}}(r) = \frac{1}{2mr^2}\ell^2 + mg(r-b)$, which has a minimum. There is then an equilibrium point. The orbit about the equilibrium point is, however, not closed.

3.7 $\phi = \cos^{-1}(2/3)$.

3.8 The frequencies for small oscillations about equilibrium are $\omega = \pm\sqrt{\omega_{\text{p}}^2 + \frac{1}{2}\omega_{\text{s}}^2 \pm \frac{1}{2}\sqrt{4\omega_{\text{p}}^4 + \omega_{\text{s}}^4}}$.

3.9 The natural frequencies are $\omega = \omega_0\sqrt{\left(2-\sqrt{2}\right)}, \omega_0\sqrt{\left(2+\sqrt{2}\right)}$, where $\omega_0 = \sqrt{\frac{g}{\ell}}$.

3.10 This is a proof.

3.11 This is a proof.

3.12 The first part is a demonstration that the proposed vector potentials are appropriate. The canonical equations are

$$\dot{p}_x = \frac{1}{2}\Omega p_y - \frac{1}{4}m\Omega^2 x,$$

$$\dot{p}_y = -\frac{1}{2}\Omega p_x - \frac{1}{4}m\Omega^2 y,$$

and

$$\dot{x} = \frac{1}{m}p_x + \frac{1}{2}\Omega y,$$

$$\dot{y} = \frac{1}{m}p_y - \frac{1}{2}\Omega x.$$

Solution is carried out in the complex plane.

3.13 Solution is in the complex plane with $Z = x + iy$ and $P_{\text{Z}} = p_{\text{x}} + ip_{\text{y}}$. The canonical equations produce the equations $\dot{P}_{\text{Z}} = -\frac{Q^2B^2}{4m}Z - \frac{QB}{2m}iP_{\text{Z}}$, and $\dot{Z} = \frac{P_{\text{Z}}}{m} - i\frac{QB}{2m}Z$ in the complex plane. The Ansatz $P_{\text{Z}} = \tilde{P}_{\text{Z}}\exp(i\Omega t)$ and $Z =$

$\tilde{Z} \exp(i\Omega t)$ results in $\Omega = -\frac{QB}{m}$. The orbit is then easily obtained from the $Z(t)$. Talking real and imaginary parts,

$$x(t) = \mathrm{Re}(Z(t)) = x_0 \cos(\Omega t)$$
$$y(t) = \mathrm{Im}(Z(t)) = -y_0 \sin(\Omega t).$$

This orbit is a circle. Looking down on the (x, y) −plane, the direction of motion is clockwise.
We may now couple this motion with uniform motion along the (original) $z-$axis. The result is a spiral motion along the $z-$axis. Small variations in the magnetic field along the $z-$axis will not disturb this general spiral motion. We may then consider the charges as "trapped" in the spiral paths along the magnetic field lines.

3.14 Solution uses the same approach as in 3.13. The result is

$$x(t) = \mathrm{Re}(Z) = R\cos(\Omega t) + \frac{E}{B}t$$
$$y(t) = \mathrm{Im}(Z) = -R\sin(\Omega t),$$

which is a cycloid.

3.15 With $p_\vartheta = \ell$ (constant) the canonical equations are

$$\ell = mr^2\left(\dot{\vartheta} + \frac{1}{2}\Omega\right)$$

and

$$\dot{p}_{\mathrm{r}} = mr\left(\frac{\ell}{mr^2} + \frac{\Omega}{2}\right)\left(\frac{\ell}{mr^2} - \frac{\Omega}{2}\right).$$

With

$$\frac{\ell}{mr^2} = \dot{\vartheta} + \frac{\Omega}{2}$$

we have

$$\dot{p}_{\mathrm{r}} = mr\left(\dot{\vartheta} + \Omega\right)\left(\dot{\vartheta}\right).$$

and

$$\dot{r} = \frac{1}{m}p_{\mathrm{r}}.$$

These are nonlinear. We can show that a circle solves these, but cannot guarantee it is the only solution.

3.16 With $\Omega = QB/m$. The canonical equations are

$$
\begin{aligned}
\dot{x} &= \frac{\partial \mathcal{H}}{\partial p_{\mathrm{x}}} = \frac{1}{m} p_{\mathrm{x}} + \frac{1}{2} \Omega y \exp(az) \\
\dot{y} &= \frac{\partial \mathcal{H}}{\partial p_{\mathrm{y}}} = \frac{1}{m} p_{\mathrm{y}} - \frac{1}{2} \Omega x \exp(az) \\
\dot{z} &= \frac{\partial \mathcal{H}}{\partial p_{\mathrm{z}}} = \frac{1}{m} p_{\mathrm{z}} \\
\dot{p}_{\mathrm{x}} &= -\frac{\partial \mathcal{H}}{\partial x} = \frac{1}{2} \Omega \left(p_{\mathrm{y}} - \frac{1}{2} m \Omega x \exp(az) \right) \exp(az) \\
\dot{p}_{\mathrm{y}} &= -\frac{\partial \mathcal{H}}{\partial y} = -\frac{1}{2} \Omega \left(p_{\mathrm{x}} + \frac{1}{2} m \Omega y \exp(az) \right) \exp(az) \\
\dot{p}_{\mathrm{z}} &= -\frac{\partial \mathcal{H}}{\partial z} = -\frac{1}{2} \Omega a \left[\left(p_{\mathrm{x}} + \frac{1}{2} m \Omega y \exp(az) \right) y \right. \\
&\quad \left. - \left(p_{\mathrm{y}} - \frac{1}{2} m \Omega x \exp(az) \right) x \right] \exp(az) .
\end{aligned}
$$

We can show that the solution in planes of constant z is a circle. But the solution shown in the statement of the exercise was obtained numerically. This solution demonstrates the magnetic bottle.

3.17 The Hamiltonian is

$$
\mathcal{H} = \frac{1}{2m} p_\rho^2 + \frac{1}{2m\rho^2} \left(p_\vartheta - Q \frac{\mu_0 M}{4\pi} \frac{1}{\rho^2} \right)^2 + \frac{Q_{\mathrm{N}} Q}{4\pi\varepsilon_0} \frac{1}{\rho}
$$

Because the Lagrangian is cyclic in ϑ we have p_ϑ = constant. If we expand this Hamiltonian for small values of M we have

$$
\mathcal{H} = \frac{1}{2m} p_\rho^2 + \frac{1}{2m\rho^2} p_\vartheta^2 - \frac{M Q \mu_0}{4\pi m} \frac{1}{\rho^4} p_\vartheta + \frac{Q_{\mathrm{N}} Q}{4\pi\varepsilon_0} \frac{1}{\rho} .
$$

The effective potential is then decreased in the neighborhood of the nucleus if the nucleus possessed a magnetic moment. The $1/\rho^4$ dependence of this potential, however, makes it a weak potential at large distances.

4.1 The two products

$$
\mathbf{R}_2 \left(\delta\vartheta_2' \right) \mathbf{R}_3 \left(\delta\vartheta_3' \right)
= \begin{bmatrix} 1 & -\delta\vartheta_3' & \delta\vartheta_2' \\ \delta\vartheta_3' & 1 & 0 \\ -\delta\vartheta_2' & \delta^2 \vartheta_2' \vartheta_3' & 1 \end{bmatrix}
$$

and $\mathbf{R}_3 \left(\delta\vartheta_3' \right) \mathbf{R}_2 \left(\delta\vartheta_2' \right)$ are equal.

4.2 With the projector

$$
P_{\mathrm{X}} = \left| X_\mu \right\rangle \left\langle X_\mu \right| = \mathbf{1},
$$

which we may introduce at any point in our expression for the kinetic energy, we have

$$\begin{aligned} T_{\text{rot}} &= \frac{1}{2} I_0 \langle \omega \,| X_\lambda \rangle \langle X_\lambda |\, x_\rho \rangle \langle x_\rho \,| X_\sigma \rangle \langle X_\sigma |\, \omega \rangle \\ &= \frac{1}{2} I_0 \langle \omega \,| X_\lambda \rangle\, \delta_{\lambda\sigma} \langle X_\sigma |\, \omega \rangle \\ &= \frac{1}{2} I_0 \langle \omega \,| X_\lambda \rangle \langle X_\lambda |\, \omega \rangle , \end{aligned}$$

which is the same form as

$$T_{\text{rot}} = \frac{1}{2} \boldsymbol{I}_0 \langle \omega \,| x_\rho \rangle \langle x_\rho |\, \omega \rangle ,$$

except the representation is in a different basis.

4.3 Rolling begins at the time t_{roll} with

$$t_{\text{roll}} = \frac{I\omega_0/\mu_{\text{k}} mag}{\left(1 + I/ma^2\right)}.$$

4.4

$$\begin{bmatrix} x \\ \vartheta \end{bmatrix} = \begin{bmatrix} gt^2 \sin\alpha \\ \left(gt^2/a\right) \sin\alpha \end{bmatrix}.$$

4.5 The canonical equations are $\dot{p}_{\text{r}} - p_\vartheta^2/\left(mr^3\right) + \lambda_1 + \lambda_2 = 0, \dot{p}_\vartheta + \lambda_3 r = 0, \dot{p}_{\text{z}} + mg - \lambda_1 \tan\alpha + \lambda_4 = 0, \dot{p}_1 - \lambda_2 \left(a \sin\alpha\right) + \lambda_3 a \sin\alpha = 0, \dot{p}_2 - \lambda_4 \left(a\cos\alpha\right) = 0, \dot{p}_3 + \lambda_3 \left(a \cos\alpha\right) = 0$, and $\dot{r} = p_{\text{r}}/m$, $\dot{\vartheta} = p_\vartheta/\left(mr^2\right)$, $\dot{z} = p/m\ \dot{\vartheta}_1 = p_1/I_0$, $\dot{\vartheta}_2 = p_2/I_0$, $\dot{\vartheta}_3 = p_3/I_0$.The constraints add the equations $0 = \frac{p_{\text{r}}}{m} - (\tan\alpha)\frac{p_{\text{z}}}{m}$, $0 = \frac{p_{\text{r}}}{m} - \frac{p_1}{I_0} a \sin\alpha$, $0 = \frac{p_\vartheta}{mr} + \frac{p_1}{I_0} a \sin\alpha + \frac{p_3}{I_0} a\cos\alpha$, $0 = \frac{p_{\text{z}}}{m} - \frac{p_2}{I_0} a \cos\alpha$. There are then 16 equations for the 12 canonical variables (coordinates and momenta) and the 4 λs. That is, the mathematical representation of the system is complete.

4.6 $\Theta = \cos^{-1}(4/7)$.

4.7 $\Theta = \cos^{-1}(10/17)$.

4.8 This is an exercise in which we show that the motion of a toy gyroscope can be understood in terms of Euler's equations. We also calculate the precessional frequency as $\omega_{\text{p}} = mg\left[\ell/\left(\omega_0 I\right)\right]$.

4.9 The ratio of the periods is $\tau_{\text{rolling}} = \sqrt{3/2}\tau_0 = 1.2247\tau_0$.

4.10 If we choose $a = b$ and $\bar{I} = I/ma^2$,

$$\begin{bmatrix} \dot{\vartheta} \\ \dot{\phi} \end{bmatrix} = \frac{1}{ma^2\left[1 - \cos^2\left(\phi - \vartheta\right) + \bar{I}\right]} \begin{bmatrix} p_\vartheta\left(\bar{I} + 1\right) - p_\phi \cos\left(\phi - \vartheta\right) \\ p_\phi - p_\vartheta \cos\left(\phi - \vartheta\right) \end{bmatrix}.$$

The eigenfrequencies for small vibrations are $\omega^2 = (g/a)\left(5/2 \pm \sqrt{17}/2\right)$. And the ratios of displacement are for $\omega^2 = (g/a)\left(5/2 + \sqrt{17}/2\right)$

$$\frac{\tilde{\vartheta}}{\tilde{\phi}} = -1.2808$$

and for $\omega^2 = (g/a)\left(5/2 - \sqrt{17}/2\right)$

$$\frac{\tilde{\vartheta}}{\tilde{\phi}} = -.78078.$$

4.11 The canonical equations are

$$\begin{aligned} \dot{p}_\mathrm{r} &= \frac{p_\vartheta^2}{mr^3} + mg\cos\vartheta - k\left(r - r_0\right), \\ \dot{p}_\vartheta &= -mgr\sin\vartheta, \end{aligned}$$

$$\begin{aligned} \dot{r} &= \frac{p_r}{m}, \\ \dot{\vartheta} &= \frac{p_\vartheta}{mr^2}. \end{aligned}$$

The Euler–Lagrange equations are

$$mr\dot{\vartheta}^2 + mg\cos\vartheta - k\left(r - r_0\right) - m\ddot{r} = 0$$

and

$$-mgr\sin\vartheta - 2mr\dot{\vartheta}\dot{r} - mr^2\ddot{\vartheta} = 0.$$

The Lagrangian does not depend explicitly on the time, so

$$\mathcal{E} = \frac{1}{2}m\left(\dot{r}^2 + r^2\dot{\vartheta}^2\right) - mgr\cos\vartheta + \frac{1}{2}k\left(r - r_0\right)^2.$$

We elected not to linearize these equations, but to perform a numerical (Runge–Kutta) integration. Graphs are presented in the text.

4.12 the cosine of the angle between the $z-$component of the angular velocity vector and the angular momentum is

$$\cos\beta_\mathrm{z} = \frac{\omega_0 I}{\sqrt{I'^2A^2 + I^2\omega_0^2}}.$$

The spectator and the player cannot see the angular momentum vector. What each of them sees is the axis of the football. If the z−axis is aligned with the angular momentum vector the football will not appear to wobble because the angular momentum vector has a fixed direction in space. This is the spiral pass. If β_z is not zero, then the angular velocity vector rotates about the angular momentum vector and the pass appears wobbly.

5.1 The transform is

$$F_2(p, P, t) = F_1(q, P, t) - \sum_i q_i \frac{\partial F_1}{\partial q_i}$$
$$= F_1 - \sum_i q_i p_i .$$

And

$$\mathrm{d}F_2(p, P, t) = \sum_i Q_i \mathrm{d}P_i - q_i \mathrm{d}p_i + \frac{\partial F}{\partial t}\mathrm{d}t.$$

5.2 This is an outline of the approach using P as the link to the final configuration.

5.3 The final momentum is $P = \tau - t$. Then, recalling that the natural frequency of the harmonic oscillator is $\omega_0 = \sqrt{k/m}$, we may also write our solution as $q = \mp\sqrt{2\mathcal{E}/k}\sin[\omega_0(\tau - t)]$ and $p = \pm\sqrt{2m\mathcal{E}}\cos[\omega_0(\tau - t)]$.

5.4

$$q = \sqrt{\frac{J_q}{m\omega_0\pi}}\cos\omega_0(t - \tau).$$

and

$$p = \sqrt{\frac{m\omega_0 J_q}{\pi}}\sin\omega_0(t - \tau).$$

5.5 This is a demonstration.

5.6 The Dirac analog yields $(q_{n+1}p_{n+1} - p_{n+1}q_{n+1}) = (-t\mathcal{H} + \mathcal{H}t) = i\hbar\mathbf{1}$. Since the magnitude of $\mathcal{H}$ is the energy $\mathcal{E}$, we then have $\Delta\mathcal{E}\Delta t \geq \frac{1}{2}\hbar$.

5.7 With

$$F_1 = -\mathcal{E}t + \ell\vartheta \pm \frac{1}{\sin\beta}\int \mathrm{d}r\sqrt{2m\mathcal{E} - \frac{1}{r^2}\ell^2 - 2m^2 g\frac{r}{\tan\beta}}$$

we have

$$P_1 = \frac{\partial F_1}{\partial \mathcal{E}} = -t \pm \frac{m}{\sin\beta}\int \mathrm{d}r/\sqrt{2m\mathcal{E} - \frac{1}{r^2}\ell^2 - 2m^2 g\frac{r}{\tan\beta}} = \beta_1$$

and

$$P_2 = \frac{\partial F_1}{\partial \ell} = \vartheta \mp \frac{\ell}{\sin\beta} \int \mathrm{d}r/r^2 \sqrt{2m\mathcal{E} - \frac{1}{r^2}\ell^2 - 2m^2 g \frac{r}{\tan\beta}} = \beta_2.$$

The canonical momentum $p_\mathrm{r} = \mathrm{d}F_\mathrm{r}(r,\alpha)/\mathrm{d}r$ is

$$p_\mathrm{r} = \pm \frac{1}{\sin\beta} \sqrt{2m\mathcal{E} - \frac{1}{r^2}\ell^2 - 2m^2 g \frac{r}{\tan\beta}}$$

with

$$\mathcal{E} = \frac{\ell^2}{2mr_0^2} + mg\frac{r_0}{\tan\beta}$$

and

$$\ell = m\,(z_0 \tan\beta)^2\, \dot{\vartheta}_0.$$

5.8 With

$$F_1 = -\mathcal{E}t + \ell\vartheta \pm \int \mathrm{d}r \sqrt{\left(2m\mathcal{E} - 2m^2 g\eta r^\mathrm{n} - \frac{1}{r^2}\ell^2\right)\left(n^2\eta^2 r^{2\mathrm{n}-2} + 1\right)}$$

we have

$$P_1 = \frac{\partial F_1}{\partial \mathcal{E}} = -t \pm m \int \mathrm{d}r \sqrt{\frac{\left(n^2\eta^2 r^{2\mathrm{n}-2} + 1\right)}{\left(2m\mathcal{E} - 2m^2 g\eta r^\mathrm{n} - \frac{1}{r^2}\ell^2\right)}} = \beta_1$$

and

$$P_2 = \frac{\partial F_1}{\partial \ell} = \vartheta \mp \ell \int \frac{\mathrm{d}r}{r^2} \sqrt{\frac{\left(n^2\eta^2 r^{2\mathrm{n}-2} + 1\right)}{\left(2m\mathcal{E} - 2m^2 g\eta r^\mathrm{n} - \frac{1}{r^2}\ell^2\right)}} = \beta_2.$$

The canonical momentum $p_\mathrm{r} = dF_\mathrm{r}(r,\alpha)/\mathrm{d}\,r$ is

$$p_\mathrm{r} = \pm \sqrt{\left(2m\mathcal{E} - 2m^2 g\eta r^\mathrm{n} - \frac{1}{r^2}\ell^2\right)\left(n^2\eta^2 r^{2\mathrm{n}-2} + 1\right)},$$

and

$$p_z = \pm\left(n\eta\left(\frac{z}{\eta}\right)^{1-1/n}\right)\sqrt{\frac{\left(2m(\mathcal{E}) - 2m^2 g\eta\left(\frac{z}{\eta}\right) - \left(\frac{z}{\eta}\right)^{-2/n}(\ell)^2\right)}{\left(n^2\eta^2\left(\frac{z}{\eta}\right)^{2-2/n} + 1\right)}}.$$

From these we get the phase plots.

5.9 The equation for the (assumed separable) generator is

$$\frac{\mathrm{d}F_t(\alpha, t)}{\mathrm{d}t} = -\frac{1}{(2-\cos^2\vartheta)}\left[\frac{1}{2ma^2}\left(\frac{\mathrm{d}}{\mathrm{d}\vartheta}F_\vartheta\right)^2 + \frac{1}{m}\left(\frac{\mathrm{d}}{\mathrm{d}z}F_z\right)^2 + \frac{1}{ma}\left(\frac{\mathrm{d}}{\mathrm{d}\vartheta}F_\vartheta\right)\left(\frac{\mathrm{d}}{\mathrm{d}z}F_z\right)\cos\vartheta\right]$$
$$+mgz - \frac{1}{2}k\left(z^2 - \ell^2\right) - \frac{1}{2}k'\vartheta^2.$$

The time function can be separated. But beyond that we cannot separate the equation. That is the Hamiltonian is not, in fact, separable. We are then stuck with a partial differential equation for the generator $F = F_{1,2}(z, \vartheta, t, \alpha)$.

$$\frac{\partial F_t(\alpha, t)}{\partial t} = -\frac{1}{(2-\cos^2\vartheta)}\left[\frac{1}{2ma^2}\left(\frac{\partial}{\partial\vartheta}F\right)^2 + \frac{1}{m}\left(\frac{\partial}{\partial z}F\right)^2 + \frac{1}{ma}\left(\frac{\partial}{\partial\vartheta}F\right)\left(\frac{\partial}{\partial z}F\right)\cos\vartheta\right]$$
$$+mgz - \frac{1}{2}k\left(z^2 - \ell^2\right) - \frac{1}{2}k'\vartheta^2.$$

The solution is then not practicable. If we define $K_1 = k' + a^2k$ and $K_2^2 = a^4k^2 + k'^2 - a^2kk'$, we find that

$$\frac{K_1^2}{K_2^2} = \frac{\left(k' + a^2k\right)^2}{a^4k^2 + k'^2 - a^2kk'} = \frac{a^4k^2 + k'^2 + 2a^2kk'}{a^4k^2 + k'^2 - a^2kk'} > 0.$$

The (eigen) values of ω are the (then) real quantities

$$\omega = \frac{1}{a\sqrt{m}}\sqrt{K_1 + \sqrt{K_2^2}}, \frac{1}{a\sqrt{m}}\sqrt{K_1 - \sqrt{K_2^2}}.$$

There is then a high and a low frequency mode, as we expected.

5.10 The generator is

$$F_1(x, y, z, t, \alpha_1, \alpha_2, \alpha_3) = -\alpha_1 t + \alpha_3 x \pm \int \mathrm{d}y\sqrt{2m\alpha_1 - \alpha_2^2 - (\alpha_3 + m\Omega y)^2} + \alpha_2 z$$

The separation constants $\alpha_1, \ldots \alpha_3$ are the energy $\mathcal{E}$ the final constant coordinates Q_2, and Q_3. The final constant momenta are designated as β_1, ... β_3. These are found from $\beta_j = -\partial F_1/\partial\alpha_j$. Then

$$\beta_1 = -\frac{\partial F_1}{\partial \alpha_1} = t \mp m \frac{1}{m\Omega} \sin^{-1} \frac{\alpha_3 + m\Omega y}{\sqrt{\left|2m\alpha_1 - \alpha_2^2\right|}},$$

$$\beta_2 = -\frac{\partial F_1}{\partial \alpha_2} = -z \pm \alpha_2 \frac{1}{m\Omega} \sin^{-1} \frac{\alpha_3 + m\Omega y}{\sqrt{\left|2m\alpha_1 - \alpha_2^2\right|}},$$

and

$$\beta_3 = -\frac{\partial F_1}{\partial \alpha_3} = -x \mp \frac{1}{m\Omega}\sqrt{2m\alpha_1 - \alpha_2^2 - (\alpha_3 + m\Omega y)^2}.$$

Then

$$(x+\beta_3)^2 + \left(y + \frac{\alpha_3}{m\Omega}\right)^2 = \frac{2m\alpha_1 - \alpha_2^2}{m^2\Omega^2}.$$

We have a complete description of the motion. The charge moves uniformly along the $z-$axis at a rate $\mathrm{d}z/\mathrm{d}t = \alpha_2/m$. The momentum along the $z-$axis is then α_2. This is the constant we have for the final coordinate Q_2. The motion in the $(x, y)-$plane is a circle centered at $(-\beta_3, -\alpha_3/m\Omega)$ with a radius $R = \sqrt{\left|2m\alpha_1 - \alpha_2^2\right|}/m\Omega$. And the $(x, y)-$motion in the circle is uniform at a frequency Ω as we see from the sinusoidal solution for y.

5.11 The generator is

$$F_1\left(r, \vartheta, z, t, \alpha_1, \ldots, \alpha_3\right)$$
$$= -\alpha_1 t \pm \int \mathrm{d}r \sqrt{2m\alpha_1 - \alpha_2^2 - \left(\frac{1}{r}\alpha_3 - r\frac{m\Omega}{2}\right)^2} + \alpha_3\vartheta + \alpha_2 z.$$

The $\alpha_1, \ldots, \alpha_3$ are the constant final coordinates and the energy (α_1). The final momenta will be the constants $\beta_1, \ldots, \beta_3$. We could integrate what we have here and obtain the generator directly. But the differentiation of the result looks formidable. So we calculate first the β's.

$$\beta_1 = -\frac{\partial F_1}{\partial \alpha_1} = t \mp m \int \mathrm{d}r \frac{1}{\sqrt{2m\alpha_1 - \alpha_2^2 - \left(\frac{1}{r}\alpha_3 - r\frac{m\Omega}{2}\right)^2}},$$

$$\beta_2 = -\frac{\partial F_1}{\partial \alpha_2} = -z \pm \alpha_2 \int \mathrm{d}r \frac{1}{\sqrt{2m\alpha_1 - \alpha_2^2 - \left(\frac{1}{r}\alpha_3 - r\frac{m\Omega}{2}\right)^2}},$$

and

$$\beta_3 = -\frac{\partial F_1}{\partial \alpha_3} = -\vartheta \pm \int \frac{\mathrm{d}r}{r} \frac{\left(\frac{1}{r}\alpha_3 - r\frac{m\Omega}{2}\right)}{\sqrt{2m\alpha_1 - \alpha_2^2 - \left(\frac{1}{r}\alpha_3 - r\frac{m\Omega}{2}\right)^2}}.$$

The integrals in these results require some care because of the form of the expression $(\alpha_3/r - rm\Omega/2)$, which causes the difficulty.

$$\beta_1 = t \pm \frac{1}{\Omega}\cos^{-1}\frac{A - r^2\frac{m\Omega}{2}}{|B|},$$

$$z + \beta_2 = \pm\alpha_2\frac{1}{m\Omega}\sin^{-1}\frac{A - r^2\frac{m\Omega}{2}}{|B|},$$

$$R^2 = r^2 + r_0^2 - 2rr_0\cos(\vartheta + \beta_3),$$

which is the general form of the cosine law. The orbit is a circle of radius R with center located by r_0, and is distinct from the origin of coordinates, which is designated by 0. The charge rotates in a clockwise fashion around the circle, which results in a negative angular momentum along the $z-$axis, which is out of the plot. The motion of the charge in the $z-$direction is uniform, and may also be stationary.

5.12 The generator as

$$F_{1,2}(q, \alpha, t) = -\alpha_1 t + \alpha_3 x \pm \int dy\sqrt{2m\alpha_1 + 2mQEy - \alpha_2^2 - (\alpha_3 + m\Omega y)^2} + \alpha_2 z.$$

Again, because the partial derivatives become complicated, we shall proceed first to the momenta $\beta_j = -\partial F_1(q, t, \alpha)/\partial\alpha_j$.

$$\beta_1 = -\frac{\partial F_1}{\partial\alpha_1} = t \mp m\int dy\frac{1}{\sqrt{2m\alpha_1 + 2mQEy - \alpha_2^2 - (\alpha_3 + m\Omega y)^2}},$$

$$\beta_2 = -\frac{\partial F_1}{\partial\alpha_2} = -z \pm \alpha_2\int dy\frac{1}{\sqrt{2m\alpha_1 + 2mQEy - \alpha_2^2 - (\alpha_3 + m\Omega y)^2}},$$

and

$$\beta_3 = -\frac{\partial F_1}{\partial\alpha_3} = -x \pm \int dy\frac{(\alpha_3 + m\Omega y)}{\sqrt{2m\alpha_1 + 2mQEy - \alpha_2^2 - (\alpha_3 + m\Omega y)^2}}.$$

Integrating

$$y = \left(\frac{QE}{m\Omega^2} - \frac{\alpha_3}{m\Omega}\right) \pm A\cos\Omega(t - \beta_1)$$

and

$$y = \left(\frac{QE}{m\Omega^2} - \frac{\alpha_3}{m\Omega}\right) \pm A\cos\frac{m\Omega}{\alpha_2}(z + \beta_2).$$

The charged particle then moves along the $z-$axis according to

$$z = \frac{\alpha_2}{m} t.$$

This may also be zero if $\alpha_2 = 0$. And performing the β_3 integration

$$x + \beta_3 = \frac{QE}{m\Omega^2} \Omega \left(t - \beta_1 \right) - A \sin \Omega \left(t - \beta_1 \right).$$

Because of the substitutions none of these integrals is easy. With

$$y + \frac{\alpha_3}{m\Omega} = \left(\frac{QE}{m\Omega^2} \right) - A \cos \Omega \left(t - \beta_1 \right),$$

we have the general form of the equation for a cycloid. The constants β_3 and $\alpha_3/m\Omega$ locate the cycloid relatively to the origin in the $(x, y)-$plane. The form of the cycloid is determined by the relative sizes of $QE/m\Omega^2$ and A. There are three general forms of the cycloid depending upon whether $QE/m\Omega^2$ is greater than, less than, or equal to A.

7.1 This exercise is a demonstration.
7.2 (a) $\mathcal{H} = \frac{1}{2m} \left(p_1^2 + p_3^2 \right) + \frac{1}{2\mu m} p_2^2 + \frac{1}{2} k \left((x_1 - x_2)^2 + (x_2 - x_3)^2 \right)$.
(b) Because of the product terms $x_1 x_2$ and $x_2 x_3$ this Hamiltonian is not separable.
(c) The eigenvalues (frequencies) are: $\pm i\omega_0$, 0, and $\pm i\omega_0 \sqrt{\left(1 + \frac{2}{\mu} \right)}$.

7.3 $x = c - ab$, $y = b - c/a$, and $z = c/a - b$.

8.1 This is a demonstration.

8.2 The differential world line is

$$\begin{aligned} ds' &= \begin{bmatrix} \gamma & -\gamma\beta & 0 & 0 \\ -\gamma\beta & \gamma & 0 & 0 \\ 0 & 0 & 1 & 0 \\ 0 & 0 & 0 & 1 \end{bmatrix} \begin{bmatrix} cdt \\ vdt \\ 0 \\ 0 \end{bmatrix} \\ &= \begin{bmatrix} \gamma cdt - \gamma\beta vdt \\ \gamma vdt - \gamma\beta cdt \\ 0 \\ 0 \end{bmatrix}. \end{aligned}$$

8.3 Choosing $\beta_x' = 1 - \kappa$ and $\beta = 1 - \lambda$ the equation $u_x = \left(u_x' + c\beta \right) / \left(1 + \beta u_x'/c \right)$ results in

$$\beta_x = \frac{\beta_x' + \beta}{1 + \beta\beta_x'} = \frac{2 - \kappa - \lambda}{2 - \kappa - \lambda + \kappa\lambda} < 1.$$

This is the argument that Einstein presented.

8.4 $h\nu' = m_0\gamma_u (h\nu)(1+\beta_u) / \left[2m_\gamma + m_0\gamma_u(1-\beta_u)\right]$.

8.5 $mc^2 = 2\sqrt{\mathcal{E}_1\mathcal{E}_2}$

8.6 $h\nu = 2(1 + m_{0e}/m_{01})\, m_{0e}c^2$. If the mass m_{01} is a nucleon (proton or neutron) then $m_{0e}/m_{01} \approx 0$ and $h\nu \approx 2m_{0e}c^2$. That is, the absolute minimum energy of the incoming photon is the sum of the rest energies of the electron and positron.

8.7 This exercise is a demonstration.

8.8 This exercise is a demonstration.

8.9 This exercise is a demonstration.

8.10 This exercise is a demonstration.

8.11 We find that we must require $\boldsymbol{f} \cdot \boldsymbol{u} = 0$, as well as $\boldsymbol{f} = m_0\gamma_u \mathrm{d}\boldsymbol{u}/\mathrm{d}t$. The latter of these is not outside of our expectations, but the former is. This (also) violates the expectation that a force, while it may not change the rest mass, in the case of a pure force, it must not be limited in such a way that it cannot change the total relativistic energy.

8.12 This exercise is a demonstration.

8.13 This exercise is a demonstration.

8.14 This exercise is a demonstration.

References

1. R. Baierlein, *Newtonian Dynamics* (McGraw-Hill Book Company, New York, 1983)
2. V. Bernard, M. Lage, U.-G. Meissner, A. Rusetsky: J. High Energy Phys. JHEP **01**, 019 (2011). (published for SISSA by Springer)
3. P.G. Bergmann, *Introduction to the Theory of Relativity* (Prentice-Hall, New York, 1942)
4. R.C. Bishop, "Chaos", The Stanford Encyclopedia of Philosophy (Fall 2009 Edition), E.N. Zalta (ed.), http://plato.stanford.edu/archives/fall2009/entries/chaos/
5. A. Alexander, *Infinitesimal* (Scientific American, New York, 2014)
6. *Encyclopaedia Britannica* (William Benton, Chicago, London 1969)
7. G. Contopoulos, Stockholms Obs. Ann. **19**(10) (1957)
8. G. Contopoulos, Stockholms Obs. Ann. **20**(5) (1958)
9. G. Contopoulos, Z. Astrophys. **49**, 273 (1960)
10. G. Contopoulos, Astron. J. **68**, 1 (1963)
11. R. Courant, D. Hilbert, *Methods of Mathematical Physics*, vol. 2 (Wiley, New York, 1989)
12. *CRC Standard Mathematical Tables 28th Edition.* (CRC Press, Boca Raton, FL 1987)
13. W.H. Cropper, *Great Physicists: the Life and Times of Leading Physicists from Galileo to Hawking* (Oxford University Press, New York, 2001)
14. P. Curd, "Presocratic Philosophy", The Stanford Encyclopedia of Philosophy (Spring 2012 Edition), E.N. Zalta (ed.), http://plato.stanford.edu/archives/spr2012/entries/presocratics/
15. J. Cushing, *Philosophical Concepts in Physics* (Cambridge University Press, Cambridge, 1998)
16. J.B. le R. d'Alembert, *Traité de dynamique*. (David, Paris 1743)
17. M.T. d'Alverny, "Translations and Translators", in *Renaissance and Renewal in the Twelfth Century* ed. by R.L. Benson, G.C Constable (Harvard University Press, Cambridge 1982)
18. P.A.M. Dirac, *The Principles of Quantum Mechanics* (Oxford University Press, London, 1958)
19. L. de Broglie: *Recherches sur la Théorie des Quanta*, Ann. de Phys., 10e série, t. III (Janvier - F évrier 1925, trans. by A.F. Kracklauer AFK 2004)
20. R. Descartes, *Principia philosophiae*, I, 51. *Oeuvres de Descartes*, ed. Charles Adam and Paul Tannery, vol. 8, p. 24 (Cerf, Paris 1905)
21. J.W. Dettman, *Mathematical Methods in Physics and Engineering* (McGraw-Hill, New York, 1962)
22. R. Dugas, *History of Mechanics* (Dover, Mineola, New York, 1988)
23. A. Einstein, Annalen der Physik **17**, 132 (1905)
24. H.A. Lorentz, A. Einstein, H. Minkowski, H. Weyl, *The Principle of Relativity* (Dover, New York, 1952)

C.S. Helrich, *Analytical Mechanics*, Undergraduate Lecture Notes in Physics, DOI 10.1007/978-3-319-44491-8

25. S.G. Eubank, J.D. Farmer, Introduction to dynamical systems, in *Introduction to Nonlinear Physics*, ed. by L. Lam (Springer, New York 1997) pp. 55–175
26. G. Farmelo, *The Strangest Man* (Basic Books, New York, 2009)
27. M. Feingold, Decline and fall: arabic science, in seventeenth century England, in *Tradition, Transformation, Transmission*, ed. by Ragep and Ragep (Brill, Leiden, 1996), pp. 441–469
28. R.H. Fowler, *Statistical Mechanics* (Cambridge University Press, London, 1966)
29. D.Y. Gao, H.D. Sherali, Canonical duality theory: connections between nonconvex mechanics and global optimization, in: *Advances in Mechanics and Mathematics*, vol III, ed. by Gao, Sherali (Spinger, New York 2006)
30. H. Geiger, E. Marsden, Proceedings of the Royal Society **82**, 495–500 (1909)
31. I.M. Gelfand, S.V. Fomin, *Calculus of Variations* (Prentice-Hall, Englewood Cliffs, 1963)
32. J.W. Gibbs, *Elementary Principles in Statistical Mechanics* (Yale University Press, New Haven, 1902)
33. O. Gingerich, *God's Planet* (Harvard University Press, Cambridge, 2014)
34. H. Goldstein, *Classical Mechanics*, 2nd edn. (Addison-Wesley, Reading, MA, 1980)
35. I.S. Gradshteyn, I.M. Ryzhik, *Table of Integrals, Series, and Products*, 4th edn. (Academic Press, San Diego, CA, 1980)
36. D.T. Greenwood, *Classical Dynamics* (Dover Mineola, New York, 1997)
37. D.J. Griffiths, *Introduction to Electrodynamics*, 3rd edn. (Addison Wesley, Upper Saddle River, NJ, 1999)
38. J. Hackett, Roger Bacon, in *The Stanford Encyclopedia of Philosophy*, ed. by E.N. Zalta (Spring 2012 Edition), http://plato.stanford.edu/archives/spr2012/entries/roger-bacon/
39. W.R. Hamilton, Philosophical Transactions of the Royal Society, part I for 1835, pp. 247–308 (1834)
40. W.R. Hamilton, Philosophical Transactions of the Royal Society, part II for 1834, pp. 95–3144 (1835)
41. L.N. Hand, J.D. Finch, *Analytical Mechanics* (Cambridge University Press, Cambridge, 1998)
42. S. Hassani, *Foundations of Mathematical Physics* (Allyn and Bacon, Needham Heights, MA, 1991)
43. M. Hénon, C. Heiles, Astron. J. **69**, 73 (1964)
44. C.S. Helrich, *Modern Thermodynamics with Statistical Mechanics* (Springer, Heidelberg, 2009)
45. C.S. Helrich, *The Classical Theory of Fields: Electromagnetism* (Springer, Heidelberg, 2012)
46. C.S. Helrich, T. Lehman, *Am. J. Phys.* **47**(4), (1979)
47. R.G. Hewlett, O.E. Anderson Jr., *The New World, 1939/1946* (University of Califormia Press, Berkeley, 1990)
48. R.C. Hilborn, *Chaos and Nonlinear Dynamics* (Oxford University Press, Oxford, 2000)
49. G. Holton, S.G. Brush, *Physics, the Human Adventure* (Rutgers University Press, New Brunswick, NJ, 2001)
50. http://www.roberthooke.org.uk/chronolo.htm, Accessed 21 Jan 2013
51. J.D. Jackson, *Classical Electrodynamics*, 2nd edn. (Wiley, New York, 1975)
52. C.G.J. Jacobi, Crelle's J. **17**, 68 (1837), http://gdz.sub.uni-goettingen.de/dms/load/img/?PPN=PPN243919689_0017
53. C.G.J. Jacobi, Crelle's J. **17**, 97 (1837), http://gdz.sub.uni-goettingen.de/dms/load/img/?PPN=PPN243919689_0017
54. C.G.J. Jacobi, *Vorlesungen Über Dynamik*. ed. by E. Lottner (Reimer, Berlin 1884)
55. M. Jammer, *Concepts of Force: A Study in the Foundations of Dynamics* (Harvard University Press, Cambridge, 1957)
56. M. Jammer, *Concepts of Mass in classical and Modern Physics* (Dover, Mineola, New York, 1997)
57. J.V. Jose, E.J. Saletan, *Classical Dynamics: A Contemporary Approach* (Cambridge University Press, Cambridge, 1998)
58. M.J. Klein: *Proceedings of the International School of Physics "Enrico Fermi" Course LVII*. ed. by C. Weiner (Academic Press, New York 1977)

59. A. Koestler, *The Watershed: A Biography of Johannes Kepler* (Anchor Books, New York, 1960)
60. N.A. Kostov, V.S. Gerdjikov, M. Mioc, arXiv:0911.42601 [nlin.SI]. Accessed 22 Nov 2009
61. L. Lam (ed.), *Introduction to Nonlinear Physics* (Springer, New York, 1997)
62. J.L. Lagrange, Essai d'une nouvelle méthode pour déterminer les maxima et les minima des formules intégrales indéfinies. Mélanges de Turin **2**, 173–95 (1760–61)
63. J.L. Lagrange, Applications de la méthode exposée dans le mémoir précédent à la solution de différents problèmes de dynamique. M élanges de Turin **2**, 196–298 (1760–61)
64. J.L. Lagrange, *Mécanique Analitique.* (Desaint, Paris, 1788)
65. C. Lanczos, *The Variational Principles of Mechanics* (Dover, Mineola, New York, 1986)
66. A.M. Legendre, L'intégration de quelques é quations aux différences Partielles, in Histoire de l'Académie Royale des Sciences (1787)
67. D.S. Lemons, *Perfect Form: Variational Principles, Methods, and Applications in Elementary Physics* (Princeton University Press, Princeton, 1997)
68. G.N. Lewis, R.C. Tolman, *Philosophical Magazine* vol. 18 (1909), p. 510
69. G.E.R. Lloyd, *Early Greek Science: Thales to Aristotle* (Norton, New York, 1970)
70. C.R. MacCluer, *Calculus of Variations: Mechanics, Control and Other Applications* (Dover, Mineola, New York, 2012)
71. E. Mach, *The Science of Mechanics Trans* T.J. McCormack (Open Court, LaSalle, IL, 1942)
72. K. Mainzer, *Thinking in Complexity* (Springer, Berlin, 1997)
73. B.B. Mandelbrot, *The Fractal Geometry of Nature* (Freeman, New York, 1982)
74. J.E. Marsden, T.S. Ratiu, *Introduction to Mechanics and Symmetry* (Springer, New York, 1999)
75. J.C. Maxwell, On the dynamical theory of gases, in *The Scientific Papers of James Clerk Maxwell, V II*, ed. by W.D. Niven (Cambridge University Press, Cambridge 1980; reprint Dover, New York 1952)
76. J. Mehra, H. Rechenberg, *The Historical Development of the Quantum Theory*, vol 1, Part 2. (Springer, New York, 1982)
77. L. Meitner, O.R. Frisch, Nature **143**, 239 (1939)
78. E. Merzbacher, *Quantum Mechanics*, 3rd edn. (Wiley, New York, 1998)
79. A.A. Michelson, E.W. Morley, Am. J. Sci. **34**, 333 (1887)
80. S.W. McCuskey, *Introduction to Advanced Dynamics* (Adison-Wesley, Reading, 1959)
81. S.W. McCuskey, *Introduction to Celestial Mechanics* (Adison-Wesley, Reading 1963)
82. W. Moore, *Schrödinger: Life and Thought* (Cambridge University Press, Cambridge, 1989)
83. P.M. Morse, H. Feshbach, *Methods of Theoretical Physics, Part I* (McGraw-Hill, New York, 1953)
84. I. Muzaffar, *Science and Islam* (Greenwood, Westport, Conn., 2007)
85. M. Nakane, C.G. Fraser, Centaurus **44**, 11–227 (2002)
86. O. Neugebauer, *A History of Ancient Mathematical Astronomy* (Springer, New York, 1975)
87. I. Newton, *Newton's Papers and Letters on Natural Philosophy*. ed. by I.B. Cohen (Harvard University Press, Cambridge, 1958)
88. I. Newton, *Principia Mathematica*, trans. and ed. by A. Motte as *Mathematical Principles of Natural Philosophy* (1729), rev. Florian Cajori (University of California Press, 1934)
89. A. Pais, *Subtle is the Lord* (Oxford University Press, Oxford, 1982)
90. M. Panza, The Origins of Analytical Mechanics in the 18th Century, in *A History of Analysis*, ed. by H.N. Jahnke (American Mathenatical Society and London Mathematical Society 2003) pp. 137–153
91. D. Park, *The How and the Why* (Princeton University Press, Princeton, 1988)
92. W. Pauli, *Theory of Relativity* (Dover, New York, NY, 1981)
93. Poincaré return map. Encyclopedia of Mathematics, http://www.encyclopediaofmath.org/index.php?title=Poincar%C3%A9_return_map&oldid=24535. Accessed 8 Mar 2014
94. J.J. O'Connor, E.F. Robertson, Charles Eugene delaunay, in *School of Mathematics and Statistics University of St Andrews, Scotland*, http://www-history.mcs.st-andrews.ac.uk/Biographies/Delaunay.html. Accessed 17 July 2013

95. J.J. O'Connor, E.F. Robertson, Carl Gustav Jacob Jacobi, in *MacTutor History of Mathematics* (2000), http://www-history.mcs.st-andrews.ac.uk/Biographies/Lagrange.html. Accessed 20 Dec 2012
96. J.J. O'Connor, E.F. Robertson, Joseph-Louis Lagrange, in *MacTutor History of Mathematics* (1999), http://www-history.mcs.st-andrews.ac.uk/Biographies/Lagrange.html. Accessed 29 Nov 2012
97. A. Ollongren, Bull. Astron. Inst. Neth. **16**, 241 (1962)
98. E. Ott, *Chaos in Dynamical Systems* (Cambridge University Press, Cambridge, UK, 1993)
99. W. Rindler, *Introduction to Special Relativity* (Oxford University Press, Oxford, 1982)
100. R. Rhodes, *The Making of the Atomic Bomb* (Simon and Schuster, New York, 1986)
101. V. Roberts, Isis **57**, 210 (1966)
102. S. Ruijsenaars, *Integrable Systems: An Overview* (2009), https://maths.leeds.ac.uk/siru/ISsurvey.pdf. Accessed 10 Feb 2014
103. E. Rutherford, Philos. Mag. **21**, 669–688 (1911)
104. H. Sagan, *Introduction to the Calculus of Variations* (Dover, Mineola, New York, 1992)
105. E. Schrödinger, Annalen der Physik **79**(4), 361–376 (1926)
106. K. Schwarzschild, Sitzungsber. der Kgl. Akad. der Wiss., **548** (1916)
107. C. Shields, Aristotle, in *The Stanford Encyclopedia of Philosophy*, ed. by E.N. Zalta (Summer 2012 Edition), http://plato.stanford.edu/archives/sum2012/entries/aristotle/
108. G. Saliba, *Islamic Science and the Making of the European Renaissance* (MIT Press, Cambridge, 2007)
109. G. Saliba, *A History of Arabic Astronomy: Planetary Theories During the Golden Age of Islam* (New York University Press, New York, 1994)
110. C.J. Scriba, Biography of C.G.J Jacobi, in *Dictionary of Scientific Biography* (New York 1970–1990), http://www.encyclopedia.com/doc/1G2-2830902153.html
111. D. Sobel, *A More Perfect Heaven* (Walker, New York, 2011)
112. I. Suhendro, Abraham Zelmanov J. **1**, xiv–xix (2008)
113. N. Sverdlow, Proc. Am. Philos. Soc. **117**(6), 423–512 (1973)
114. J.J. Sylvester, Philosophical Magazine, 4th series, vol. 2 (1851), pp. 391–410
115. K. Symon, *Mechanics* (Addison-Wesley, Reading, MA, 1977)
116. J.J. Thomson, Camb. Lit. Philos. Soc. xv pt. 5 (1910)
117. S.T. Thornton, A. Rex, *Modern Physics for Scientists and Engineers*, 4th edn. (Brooks/Cole, Boston, 2013)
118. G. Toomer, *Ptolemy's Almagest* (Springer, New York, 1984)
119. J.R. Voelkel, *Johannes Kepler and the New Astronomy* (Oxford University Press, New York, 1999)
120. C.E. Wayne: *An Introduction to KAM Theory* (2008), https://math.bu.edu/people/cew/preprints/introkam.pdf. Accessed 13 Feb 2014
121. R.S. Westfall, *Never at Rest: A Biography of Isaac Newton* (Cambridge University Press, Cambridge, 1983)
122. R.S. Westfall, *The Life of Isaac Newton* (Cambridge University Press, Cambridge, 1993)
123. H. Weyl, *Space—Time—Matter* (Dover, New York, 1950)
124. M. White, *Isaac Newton: The Last Sorcerer* (Perseus, Reading, Massachusetts, 1997)
125. E.T. Whittaker, *A Treatise on the Analytical Dynamics of Particles and Rigid Bodies* (Cambridge University Press, New York, 1964)
126. E.T. Whittaker, *A History of the Theories of Aether and Electricity from the Age of Descartes to the Close of the Nineteenth Century* (Longmans Green, London, 1910)
127. E.P. Wigner, Ann. Math. **40**, 149 (1939)
128. W. Yourgrau, S. Mandelstam, *Variational Principles in Dynamics and Quantum Theory* (Dover, New York, 1979)
129. R. Zach, Hilbert's Program, The Stanford Encyclopedia of Philosophy (Summer 2015 Edition), E.N. Zalta (ed.), http://plato.stanford.edu/archives/sum2015/entries/hilbert-program/
130. J. Zupko, John Buridan, The Stanford Encyclopedia of Philosophy (Fall 2011 Edition), E.N. Zalta (ed.), http://plato.stanford.edu/archives/fall2011/entries/buridan/

Index

C.S. Helrich, *Analytical Mechanics*, Undergraduate Lecture Notes in Physics, DOI 10.1007/978-3-319-44491-8

Printed in the United States
By Bookmasters